Genome Editing and Global Food Security

With the rapid increase in the global population and changing climatic impacts on agriculture, this book demonstrates how genome editing will be an indispensable technique to overcome ongoing and prospective agricultural challenges.

This book examines the role of genome editing in improving crop yields and contributing to global food security. It summarizes a range of genome editing techniques and discusses the roles they can play in producing a new generation of high-yielding, climate-ready crops. This includes site-specific nucleases, precision genome engineering, clustered regularly interspaced short palindromic repeats, and bioinformatics. It showcases how these gene editing techniques can tailor plants to not only increase yield-related traits but to also make them better suited to their environment and to be resistant to pests and extreme climatic events, such as droughts. The book also examines genome editing regulations and policies, the commercialization of genome-edited crops, and biosafety and biosecurity concerns. Overall, this book reveals and showcases how genome editing can improve crop resilience and production to address current and future agricultural challenges and alleviation of global food security concerns.

This book will be of great interest to students and scholars of agricultural science, crop and plant science, genome editing, sustainable agriculture, biotechnology, and food security.

Zeba Khan is a faculty member at the Center for Agricultural Education, Department of Agricultural Sciences at the Aligarh Muslim University, India. She holds a PhD from Aligarh Muslim University.

Durre Shahwar is a post-doctoral fellow at the Department of horticultural biosciences, Pusan National University, South Korea. She holds a PhD from the Aligarh Muslim University, India.

Yasmin Heikal is an assistant professor in the Department of Botany at Mansoura University, Egypt. She holds a PhD in genetic diversity from Mansoura University.

Earthscan Food and Agriculture

For more information about this series, please visit: www.routledge.com/books/series/ECEFA/

Genome Editing and Global Food Security

Molecular Engineering Technologies for Sustainable Agriculture

**Edited by Zeba Khan,
Durre Shahwar, and
Yasmin Heikal**

Routledge
Taylor & Francis Group
LONDON AND NEW YORK

earthscan
from Routledge

First published 2024
by Routledge
4 Park Square, Milton Park, Abingdon, Oxon OX14 4RN

and by Routledge
605 Third Avenue, New York, NY 10158

Routledge is an imprint of the Taylor & Francis Group, an informa business

British Library Cataloguing-in-Publication Data
A catalogue record for this book is available from the British Library

ISBN: 978-1-032-46524-1 (hbk)
ISBN: 978-1-032-46526-5 (pbk)
ISBN: 978-1-003-38210-2 (ebk)

DOI: 10.4324/9781003382102

Typeset in Times New Roman
by MPS Limited, Dehradun

Contents

Contributors

Mohamed Farah Abdulla is a PhD candidate at Plant Biotechnology Division, Department of Agricultural Biotechnology, Faculty of Agriculture, Ondokuz Mayıs University, Kurupelit campus, Samsun, Turkey.

Syed Mohammad Bashir Ali is a student of agricultural sciences at Aligarh Muslim University. In his academic journey thus far, he has developed his interest in crop improvement using genomics, bioinformatics, and biotechnology. Specifically, he is passionate about studying functional genomics, transcriptomics, and genetic engineering.

Mohammed Ali is an assistant professor of Crop Genetic and Breeding at Desert Research Centre (DRC) and he received his PhD in 2018 from Huazhong Agricultural University, China. From 2008 to 2023, he worked as a researcher in DRC. He He has been working in Plant Biotechnology in the fields of Plant Functional Genomics and Secondary Metabolism for 15 years; his current research involves the usage of different plant metabolites databases and new techniques such as RNA-Seq as a method to identify and discover the genes involved in the biosynthesis of various primary and secondary metabolite pathways in model and non-model plants with unknown genomic sequences.

M.A. AL-KORDY is a professor of plant genetics & amp; biotechnology at Genetics & amp; Cytology Department, Biotechnology Research Institute, National Research Centre (NRC), Egypt. He obtained his PhD in 1997 from the Department of Genetics, Faculty of Agriculture, Ain Shams University, Egypt. He is working on the genetic diversity of wild, medicinal, and aromatic plants and the functional analysis of secondary metabolites genes under biotic and abiotic stress.

Ahmad M. Alqudah is an assistant professor in Plant Molecular Genetics at Qatar University. He received BSc in 2005 and then MSc in plant science from Jordan University of Science and Technology, Jordan, in 2007. In 2015, he completed his PhD in Plant Breeding and Genetics from Martin Luther University Halle-Wittenberg, Germany. Then, he worked as a full-time postdoctoral researcher and scientist in different research groups specializing in Cereal Genetics in Germany. His projects aimed to understand the underlying molecular genetic factors of agronomic, developmental, adaptive, and grain yield-related traits in wheat and barley. In 2021, he was awarded a grant for his project to do gene editing using the latest plant molecular technologies including CRISPR-Cas from Novo Nordisk Foundation. The project was hosted by Aarhus University, Denmark's crop genetics, and biotechnology section. Soon after that, he was appointed as tenure track assistant professor in crop genetics and biotechnology at Aarhus University. Recently, he became an assistant professor in plant molecular genetics at Qatar University.

Mohd Yunus Khalil Ansari worked as a professor in the Department of Botany, A.M.U., Aligarh, and retired as a chairperson on August 20, 2018. He has been associated with teaching cytogenetics and molecular biology. His research areas are cytogenetics, mutation breeding, and biotechnology. He produced several important mutants in more than 30 plants. He has been actively involved in research and published around 70 scientific journals and was the co-author of 4 books. He has produced 15 PhD and 7 M.Phil. students. He was a co-principal investigator in the Department of Biotechnology sponsored project "In-vitro propagation of Pterocarpus sp." and worked as a co-investigator in a project sponsored by the Ministry of Environment and Forest on preserving 30 endangered plants.

Abdelfattah Badr is a professor of Genetics and Plant Biodiversity, Faculty of Science at Helwan University. He has a BSc in Botany from Assiut University, Egypt, in 1972, and a PhD in Plant Genetics from Sheffield University, UK, in 1977. He has been a professor since 1986, and was awarded postdoctoral scholarships by the British Council in 1980, Fulbright Foundation in 2001 and 2011, and a DAAD Science Tour in 2012. He has been Alexander von Humboldt's Fellow since 1990. He was awarded the Egyptian State Appreciation Prize for Excelling Research in Biological Sciences in 1996 and Tanta University Appreciation Prize for Merit in Basic Sciences in 2008. He is currently the Chairperson of the Environmental Research Council at the Academy of Science and Technology, a member of the Experts Committee for nominating Deans of University Colleges, and the Promotion Committee of Professor and Associate Professors in the Supreme Council of Egyptian Universities.

Priya Brata Bhoi is currently working as an assistant professor at Punjab Agriculture University, Ludhiana. Worked on environmental impact assessment and socio-economic issues which required designing questionnaires, and collecting and tidying up unstructured data.

Nourhan Fouad is currently a senior-level research assistant in the Department of Genetic Innovation at the International Centre for Agricultural Research in the Dry Areas. She obtained her BSc from the Faculty of Science, Cairo University, and her MSc on Biotechnology Biomolecular entitled:

"Development of Molecular Markers Linked to Fusarium Wilt Resistant Gene(s) in Chickpea (*Cicer arietinum*)," from the Faculty of Science, Cairo University. Presently, she is enrolled for her PhD, working on plant metagenomics and microbiome on bread wheat, at the Biotechnology Department, Faculty of Science, Cairo University, Giza, Egypt.

Ashutosh Sawarkar Ganpatrao is currently working as an assistant professor (Genetics and Plant Breeding) at Ramakrishna Mission Vivekananda Educational and Research Institute, Kolkata, India. He is having more than four years of teaching experience.

Rama Krishna Satyaraj Guru is a PhD research scholar at Indira Gandhi Krishi Viswavidyalaya (IGKV), Raipur. He has completed his master's degree from IGKV, Raipur.

Aladdin Hamwieh is a scientist working as a chickpea breeder and biotechnologist at the International Center for Agricultural Research in Dry Areas. He has dedicated his career to developing and implementing innovative breeding strategies for improving chickpea production in dry areas. His research focuses on enhancing chickpea tolerance to abiotic and biotic stresses, such as drought, salinity, and disease. He has made significant contributions to the identification and characterization of genomic regions associated with traits such as Ascochyta blight resistance and yield under water-limited conditions. He is also known for his work on developing high-throughput phenotyping and genotyping techniques to accelerate the breeding process and improve accuracy in selecting desirable traits. His publications cover a wide range of topics related to chickpea breeding, including the development of breeding strategies, the identification of genomic regions associated with important traits, and the use of high-throughput phenotyping techniques. His work has not only advanced the field of chickpea breeding but has also contributed to food security and sustainability in dry regions of the world.

Hanaa Hegazy Ali Ibrahim is a professor of Genetics at the Faculty of Education, Ain Shams University. She has a BSc in Botany from Ain Shams University, Egypt, in 1981, and PhD in Plant Genetics from Ain Shams University in 1995. Pressors at Ain Shams University, in 2007. She was a visiting scientist at the Max Planck Institute of Plant Breeding in Cologne Germany in 1997 and 1998 and at Miami University, Oxford, Ohio, in 2001. Recently she was a visiting professor at the German Gene Bank in 2018 and 2019.

Bushra khatoon is a budding researcher in the field of parasitology, with a special emphasis on parasite physiology, biochemistry, and enzymology. She completed both her bachelor's and master's degrees in Zoology with a specialization in Parasitology from Aligarh Muslim University. During her postgraduate studies, she conducted an in-depth research project on "The

Antioxidant Enzymes in the Liver Infected by Echinococcus granulosus," which was supervised by Professor Malik Irshadullah. Her work has important implications for the development of new treatments for parasitic infections.

Shanta Karki is a Joint Secretary working for the Ministry of Agriculture and Livestock Development, Government of Nepal. She has a bachelor's degree from the Institute of Agriculture and Animal Science, Tribhuvan University, Nepal; Master's and PhD degrees from Kyoto University, Japan, with a major in Plant Breeding and Molecular Biology. She worked as a molecular biologist at the C4 Rice Center (Genetic Engineering Team), International Rice Research Institution, Philippines, from 2009 to 2015. She has authored and co-authored several research articles, book chapters, conference papers, annual reports, and opinion columns.

Musa Kavas is an associate professor at the Plant Biotechnology Division, Department of Agricultural Biotechnology, Faculty of Agriculture, Ondokuz Mayıs University, Kurupelit campus, 55200, Samsun, Turkey.

Zeba Khan is presently engaged in academic teaching at the Faculty of Agricultural Sciences, Aligarh Muslim University, Aligarh, India. She obtained her master's degree and PhD from Aligarh Muslim University. She worked as SPO at the National gene bank of India, NBPGR, New Delhi, India. She also worked as a postdoctoral research associate under DST-PURSE-funded Project at Aligarh Muslim University. She has published books, impacted research articles, invited chapters in both nationally and internationally reputed books/journals, and presented papers at various national and international conferences, seminars, and symposiums. She has been awarded by various international scientific organizations for her outstanding performance in research.

Atul Pradhan Madhao recently joined Sanjeev Agrawal Global Educational University, Bhopal as an assistant professor (Genetics and Plant Breeding). Earlier, he worked as a post-doctoral fellow at Ramakrishna Mission Vivekananda Educational and Research Institute, Kolkata.

Purandar Mandal is currently working as a farm superintendent/assistant professor at Odisha University of Agriculture & Technology. He holds a PhD in Horticulture (Fruit Science). He is having more than 15 years of teaching experience.

Elena V. Mikhaylova is a PhD in plant science, currently working as a senior researcher at the Institute of Biochemistry and Genetics of the Ufa Federal Research Center and as an associate professor at the Ufa State Petroleum Technological University and Bashkir State Pedagogical University. She is the leader of a research group focused on plant genetic transformation (including CRISPR) and the development of stress-resistant varieties.

Ayesha Mohanty is currently working as an assistant professor at the Odisha University of Agriculture & Technology. She is the recipient of the DST-INSPIRE fellowship during her PhD on Tenures hip. Her area of interest is soil microbiology. She is having nine years of teaching experience at the Odisha University of Agriculture & Technology.

Karam Mostafa works in The Central Laboratory for Date Palm Research and Development, Agricultural Research Center (ARC), Giza 12619, Egypt, and is a PhD candidate at Plant Biotechnology Division, Department of Agricultural Biotechnology, Faculty of Agriculture, Ondokuz Mayıs University, Kurupelit campus, Samsun, Turkey.

Kaushik Kumar Panigrahi is currently working as an assistant professor at Odisha University of Agriculture & Technology. He is having nine years of teaching experience in the discipline of Genetics and Plant Breeding as well as Agricultural Biotechnology. He has visited IRRI, Manila, twice as a visiting scientist and is having a short stint as a research trainee in Israel.

Aurel Popescu is an associate professor at the Department of Natural Sciences, University of Piteşti, Romania. In 1998, he received his PhD in Biology from the University of Bucharest for research on the genetic manipulation of Fragaria genotypes by in vitro culture techniques. He is the author/co-author of 9 books and 5 book chapters published by national/international publishing houses; more than 90 articles indexed in various databases and/or presented at national and international congresses, conferences, and symposia; manager/responsible for many research projects in the field of horticultural genetics and biotechnology; and member of the team working on 13 national and international research projects for which he was given several awards.

Smruti Ranjan Padhan is currently pursuing his PhD at Indian Agriculture Research Institute, New Delhi. He has completed his master's degree from IARI, New Delhi.

Rojalin Pradhan is currently working as an assistant professor at the Institute of Agriculture Science, Siksha O Anusandhan University, Bhubaneswar, India. She holds a PhD in Horticulture (Vegetable Science). Her area of interest is the genetic improvement of vegetable crops.

Khaled Radwan graduated from Alexandria University with a major in Plant Biodiversity, specializing in the Egyptian flora, and participated in the research concerning the regional environmental management of the Mediterranean desert ecosystems of Egypt. With the emergence of the field of genetic engineering in Egypt by 1992, he joined the Agricultural Genetic Engineering Research Institute, ARC, where he was able to conduct research connecting both sciences. He has been a science communicator at the Egyptian Biotechnology Information Center, one of the International Service for the Acquisition of Agri-Biotech Applications Knowledge

Centers, since 2011. He has been a member of the National Biotechnology Network for Expertise Board of Directors, one of the Egyptian networks founded by the Academy of Scientific Research and Technology, since 2016.

Govinda Rizal teaches Plant Breeding, Molecular Genetics, Population Genetics, Biostatistics, and Technical Writing at the Institute of Agriculture and Animal Science, Tribhuvan University, Nepal. He had his bachelor's degree from the same institution, and master's and PhD in Plant Breeding from Kyoto University, Japan, and worked as a molecular biologist at the C4 Rice Project at International Rice Research Institution in the Philippines. He is affiliated with non-governmental organizations in Nepal and works as a consultant to International Center for Agricultural Research in the Dry Areas. He has published books, research articles, and opinion columns which can be accessed from orcid.org/0000- 0002-6245-1996.

Durre Shahwar received her PhD in Botany from Aligarh Muslim University, Aligarh, India. She is the recipient of the Maulana Azad National Fellowship and National Fellowship from the University Grant Commission, New Delhi. She has been awarded the Junior Scientist of the Year Award (2018) by International Foundation for Environment and Ecology, Kolkata, at the 5th International Conference on Environment and Ecology, University Gold Medal for securing first rank in MSc (Botany) in 2014. She has published several research articles and book chapters in peer-reviewed national and international journals. She has also participated in various national and international conferences and received life membership of scientific bodies in India. Her research interests are cytogenetics, plant breeding, and molecular biology (proteomics and genomics). She is actively engaged in mutation breeding for the genetic improvement of legume crops.

Nicoleta Anca Şuţan is an associate professor at the Department of Natural Sciences, University of Piteşti. In 2010, she received her PhD in Biology from Babeş-Bolyai University for the research on *in vitro* micropropagation capacity and genetic stability of some intergeneric hybrids *Fragaria* x *Potentilla*. More recently, the focus has been on the evaluation of the genotoxicity and cytotoxicity of vegetable extracts, including photosynthesis of metallic nanoparticles. She is the author/co-author of more than 50 articles indexed in Web of Science, Scopus, and other databases and/or presented at international conferences; of 5 books/chapters published at national/international publishing houses; manager/responsible for 2 research projects; member of the team working on 6 national projects; and has received 22 national and 1 international prize.

Ranjan Kumar Tarai is an associate professor (Fruit Science) at Odisha University of Agriculture & Technology. His area of interest is Fruit Breeding. He is having 18 years of teaching experience.

Mohd Hadi Yunus obtained his undergraduate studies in chemistry from Aligarh Muslim University, Aligarh, and post-graduate studies in biotechnology at Jamia Hamdard, New Delhi. He has published several review articles and a book chapter in reputed journals and books. He has also participated in various national and international conferences/ seminars. His research interests are microbiology, genome engineering, and biotechnology.

1 Genome editing in plants via CRISPR/Cas9

A genomic scissor borrowed from bacterial immune system

Zeba Khan and Durre Shahwar

1.1 Introduction

Conventional crop improvement methods rely on time-consuming and labor-intensive breeding technologies. With the discovery of versatile genomic editing tools and rapid increase in sequencing technologies, an increase in genomic information on a broad spectrum of cells line and organisms is becoming available. With the emergence of gene editing (GE) techniques, genes can be edited with incredible precision and can be used as an effective tool in crop improvement. The quantity of GE-based crop improvement cases has significantly increased. Crop quality is one of the top priorities among the numerous target traits for crop improvement. The technology utilizes a specific nuclease sequence to initiate a double-stranded break (DSB) at a particular site in DNA. The repair process occurs either through the homology-directed repair (HDR) pathway or non-homologous end joining (NHEJ) and induces mutation. With the least chance of going off-target and no integration of exogenous gene sequences, genome editing can produce predictable and inheritable mutations in particular locations of the genome. Deletions, insertions, single-nucleotide substitutions (SNPs), and large fragment substitutions are all examples of GE-mediated DNA changes. A method for nucleotide excision involves four families of site-directed nucleases (SDNs): Zinc-finger nucleases (ZFNs), transcription activator-like effecter nucleases (TALENs), Homing endonucleases or mega-nucleases (HEs) meganucleases and clustered regularly interspaced short palindromic repeat (CRISPR)/Cas 9 nucleases (Figure 1.1) which can cause DNA alterations and induce mutations. Zinc-finger nucleases are the most common nucleases found in eukaryotes having a target site of 18–36 base pair, identifying 4–6 DNA triplets. ZNFs bind to nucleases which function as dimers therefore pairs of ZNFs are needed for targeting any specific sequences: one that recognizes the sequence upstream and the other that identifies the sequence downstream of the site to be altered. TALENs have target binding site of 24–38 base pairs and show similarity to ZNFs in that they utilize a specific targeting DNA binding site and a FokI nuclease to initiate double-stranded break at the specified site, rather than identifying nucleotide triplets, each domain identifies a single nucleotide.

DOI: 10.4324/9781003382102-1

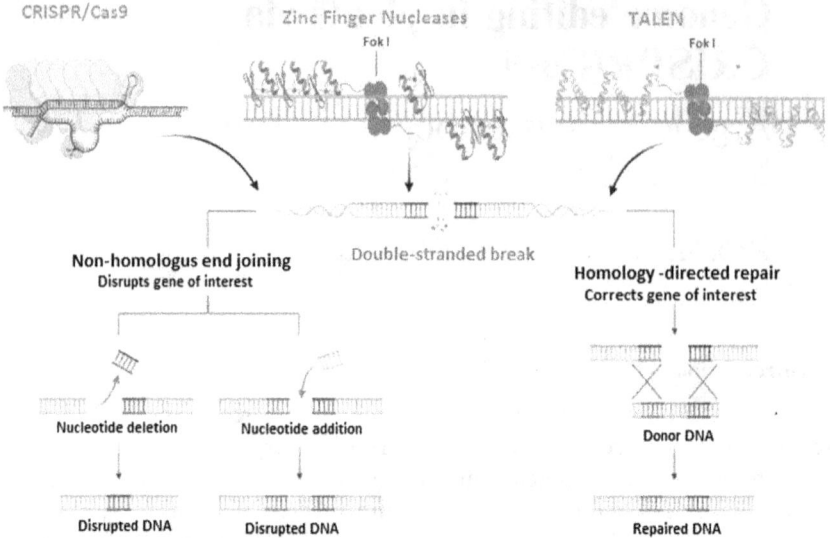

Figure 1.1 Major genome editing technologies and their basic working principle.

TALENs and ZFNs activation of the FokI nuclease domain requires dimerization. Therefore, their construction requires complex protein engineering, which is expensive and thus limits their applicability (Wada et al. 2020). CRISPR are short DNA sequences present in bacteria that provide resistance against viral DNA by giving them a kind of acquired immunity. CRISPR arrays are characterized by four factors: they are found in intergenic regions; they contain numerous short repeats with minimal variation in the sequence; these repeating patterns are separated from one another by distinct "spacer" sequences; and one side of the array contains a common leader sequence spanning hundreds of bases. These arrays were found near a group of genes known as CRISPR-associated or cas genes.

Recent advancements in CRISPR (clustered regularly interspaced short palindromic repeats)-Cas9 (CRISPR-associated protein)-based technologies have revolutionized agricultural science by demonstrating their ability to alter plant species' genomes and opening up new opportunities. The current standard technology for resistance breeding, CRISPR/Cas, can be used to improve crops in a specific way and drastically reduce breeding. It has evolved as a flexible tool that has been used to enhance crucial crop traits for agriculture, including quality, resistance to disease, and herbicide tolerance. The most current developments in CRISPR/Cas9-mediated crop quality improvement are summarized in this chapter.

1.2 History of CRISPR/Cas9

The technique was primarily discovered in bacteria as an adaptive immune response to defend against invasive viral DNA or plasmids (Pourcel et al. 2005). Initially, the protein, Cas, was thought to be involved in the DNA repair mechanism (Makarova et al. 2002). The discovery of spacer sequences being complementary to invading DNA led to the speculation that possibly CRISPR-Cas complex is associated with the immune response in bacteria against invading phage DNA (Mojica et al. 2005). However, with advancements in technology this powerful molecular scissor is not just limited to bacteria, it is now efficiently used in almost every cell type, from archaebacteria and bacteria to plants and animals including humans. CRISPR array occurs on chromosomal as well as plasmid DNA, confirming the fact that they are inclusions of the anti-virus system (Mojica et al. 2005). CRISPR array consists of spacers derived from the viral genome and plasmids, which are used as constituents for recognizing matching viral nucleic acid and neutralizing it (Rath et al. 2015). For activation of CRISPR system, Cas (CRISPR-associated) genes are required that code for the proteins essential for immune response. CRISPR-Cas defense involves three distinct stages (Figure 1.2). The first stage involves the addition of new spacers in CRISPR array system followed by the expression of cas genes, which undergoes transcription to produce a pre-crRNA which is further processed into mature crRNA. Interference is the last stage, where crRNA and Cas complex neutralizes viral DNA.

The use of CRISPR/Cas9 technique can generate transgene-free gene-edited plants that do not undergo any regulatory scrutiny in a limited time period (Ishii and Araki, 2017). The molecular scissors used for editing the genome is Cas9 protein which works by using guide RNA with short sequences complementary to the specific genetic target to break the DNA strand. The technology precisely modifies the genomic sequence by exploiting designer nucleases and cellular DNA repair systems and can precisely improve a plant without the need to incorporate DNA from other species (Voytas, 2013).

1.3 CRISPR/Cas Systems

The major components of CRISPR-Cas9 technology are: Cas9, a protein component and an RNA-dependent DNA endonuclease along with gRNA (guide RNA). The gRNA is an array of CRISPR RNA (crRNA) and trans-activating crRNA (tracrRNA) having 20 nucleotides complementary to the sequence adjoining to a protospacer-adjacent motif (PAM), 5'-NGG-3' (Pellagatti et al. 2015). Processing of crRNA and tracrRNA is linked and requires an endogenous factor RNase III. The complex crRNA: tracrRNA: Cas9 ensemble together to form crRNA-guided endonuclease (Chylinski et al. 2014). Cas9 nuclease generates a double-stranded break under the

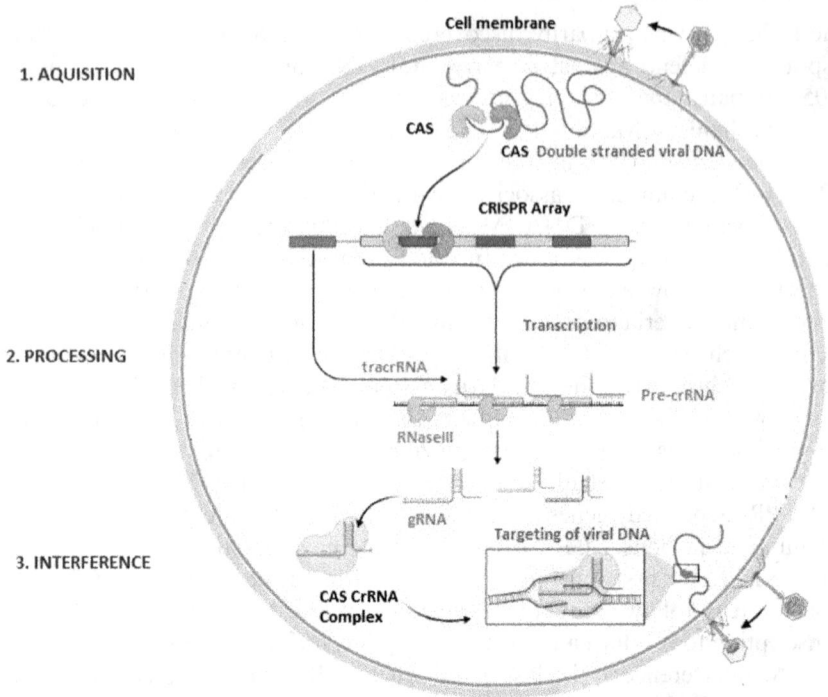

Figure 1.2 Bacterial system of defense against invading phage genome.

influence of gRNA, at ~3 bp upstream of the PAM which can be repaired either by error-prone non-homologous end joining (NHEJ), which usually culminates in gene knockout or, via homology-directed repair (HDR), where an external template DNA is supplemented followed by insertion of the repair template into a target sequence (Sun et al. 2016). The technique relies on DNA-RNA interaction in contrast to other gene-editing technologies (ZFNs and TALENs), where DNA-protein interacts for sequence recognition of target DNA (Wada et al. 2020). The mutants generated via CRISPR-Cas9 engineered tools are more précised with the magnitude of one base (Lu and Zhu, 2017) having a greater efficiency than TALEN and ZFN. Moreover, the technique has no ethical issues; hence it has greater scope for practical research.

CRISPR/Cas complexes (Figure 1.3) are characterized mainly into three types on the basis of associated Cas proteins (Wang, 2019). Type I and Type III CRISPR complex utilizes a large multi-Cas protein complex. In Type I systems, a multi-protein CRISPR RNA (crRNA) complex is utilized for binding and targeting. The crRNA complex scans the invading DNA and is then cleaved by Cas3. In Type III CRISPR system, Cas10 renders into a Cascade-like complex that identifies and cleaves the target DNA (Christopher et al. 2018). Unlike

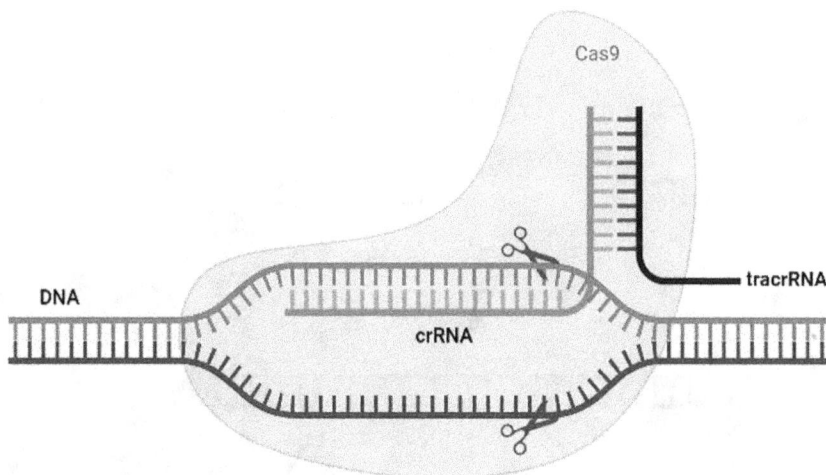

Figure 1.3 A CRISPR/Cas9 complex.
tracrRNA: trans-activating crRNA; crRNA: CRISPR RNA.

Type I and III, Type II requires only Cas9 protein for DNA interference. Both domains of Cas 9 nuclease, Ruvc and HNH are utilized for the identification, binding, and cleavage of the target DNA (Gasiunas et al. 2012).

1.4 Mechanism of CRISPR-Cas9

Generally, gene editing through CRISPR-Cas9 involves the following three steps. First, an invading foreign DNA is recognized, and an immunological memory is established by inserting short segments of sequences (26–72) known as spacer sequences into the host CRISPR assembly, having homology to the target DNA (Wang et al. 2017). Next, Cas9 protein is expressed, and transcription of inserted DNA occurs to create a precursor transcript known as precursor CRISPR RNA (pre-crRNA), which forms a CRISPR/Cas complex along with non-coding trans-activating crRNA (tracrRNA) and Cas9 endonucleases. The RNA complex of pre-crRNA and tracrRNA, called guide RNA (gRNA) confers specificity to the targeting DNA (Fujita and Fujii, 2017). The mature crRNA forms CRISPR/Cas complex along with gRNA guides the Cas9 nuclease to identify the target DNA having complementarities with guide RNA, causing its cleavage and breakdown of Cas9. Repair of blunt cleaved end induced by Cas9 protein is either done by error-prone non-homologous end joining (NHEJ) or highly précised homology-directed repair (HDR) (Figure 1.4). In NHEJ mechanism, ligase IV joins to a double-stranded break which often results in nucleotide insertions/deletion of base pairs can occur, which causes frame-shift mutation and/or gene knockout (Voytas, 2013). In HDR repair mechanism, either an

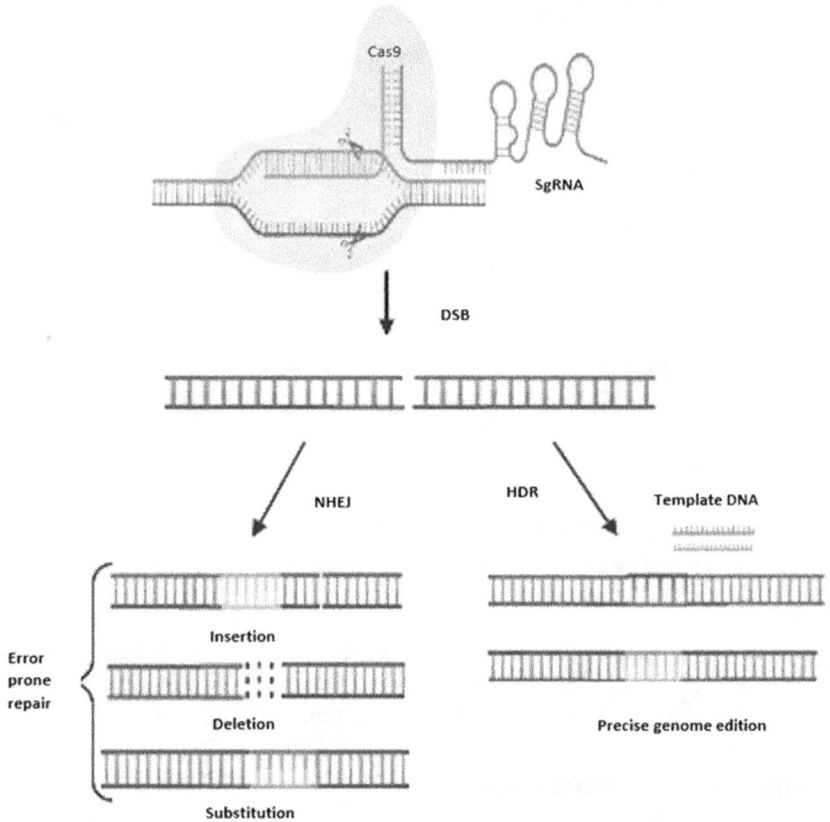

Figure 1.4 CRISPR-Cas9 complex: Cas9 protein and guide RNA (gRNA). This complex generates a double-stranded break, at the highly discrete and targeted site. Repair of blunt cleaved end by Cas9 protein is done either by non-homologous end joining (NHEJ) or highly precise homology-directed repair (HDR).

exogenous or an endogenous repair template DNA (single or double-stranded) is required to facilitate precise gene editing (Hsu et al. 2014). Out of these two, the most common repair mechanism is NHEJ as it is error-prone and can induce several mutations during the process. Thus, gene editing can be achieved with precision through CRISPR/Cas9 technology with limited effects on the genome and avoiding double-stranded breaks (Wang et al. 2016).

1.5 CRISPR-Cas9 in crop improvement

Genomic editing techniques have been found to cause multiple gene knock-outs and thus can prove an extensive tool for efficient crop management in terms of yield, nutrient composition, and plant quality and accelerate crop breeding (Jaganathan et al. 2018). According to Ricroch et al. (2017), gene editing via CRISPR/Cas9 technology has been successfully done in more than 20 crop species for yield and stress management. The purpose of its application includes improving resistance to biotic or abiotic stress, engineering metabolic pathways, and boosting grain production. Genome editing can be used for plant study and the creation of beneficial plants because the introduced mutations are transmitted down to the following generation of plants. Table 1.1 displays instances of crop genes that have been altered by the CRISPR/Cas9 techniques toward crop improvement.

1.5.1 CRISPR-Cas9 in yield-related traits

Crop improvement in rice lines Indica, IR58025B via CRISPR sgRNAs technique was done by Wang et al. (2017) by deleting large fragments of yield-related erect and dense panicle1 (DEP1) gene and created gain-of-function mutants dep1. The mutant alleles conferred improved yield-related traits, such as dense and erect panicles, reduced plant height, etc. In transgenic rice, LAZY1 gene disruption by CRISPR/Cas9 exhibited a pronounced tiller-spreading trait that was observed after tillering stage, which could enhance the yield under certain conditions (Miao et al. 2013). Gn1a and DEP1, yield-regulating genes in rice were edited to produce superior mutant plants having higher yields than natural high-yield alleles (Huang et al. 2018). Cai (2015) utilized CRISPR/Cas9 gene-editing machinery to induce targeted mutagenesis of GmFT2a in soybean, an integrator in the photoperiod flowering pathway. Mutants of GmFT2a exhibited delayed flowering phenotypes. Gene editing with the help of CRISPR/Cas9 improves yield traits in tomato breeding (Soyk et al. 2017). CRISPR-Cas9 generates mutations in SELF-PRUNING5G (SP5G) gene causing flowering suppression and thereby manipulating the photoperiod responses. Thus, engineered mutations in SP5G cause a quick suppression in flower production and early yield. CRISPR/Cas9 targeted genome editing has been used by Dupont Pioneer 2016 to produce and commercialize elite waxy corn hybrids by eliminating maize waxy gene Wx1, encoding for granule-bound starch synthase (GBSS) gene-producing amylose. In hybrids, amylose was not synthesized due to the absence of GBSS gene and produced waxy maize (high amylopectin) with improved digestibility. Thus, new variants so produced would be helpful in maize breeding programs.

1.5.2 CRISPR/Cas9 in improving crop quality

For breeders and consumers, crop quality has always been the most critical concern. Crop quality, however, is an intricate trait influenced by both genetic systems and environmental variables. The market worth of crops has largely been determined by crop quality. The area of crop quality improvement has

Table 1.1 List of crops improved by Crispr/Cas9 gene-editing techniques

Crop	Gene	Trait	Reference
Rice	DEP1	yield-related traits	Wang et al. 2017
Rice	Gn1a and DEP1	Higher yield	Huang et al. 2018
Rice	LAZY1	Tiller-spreading	Miao et al. 2013
Rice	OsERF922	Enhanced rice blast resistance	Xie and Yang, 2013
Rice	DROUGHT AND SALT TOLERANCE (OsDST)	Salinity tolerance	Kumar et al. 2020
Rice	OsmiR535	Salinity tolerance	Yue et al. 2020
Rice	ACETOLACTATE SYNTHASE	Herbicide resistance	Li et al. 2019
Wheat	DEHYDRATION RESPONSIVE ELEMENT BINDING PROTEIN 2 (TaDREB2)	Drought tolerance	Kim et al. 2018
Wheat	ETHYLENE-RESPONSE FACTOR 3 (TaERF3)	Drought tolerance	Kim et al. 2018
Orange	CsLOB1	Resistance to citrus canker	Peng et al. 2017
Maize	ARGOS8	Drought tolerance	Shi et al. 2017
Tomato	SlMAPK3	Tolerance to high-temperature stress	Yu et al. 2019
Tomato	JASMONATE ZIM-DOMAIN 2 (SlJAZ2)	Resistance to Bacterial leaf spot disease	Ortigosa et al. 2019
Cucumber	EUKARYOTIC TRANSLATION INITIATION FACTOR 4E (eIF4E)	Resistance to Cucumber vein yellowing virus	Chandrasekaran et al. 2016
Cucumber	eIF4E	Zucchini yellow mosaic virus	Chandrasekaran et al. 2016
Cucumber	eIF4E	Papaya ring spot mosaic virus-W	Chandrasekaran et al. 2016
Soybean	ALS	Herbicide resistance	Cai, 2015
Soybean	CALCIUM-DEPENDENT PROTEIN KINASE 38 (GmCDPK38)	Resistance to common cutworm	Li et al. 2022.
Wheat	EDR1	Powdery mildew resistance	Zhang et al. 2017
Tomato	SP5G	Early yield	Soyk et al. 2017
Tomato	PECTATE LYASE (SlPL)	Resistance to Grey mold	Silva et al. 2021
Tomato	MILDEW RESISTANT LOCUS O (SlMLO)	Resistance to powdery mildew	Nekrasov et al. 2017
Flax	EPSPS	Herbicide resistance	Sauer et al. 2017
Mushroom	PPO	Anti-browning phenotype	Waltz, 2016
Grapevine	VvMOL3	Resistance to powdery mildew	Wan et al. 2020
Apple	DIPM-1, DIPM-2, DIPM-4	Resistance against fire blight disease	Malnoy et al. 2016

undergone a revolution owing to the CRISPR/Cas9 genome editing technology, which allows for effectively targeted modification. Rice grains can have their amylose content reduced while also having their nutritional worth and flavor enhanced using the CRISPR/Cas9 system. By effectively knocking out the WAXY (Wx) gene, Ma et al. (2015) were able to produce waxy rice mutant plants, which successfully reduced the amylose content of rice grains from 14.6% to 2.6%, these grains with significantly low amylase contents are favored because of their nutritional and culinary value. In order to add aromas to an elite non-aromatic rice cultivar, Ashok Kumar et al. (2020) used the CRISPR/Cas9 technology to target the 7th exon of OsBADH2. Similarly to this, Hui et al. (2021) improved the three-line hybrid rice's grain aroma by focusing on the 7th exon of OsBADH2 in no-fragrant japonica and indica types.

1.5.3 *CRISPR/Cas9 in disease resistance*

It has been estimated that 20–40% of the losses in agricultural production worldwide are attributable to biotic stresses like bacterial, fungal, and viral infections (Walker, 1984). Providing pathogen control is necessary to handle the food crisis. The effect of disease on crop productivity can be reduced by host plant resistance (Borrelli et al. 2018). Through CRISPR/Cas9 knockout technology, researchers have so far produced plants that are extremely immune to bacterial, viral, fungal, and insect diseases (Chen et al. 2019). CRISPR/Cas9 mediated gene engineering was used to develop resistance against the virus in cucumbers (Chandrasekaran et al. 2016). Immunity toward cucumber vein yellowing virus (CVYV), papaya ringspot mosaic virus type-W (PRSV-W), and zucchini yellow mosaic virus (ZYMV) was conferred due to targeted mutation in the eukaryotic translation initiation factor 4E (eIF4E). Peng et al. observed resistance against citrus canker in promoter-disrupted lateral organ boundaries (CsLOB1) gene that targets the effector binding element (EBEPthA4). Deletion of this EBEPthA4 sequence from CsLOB1 alleles in Wanjincheng orange amplifies resistance against canker (Peng et al. 2017). Studies showed that editing two copies of Gh14-3-3d gene grants increased transgene-clean plant resistance to *Verticillium dahliae* infestation in allotetraploid upland cotton (Zhang et al. 2017). Transgenic tomato plants produced via CRISPR-Cas9 mediated gene engineering showed resistance against tomato yellow leaf curl viral interference by targeting replicase loci and viral coat protein (Tashkandi et al. 2018). Zhang et al. (2017) produced powdery mildew-resistant wheat plants by using CRISPR/Cas9 technology to generate Taedr1 wheat plants showing resistance to powdery mildew by concurrent modification of the three homoeologs of wheat-enhanced disease resistance1 (EDR1). Recently developed stable rice lines where *ACETOLACTATE SYNTHASE (OsALS)* gene was replaced by a mutated gene having two specific mutations which confer herbicide resistance (Li et al. 2019)

1.5.4 CRISPR/Cas9 in abiotic stress tolerance

Abiotic stresses that influence plant growth and development, such as salinity, drought, temperature extremes, and heavy metals, can reduce crop output by 50%. (Liu et al. 2022) Overexpression of the auxin-regulated gene involved in organ size (ARGOS) gene, responsible for modulating ethylene signal and down-regulator of ethylene response, was found to improve tolerance against drought in transgenic maize plants. Shi et al. (2017) created two variants (ARGOS8-v1 and ARGOS8-v2) utilizing CRISPR/Cas9 gene editing technique to produce maize hybrid. In *Oryza sativa* stress-related RING finger protein 1 (OsSRFP1) is a negative regulator for a number of abiotic stresses by modulating H_2O_2 level. Knockdown of *OsSRFP1* disrupts H2O2 and increases plant tolerance to abiotic stresses. (Fang et al. 2015). In tomatoes, heat stress tolerance has been achieved by CRISPR/Cas9 tools by targeting SlAGAMOUS-LIKE 6 (SlAGL6) gene, the mutant plants showed improved fruit sets under heat stress conditions (Klap et al. 2017). By altering protein expression patterns and scavenging reactive oxygen species, CRISPR/Cas9 caused SRL1 and SRL2 gene mutations in rice to produce the phenotype of curled leaves and drought tolerance (Liao et al. 2019). In order to create the mutant of the indica mega rice cultivar MTU1010 with wider leaves, reduced stomatal density, and improved leaf water retention capacity under drought stress, Santosh Kumar et al. (2020) edited the OsDST gene using CRISPR/Cas9 (Santosh Kumar et al. 2020). CRISPR/Cas9-mediated knockdown of OsNramp5 and OsLCT1 lowers cadmium (Cd) deposition in rice (Songmei et al. 2019). Rice grains with OsNRAMP1 knocked out by CRISPR/Cas9 had reduced levels of Cd and Pb (Chang et al. 2020).

1.5.5 CRISPR/Cas 9 in moderating self-incompatibility and Hybrid breeding

CRISPR/Cas9-mediated gene engineering can be efficiently exploited for manipulating male sterility in hybrid seed production technology. In hexaploidy wheat, CRISPR/Cas 9 has been exploited for the generation of nuclear male sterility. The *Ms1* knockout mutants produced through the introduction of targeted biallelic frameshift mutations, where male sterile wheat lines thatare considered as an agronomically important trait in hybrid seed production technology (Okada et al. 2019). In the tomato hybrid seed production program, male sterile lines were produced using this technology thereby reducing its cost and ensuring high purity of tomato varieties. Jung et al. (2020) successfully produced male-sterile tomato plants by modifying SlMS10 gene through CRISPR/Cas9 engineering. Self-compatibility was achieved in diploid potato lines via CRISPR/Cas9 tools by utilizing dual single-guide RNA (sgRNA) for targeted biallelic frameshift mutations in the regions of S-RNase gene to create targeted knockouts (KOs). Self-compatibility was attained in nine S-RNase KO T_0 lines and transmitted to T_1 progeny (Enciso-Rodriguez et al. 2019). By deleting TMS5 in rice, thermogenic male sterility has recently been created using the CRISPR-Cas system (Barman et al. 2019). Similarly, TaNP1 gene, which controls the growth of the tapetum and the production of pollen exine in wheat, was silenced at three homologus loci, resulting in male-sterility in wheat plants (Li 2020).

1.6 Conclusion and future prospects

After being discovered as a component of the bacterial immune system CRISPR/Cas9 array is now recognized as a potent machinery for gene editing. The technique has been successfully utilized to create crop varieties with desired traits and promising agronomic performance. It is a gene engineering technology through which scientists can insert or delete (indel) targeted DNA into cells with great precision. CRISPR/Cas9 generates a DSB in the DNA and a precise change can be made in the DNA by inserting tiny bits of desired DNA and genetic variability can be induced in an organism. The future of genome editing is more likely to depend on the CRISPR-Cas9 tool. The technology is no longer just limited as a molecular scissor; rather it has turned out as an ultra-modern tool for the transition of one nucleotide to another and thus can alter the epigenetic modification of the targeted DNA. Regulatory policies that label gene-edited products as GMOs may forbid their utilization in some countries, despite the overwhelming advantages of the CRISPR/Cas system for crop development (Callaway, 2018). However, after removing the transgenic label, the mutants produced by CRISPR/Cas are precisely the same as those produced by natural mutation or conventional mutagenesis. The future of genome editing with CRISPER/Cas9 will be more promising if additionally, new delivery systems would be implicated such as carbon nanotubes (CNT) for delivering CRISPR/Cas 9 system in different organisms, especially plants where they can enter the cells without much mechanical support and can cause least cellular injury, low toxicity, and yield high transformation efficiencies (Zhu et al. 2020). Vertical arrays of nanowires or nanotubes can provide a new approach for gene delivery into hard-to-transfect cells such as stem cells and primary cells. Other nano-derivatives such as layered double hydroxides and mesoporous silica NPs also can be explored as delivery vehicles. A novel delivery method "pollen magnetofection", for nucleic acid transport, using magnetic NPs was done by (Zhang et al. 2019) for exogenous delivery of DNA into pollen grains in cotton plants, whereby it was successfully incorporated into the genome and was expressed stably in succeeding generations. This smart delivery approach can be utilized in CRISPR/Cas9 cargoes to create gene modification without utilizing tissue culture techniques (Lew et al. 2020).

Acknowledgment

The author wants to express sincere thanks to the Dean, Faculty of Agricultural Sciences, Aligarh Muslim University, Aligarh for furnishing all necessary facilities required to perform the above research.

Funding: This research did not receive any specific funding.

The author declares there is no conflict of interest.

References

Ashok kumar S, Jaganathan D, Ramanathan V, Rahman H, Palaniswamy R, Kambale R, et al., 2020. Creation of novel alleles of fragrance gene *OsBADH2* in rice through CRISPR/Cas9 mediated gene editing. *PLoS One* 15, e0237018. doi: 10.1371/journal.pone.0237018

Barman HN, Sheng Z, Fiaz S, Zhong M, Wu Y, Cai Y, Wang W, Jiao G, Tang S, Wei X, Hu P, 2019. Generation of a new thermo-sensitive genic male sterile rice line by targeted mutagenesis of TMS5 gene through CRISPR/Cas9 system. *BMC Plant Biol.* 19, 109. doi: 10.1186/s12870-019-1715-0

Borrelli VMG, Brambilla V, Rogowsky P, Marocco A, Lanubile A, 2018. The enhancement of plant disease resistance using CRISPR/ Cas9 technology. *Front. Plant Sci.* 9, 1245. doi:10.3389/fpls.2018.01245

Butler NM, Baltes NJ, Voytas DF, Douches DS, 2016.Geminivirus-mediated genome editing in potato (*Solanum tuberosum* L.) using sequence-specific nucleases. *Front. Plant Sci.* 7, 1045.

Cai Y, 2015. CRISPR/Cas9-mediated genome editing in soybean hairy roots. *PLoS One* 10 (8), e0136064.

Callaway E, 2018. CRISPR plants now subject to tough GM laws in European Union. *Nature* 560 (7716), 16. doi: 10.1038/d41586-018-05814-6

Chandrasekaran J, Brumin M, Wol D, Leibman D, Klap C, Pearlsman M, et al., 2016. Development of broad virus resistance in non-transgenic cucumber using CRISPR/Cas9 technology. *Mol. Plant Pathol.* 17, 1140–1153. doi: 10.1111/mpp.12375

Chang JD, Huang S, Yamaji N, Zhang W, Ma JF, Zhao FJ, 2020. OsNRAMP1 transporter contributes to cadmium and manganese uptake in rice. *Plant Cell Environ.* 43(10), 2476–2491. doi:10.1111/pce.13843

Chen K, Wang Y, Zhang R, Zhang H, Gao C, 2019. CRISPR/Cas genome editing and precision plant breeding in agriculture. *Annu. Rev. Plant Biol.* 70, 667–697. doi:10.1146/annurev-arplant-050718-100049

Christopher AL, Jason CH, James PC, Jerilyn AT, 2018. Delivering CRISPR: a review of the challenges and approaches. *Drug Deliv.* 25(1), 1234–1257.

Chylinski K, Makarova KS, Charpentier E, Koonin EV, 2014. Classification and evolution of type II CRISPR-Cas systems. *Nucleic Acids Res.* 42, 6091–6105.

Enciso-Rodriguez, F, Manrique-Carpintero, NC, Nadakuduti, SS, Buell, CR, Zarka, D, Douches, D, 2019. Overcoming self-incompatibility in diploid potato using CRISPR-Cas9. *Front. Plant Sci.* 10, 376.

Fang H, Meng Q, Xu J, Tang H, Tang S, Zhang H, Huang J, 2015. Knock-down of stress inducible OsSRFP1 encoding an E3 ubiquitin ligase with transcriptional activation activity confers abiotic stress tolerance through enhancing antioxidant protection in rice. *Plant Mol. Biol.* 87, 441–458.

Fujita T, Fujii, H, 2017. New Directions for Epigenetics: Application of Engineered DNA-Binding Molecules to Locus-Specific Epigenetic Research, Eds: Trygve O

Tollefsbol, *Handbook of Epigenetics (Second Edition)*, Academic Press, 635–652, ISBN 9780128053881

Gasiunas G, Barrangou R, Horvath P, Siksnys V, 2012. Cas9-crRNA ribonucleoprotein complex mediates specific DNA cleavage for adaptive immunity in bacteria. *Proc. Natl. Acad. Sci. USA* 109, E2579–E2586.

Hsu PD, Lander ES, Zhang F, 2014. Development and applications of CRISPR-Cas9 for genome engineering. *Cell* 157, 1262–1278.

Huang L, Zhang R, Huang G, Li Y, Melaku G, Zhang S, et al., 2018. Developing superior alleles of yield genes in rice by artificial mutagenesis using the CRISPR/Cas9 system. *Crop J.* 6, 475–481.

Hui S, Li H, Mawia AM, Zhou L, Cai J, Ahmad S, et al., 2021. Production of aromatic three-line hybrid rice using novel alleles of BADH2. *Plant Biotechnol. J.* 20, 59–74. doi: 10.1111/pbi.13695

Ishii T, Araki M, 2017. A future scenario of the global regulatory landscape regarding genome-edited crops. *GM Crop. Food*, 8, 44–56.

Jaganathan D, Ramasamy K, Sellamuthu G, Jayabalan S, Venkataraman G, 2018. CRISPR for crop improvement: an update review. *Front. Plant Sci.*, 9, 985.

Jung JY, Kim DH, Lee JH, 2020. Knockout of SlMS10 gene (Solyc02g079810) encoding bHLH transcription factor using CRISPR/Cas9 system confers male sterility phenotype in Plants. *Plants* 9 (9), 1189.

Kim D, Alptekin B, Budak H, 2018. CRISPR/Cas9 genome editing in wheat. *Funct. Integr. Genomics* 18 (1), 31–41. doi: 10.1007/s10142-017-0572-x

Klap C, et al., 2017. Tomato facultative parthenocarpy results from SlAGAMOUS-LIKE 6 loss of function. *Plant Biotechnol. J.* 15, 634–647.

Kumar VV Santosh, Verma RK, Yadav SK, Yadav P, Watts A, Rao MV, Chinnusamy V, 2020. CRISPR-Cas9 mediated genome editing of drought and salt tolerance (OsDST) gene in indica mega rice cultivar MTU1010. *Physiol. Mol. Biol. Plants* 26, 1099–1110.

Lew TTS, Park M, Wang Y, Gordiichuk P, Yeap WC, Mohd Rais SK, Kulaveerasingam H, Strano MS, 2020. Nanocarriers for transgene expression in pollen as a plant biotechnology tool. *ACS Mater. Lett.* 2, 1057–1066.

Li J, Wang Z, He G, Ma L, Deng XW, 2020. CRISPR/Cas9-mediated disruption of TaNP1 genes results in complete male sterility in bread wheat. *J. Genet. Genomics* 47 (5), 263–272. doi: 10.1016/j.jgg.2020.05.004

Li S, Li J, He Y, Xu M, Zhang J, Du W, et al. 2019. Precise gene replacement in rice by RNA transcript-templated homologous recombination. *Nat. Biotechnol.* 37, 445–450.

Li X, Hu D, Cai L, Wang H, Liu X, Du H, et al. 2022. Calcium dependent protein kinase38 regulates flowering time and common cutworm resistance in soybean. *Plant Physiol.* 190, 480–499. doi: 10.1093/plphys/kiac260

Liao S, Qin X, Luo L, Han Y, Wang X, Usman B, et al., 2019. CRISPR/ Cas9-Induced mutagenesis of semi-rolled leaf1, 2 confers curled leaf phenotype and drought tolerance by influencing protein expression patterns and ROS scavenging in rice (*Oryza sativa* L.). *Agronomy* 9 (11), 728. doi: 10.3390/agronomy9110728

Liu, Z, Ma, C, Hou, L, Wu, X, Wang, D, Zhang, L, & Liu, P (2022). Exogenous SA Affects Rice Seed Germination under Salt Stress by Regulating Na+/K+ Balance and Endogenous GAs and ABA Homeostasis. *International Journal of Molecular Sciences*, 23 (6), 3293. doi: 10.3390/ijms23063293

Lu Y, Zhu JK, 2017. Precise editing of a target base in the rice genome using a modified CRISPR/Cas9 system. *Mol. Plant* 10, 523–525.

Ma X, Zhang Q, Zhu Q, Liu W, Chen Y, Qiu R, Wang B, Yang Z, Li H, Lin Y, et al., 2015. A robust CRISPR/Cas9 system for convenient, high-efficiency multiplex genome editing in monocot and dicot plants. *Mol. Plant* 8, 1274–1284.

Makarova KS, Aravind L, Grishin NV, Rogozin IB, Koonin EV 2002. A DNA repair system specific for thermophilic Archaea and bacteria predicted by genomic context analysis. *Nucleic Acids Res.* 30, 482–496.

Malnoy M, Viola R, Jung MH, Koo OJ, Kim S, Kim JS, Velasco R, Nagamangala, KC 2016. DNA-free genetically edited grapevine and apple protoplast using CRISPR/Cas9 ribonucleoproteins. *Front. Plant Sci.* 7, 1904.

Miao J, Guo D, Zhang J, Huang Q, Qin G, Zhang X, et al., 2013. Targeted mutagenesis in rice using CRISPR-Cas system. *Cell Res.* 23, 1233–1236.

Mojica FJ, Díez-Villaseñor C, García-Martínez, J, et al. 2005. Intervening sequences of regularly spaced prokaryotic repeats derive from foreign genetic elements. *J. Mol. Evol.* 60, 174–182.

Nekrasov V, Wang C, Win J, Lanz C, Weigel D, Kamoun S 2017. Rapid generation of a transgene-free powdery mildew resistant tomato by genome deletion. *Sci. Rep.* 7 (1), 482. doi: 10.1038/s41598-017-00578-x

Okada A, Arndell T, Borisjuk N, et al., 2019. CRISPR/Cas9-mediated knockout of Ms1 enables the rapid generation of male-sterile hexaploid wheat lines for use in hybrid seed production. *Plant Biotechnol. J.* 17 (10), 1905–1913.

Ortigosa A, Gimenez-Ibanez S, Leonhardt N, Solano, R, 2019. Design of a bacterial speck resistant tomato by CRISPR/Cas9-mediated editing of SlJAZ2. *Plant Biotechnol. J.* 17 (3), 665–673. doi: 10.1111/pbi.13006

Pellagatti A, Dolatshad H, Valletta S, Boultwood J, 2015. Application of CRISPR/ Cas9 genome editing to the study and treatment of disease. *Arch. Toxicol.* 89, 1023–1034. doi: 10.1007/s00204-015-1504-y

Peng A, Chen S, Lei T, Xu L, He Y, Wu L, et al. 2017. Engineering canker-resistant plants through CRISPR/Cas9-targeted editing of the susceptibility gene CsLOB1 promoter in citrus. *Plant Biotechnol. J.* 15, 1509–1519.

Pioneer, DuPont Announces Intentions to Commercialize First CRISPR-Cas Product. https://www.prweb.com/releases/dupont-pioneer-seed/crispr-cas-corn/prweb13349828.htm. Press Release. 18 April 2016. https://www.prweb.com/releases/dupont-pioneer-seed/crispr-cas-corn/prweb13349828.htm

Pourcel C, Salvignol G, Vergnaud G, 2005 CRISPR elements in Yersinia pestis acquire new repeats by preferential uptake of bacteriophage DNA and provide additional tools for evolutionary studies. *Microbiology* 151,653–663.

Rath D, Amlinger L, Rath A, Lundgren, 2015. The CRISPR-Cas immune system: biology, mechanism and applications. *Biochimie* 117, 119–128.

Ricroch, A, Clairand P, Harwood W, 2017. Use of CRISPR systems in plant genome editing: toward new opportunities in agriculture. *Emerg. Top. Life Sci.* 1, 169–182. doi: 10.1042/etls20170085

Sauer N, Narvaez-Vasquez J, Mozoruk J, Miller R, Warburg Z, Woodward M, 2017. Oligonucleotide-mediated genome editing provides precision and function to engineered nucleases and antibiotics in plants. *Plant Physiol.* 170(4), 1917–1928.

Shi J, Gao H, Wang H, Lafitte HR, Archibald RL, Yang M, et al. 2017. ARGOS8 variants generated by CRISPR-Cas9 improve maize grain yield under field drought stress conditions. *Plant Biotechnol. J.* 15, 207–216.

Silva CJ, Van den Abeele C, Ortega-Salazar I, Papin V, Adaskaveg JA, Wang D, et al., 2021. Host susceptibility factors render ripe tomato fruit vulnerable to fungal disease despite active immune responses. *J. Exp. Bot.* 72(7), 2696–2709. doi:10.1093/jxb/eraa601

Songmei L, Jie J, Yang L, Jun M, Shouling X, Yuanyuan T, et al., 2019. Characterization and evaluation of OsLCT1 and OsNramp5 mutants generated through CRISPR/Cas9-mediated mutagenesis for breeding low Cd rice. *Rice Sci.* 26, 88–97. doi: 10.1016/j.rsci.2019.01.002

Soyk S, Müller NA, Park SJ, Schmalenbach I, Jiang K, Hayama R, Zhang L, Van Eck J, Jiménez-Gómez JM, Lippman ZB, 2017. Variation in the flowering gene SELF PRUNING 5G promotes day-neutrality and early yield in tomato. *Nature Genet.* 49 (1), 162–168.

Sun Y, Zhang X, Wu C, He Y, Ma Y, Hou H, Guo X, Du W, Zhao Y, Xia L, 2016. Engineering herbicide-resistant rice plants through CRISPR/Cas9-mediated homologous recombination of acetolactate synthase. *Mol. Plant* 9, 628–631.

Tashkandi M, Ali Z, Aljedaani F, Shami A, Mahfouz MM, 2018. Engineering resistance against Tomato yellow leaf curl virus via the CRISPR/Cas9 system in tomato. *Plant Signal. Behav.* 13, e1525996

Voytas DF, 2013. Plant genome engineering with sequence-specific nucleases. *Annu. Rev. Plant. Biol.* 64,327–350.

Wada N, Ueta R, Osakabe Y, et al., 2020. Precision genome editing in plants: state-of-the-art in CRISPR/Cas9-based genome engineering. *BMC Plant Biol* 20, 234.

Walker PT, 1984. "Quantification and economic assessment of crop losses due to pests, diseases and weeds", in: Advancing Agricultural Production in Africa: Cabs First Scientific Conference, Arusha, Tanzania, 12–18 February.

Waltz E, 2016. Gene-edited CRISPR mushroom escapes US regulation. *Nat. News* 532(7599), 293.

Wan DY, Guo Y, Cheng Y, Hu Y, Xiao S, Wang Y, et al., 2020. CRISPR/Cas9-mediated mutagenesis of VvMLO3 results in enhanced resistance to powdery mildew in grapevine (*Vitis vinifera*). *Hortic. Res.* 7, 116. doi: 10.1038/s41438-020-0339-8

Wang H, La Russa M, Qi LS, 2016. CRISPR/Cas9 in genome editing and beyond. *Annu. Rev. Biochem.* 85, 227–264.

Wang T, Zhang H, Zhu H, 2019. CRISPR technology is revolutionizing the improvement of tomato and other fruit crops. *Hortic. Res.* 6, 77.

Wang Y, Geng L, Yuan M, Wei J, Jin C, Li M, Yu K, Zhang Y, Jin H, Wang E, Chai Z, Fu X, Li X, 2017. Deletion of a target gene in Indica rice via CRISPR/Cas9. *Plant Cell Rep.* 36(8), 1333–1343.

Xie K, Yang Y, 2013. RNA-guided genome editing in plants using a CRISPR–Cas system. *Mol. Plant,* 6(6), 1975–1983.

Yu W, Wang L, Zhao R, Sheng J, Zhang S, Li R, et al., 2019. Knockout of SlMAPK3 enhances tolerance to heat stress involving ROS homeostasis in tomato plants. *BMC Plant Biol.* 19(1), 354. doi: 10.1186/s12870-019-1939-z

Yue E, Cao H, Liu B. 2020. OsmiR535, a potential genetic editing target for drought and salinity stress tolerance in *Oryza sativa. Plants (Basel)* 9(10), E1337. doi: 10.3390/plants9101337

Zhang R, Meng Z, Abid MA, Zhao X, 2019. Novel pollen magnetofection system for transformation of cotton plant with magnetic nanoparticles as gene carriers. *Methods Mol. Biol.* 1902, 47–54.

Zhang Y, Bai Y, Wu G, Zou S, Chen Y, Gao C, Tang D, 2017. Simultaneous modification of three homoeologs of TaEDR1 by genome editing enhances powdery mildew resistance in wheat. *Plant J.* 91, 714.

Zhang Z, Ge X, Luo, Wang P, Fan Q, Hu G, et al., 2018. Simultaneous editing of two copies of Gh14-3-3d confers enhanced transgene-clean plant defense against *Verticillium dahliae* in allotetraploid upland cotton. *Front. Plant Sci.* 842.

Zhu H, Li C & Gao C, 2020. Applications of CRISPR–Cas in agriculture and plant biotechnology. *Nat. Rev. Mol. Cell. Biol.* 21, 661–677.

Abbreviation

Cas	CRISPR-associated protein
CRISPR/Cas	Clustered regularly interspaced short palindromic repeat/ CRISPR associated proteins
crRNA	CRISPR RNA
DNA	Deoxyribonucleic acid
DSB	Double-strand break
dsDNA	Double-stranded DNA
gRNA	Guide RNA
NHEJ	Non-homologous end joining
GBSS	Granule bound starch synthase
sgRNA	Single guide RNA
HDR	Homology-directed repair
PAM	Protospacer adjacent motif
TALENs	Transcription activator like effector nucleases
ZFNs	Zinc-finger nuclease
tracrRNA	Trans-activating CRISPR RNA
ZYMV	Zucchini yellow mosaic virus
PRSV-W	Papaya ringspot mosaic virus type-W
CVYV	Cucumber vein yellowing virus

2 Genome editing by site-directed nucleases and its applications in producing climate change resilient crop plants

Abdelfattah Badr and Hanaa H. El-Shazly

2.1 Introduction

2.1.1 Climate change definition and causes

Climate change is regarded as extreme changes in weather patterns such as temperature and rainfall. These changes have occurred naturally due to variations in the solar cycle in Earth's long history. However, human activities have become a major cause of climate change in recent centuries (www.un.org/en/climatechange/what-is-climate-change). Climate changes are due to an increase in the rate of emissions of greenhouse gases (GHGs) that lead to global warming and a global rise in the Earth's temperature because of an increase in the rate of absorption of infrared radiation. The names and relative proportions of these gases are shown in Figure 2.1A, although their proportion varies in different estimates (Morice et al. 2012). However, all estimates agree that CO_2 is the major GHG followed by methane, nitrous acid, and fluorocarbon gases. The main causes of global warming GHGs include electricity and heat, agriculture, transportation, industry, and buildings (see Figure 2.1B).

- Fossil fuels derived from coal, petroleum, and gases are the most important sources of GHGs. Carbon dioxide (CO_2) in fossil fuels and land, nitrous oxide (N_2O), and methane (CH_4) are the major GHGs produced in the atmosphere.
- Cutting down forests reduces the storing of CO_2 in the trees as carbohydrates in the forest's vegetation, leading to increases in its amount in the air.
- Agriculture is the second most important source of GHGs. Fertilizers containing nitrogen produce nitrous oxide emissions, and livestock farming produces a large amount of methane.
- Fluorinated gases emitted from equipment and transport vehicles have a strong warming effect.
- Black carbon sources such as forests, savanna fires, and volcanoes have been increasingly regarded as a major cause of GHGs immersion and contribute to global warming.

DOI: 10.4324/9781003382102-2

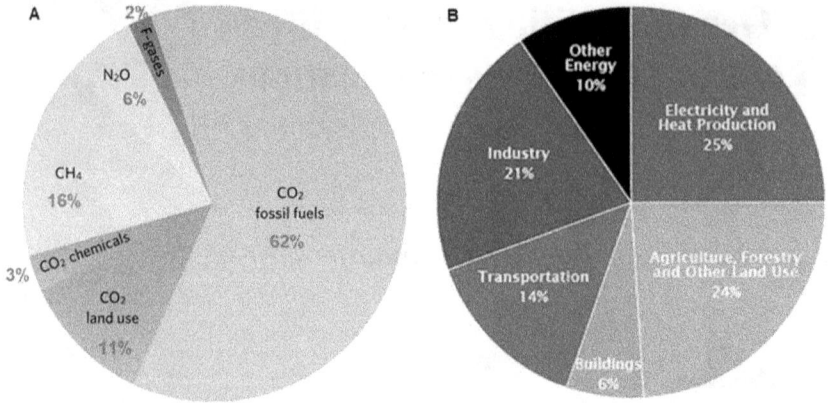

Figure 2.1 (A) Graph illustrating the global average concentrations of GHGs CO_2, NH_4, and NO_2. (B) Graph illustrating the causes of global warming GHGs including electricity and heat, agriculture, transportation, industry, and buildings.

The increase in the temperature of Earth's atmosphere and oceans is generally projected as global warming. Historical estimates indicate that the temperature rise is correlated with the changes in CO_2 concentration (Lacis et al. 2010) which represents more than 75% of GHGs. Changes in global average CO_2 concentration and Antarctic temperature over the past 800,000 years are shown in Figure 2.2. However, a lag between the concentration of CO_2 and the final temperature rise was reported, indicating that when the global and regional atmospheric CO_2 concentration stabilized, a global temperature response may be delayed after emission mitigation (Mitchell et al. 2000).

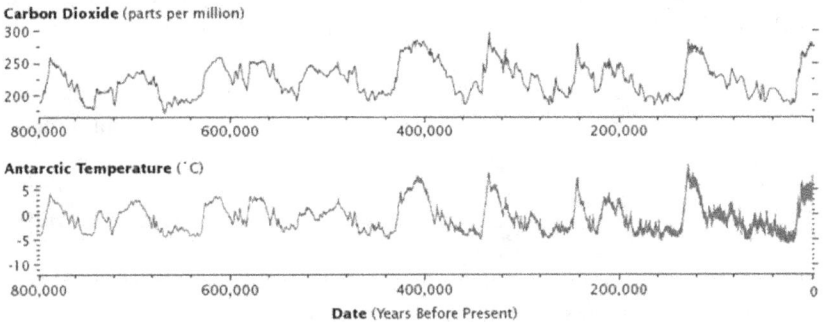

Figure 2.2 Graph illustrating the changes in global average concentrations of CO_2 in the atmosphere and the Antarctic temperature over the past 800,000 years. (Graphs by Robert Simmon, using data from Lüthi et al. 2008 and Jouzel et al. 2007). https://earthobservatory.nasa.gov/features/CarbonCycle/page4.php.

Limiting Earth's warming to about 1.5°C may be out of reach unless a deep reduction in GHGs emissions is controlled (Samset et al. 2020).

2.1.2 Consequences of climate change

Problems associated with climate change have several consequences including temperature rise, drought, sea level rise, ocean acidification, and threats to safe access to water sources and biodiversity. An annual increase of about 3 mm increase in sea level per year has been observed from 2000 to 2020 (Schleussner et al. 2016). However, a change in the sea levels with a local variation in sea level rise is expected (Magnan et al. 2016). However, there are uncertainties regarding the future rise in the global average sea level. Depending on the methodology, future projections about sea level rise range from 52 to 190 cm by 2100. These uncertainties will greatly affect the rise in the level of the Mediterranean Sea due to its connection to the Atlantic Ocean through the Strait of Gibraltar (Meyssignac et al. 2011). Freshwater resources are expected to witness a significant decrease at a rate of between 2% and 15% at an increase in temperature of 2.0°C, which is considered one of the largest rates of decline worldwide and could lead to a reduction in available water resources and insufficient drinking water (Meier et al. 2014). Sea level rise also affects the water cycle under the surface of the coastal areas, which may decrease the freshwater areas due to the reduction of freshwater (Marchane et al. 2017).

International efforts to control the impacts of climate change have adopted a global agreement by 196 United Nations Parties on the Convention Framework on Climate Change at the COP 2015 meeting in Paris, France. The Paris Accords aim to reduce global warming to 1.5°C compared to the 18th-century levels. In agreement with these accords, the UN Intergovernmental Panel on Climate Change (IPCC) in 2018 considered a temperature rise of 1.5°C as the committed temperature goal, which must be achieved by widespread action to prevent global warming and to achieve no emissions by 2050. A special report released by the IPCC in 2019 focused on climate change connections with food security and sustainable land management (Smith et al. 2019). The IPCC report also dealt with the impacts of climate change on different ecosystems including coastal regions, oceans, and mountains, linking the well-being of these ecosystems and the human communities that depend on these ecosystems. The limited land, water resources, and high rates of the human population increase may reduce the domestic food supply systems, making them insufficient to the demand for the dimensions of food security, through access, (nutrition), utilization, and availability (security).

2.2 Introduction to the genome, genes, and gene editing

The word genome refers to the DNA of all an organism's genes whereas genes are parts of the genome that control the traits which may be visible as external features or internal structures or molecular pathways. The genome may be regarded as a word text and the gene can be considered as a sentence in the whole text. Genes determine heritable characteristics of the features of the organism. Spontaneous changes to the genes through natural selection, hybridization, or mutation are the basis of evolution. Gene editing is a new technology recently developed to modify DNA precisely and efficiently within cells, by making changes to specific DNA sequences by adding, removing, or substitution of DNA bases in the genome that leads to changes in the characteristics of a cell or an organism (Gaj et al. 2013). Gene editing differs from genetic engineering methods which involve introducing DNA from one species to another to produce genetically modified organisms (GMOs). Unlike genetic engineering techniques, gene editing does not involve the insertion of foreign genes from other species. Gene editing involves modifying genes in a way similar to what happens in nature by spontaneous mutations but much faster (Grohmann et al. 2019; Gu et al. 2021). The genome-editing changes generate small insertions of DNA bases (indels) and base pair and specific short-sequence changes by homology-directed repairs (HDR). These resemble natural genome variants (Grohmann et al. 2019) and produce mutants that cannot be considered GMOs and may not be controlled by GMOs regulations in many countries (Kim and Kim, 2016; Turnbull et al. 2021).

2.3 Gene editing systems

Genome editing is achieved by nuclease enzymes known as site-directed nucleases that bind to DNA domains by RNA molecules at a target site and induce DNA double-strand breaks (DSBs) by specific endonucleases that recognize genomic sites exploiting the DSBs repair mechanisms (Carroll, 2014). The most common zinc-finger nucleases (ZFNs) were described by Urnov et al. (2005), transcription activator-like effector nucleases (TALENs) were described by Christian et al. (2010) and Li et al. (2011), and clustered regularly interspaced short palindromic repeats (CRISPR) was described by Jinek et al. (2012), Cong et al. (2013), and Mali et al. (2013). Variants of nucleases can also be engineered for genome editing by DNA single-strand breaks (Rees and Liu, 2018) or epigenome action involving no DNA breakage and repair (Holtzman and Gersbach, 2018). In addition, the direct delivery of Cas9-gRNA RNPs was reported as a genome-editing system by Toda et al. (2019), and a transgene-free gene editing system involving three major strategies was described by Gu et al. (2021).

2.3.1 The ZFN system

The ZFNs are ZF proteins combined with the FokI restriction endonuclease and non-specific DNA cleavage domain. This enzyme induces DSBs to stimulate DNA damage response pathways at a specific genomic site (Gaj et al. 2013). The ZFNs were adopted as precise gene replacement via directed

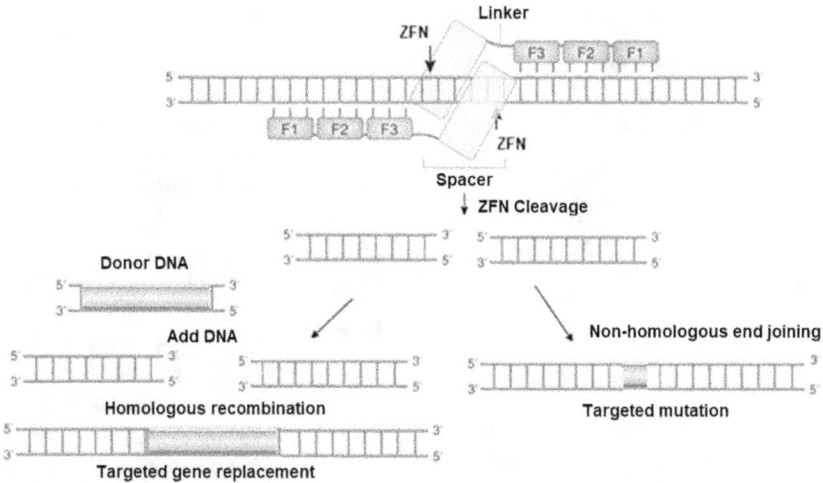

Figure 2.3 Illustration of ZFN dimers induced targeted DNA DSBs that stimulate DNA damage response. The binding specificity of the designed zinc-finger domain directs the ZFN to a specific genomic site. The repair can proceed by HR using the donor DNA as a template leading to target gene replacement. Alternatively, the break can be repaired by NHEJ, leading to mutations at the cleavage site. These may be deletions, insertions, and substitutions.

repair homology (Chen et al. 2019). Figure 2.3 illustrates how a pair of ZFNs induces a DSB, using a donor DNA template, targeting genes by homologous recombination (HR). The distance from the original break determines the length of inserted donor sequence. Alternatively, non-homologous end joining (NHEJ) can repair the break mutations caused by deletions, insertions, and base substitutions. Remarkably high frequencies of ZFN-induced DSB and DNA repair lead to targeted mutagenesis and replacement of selected genes (Carroll, 2011). Despite the early progress of ZFN nucleases in DNA repair and replacement, gene editing using ZFN-initiated events may have limited applications in gene editing in the future.

2.3.2 The TALENs system

The TALENs are fusions of DNA domains derived from TALE proteins combined with the FokI enzyme cleavage domain. They make possible the use of the transcription activator-like effector for DNA targeting. TALEN contains 33–35 amino acid multiple repeats, and each can recognize a single base pair (Boch et al. 2009; Moscou and Bogdanove 2009). TALENs induce DSBs that cause pathways alterations that can make changes to specific sequences of DNA (Gaj et al. 2013). However, the TALENs use a simple DNA-binding TALE repeat domain to target individual bases and make

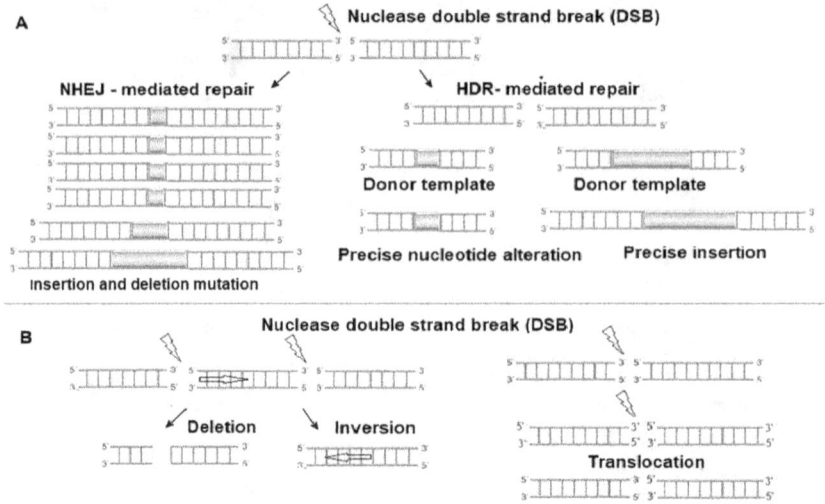

Figure 2.4 Illustrative diagram of the genome editing using TALENs: The nuclease of the FokI enzyme and Tale proteins induce DSBs in a gene locus. These breaks can be repaired by the mediation of either NHEJ or HDR. The NHEJ leads to variable lengths of insertion and deletion mutations while the HDR guides double-stranded DNA 'donor templates' which can lead to the introduction of precise nucleotide alteration or insertions. B. Introduction of two nuclease-induced DSBs in the cis position on the same chromosome can lead to deletion or inversion and the nuclease-induced DSBs on two chromosomes can lead to translocations.

them more attractive because they can be easily and rapidly recognized. Cong et al. (2012) refined the TALEN framework by proposing that the nuclease activities can be enhanced by truncating the carboxy-terminal derived by TALE-derived sequence. The high rates of cleavage activity of TALENs made them attractive in gene editing and were further encouraged by non-specialist researchers (Joung and Sander, 2013). The steps of using TALENs in genome editing are done through nuclease DSBs in a gene locus and its mediated repair by either NHEJ or HDR as illustrated in Figure 2.4.

2.3.3 *The CRISPR-Cas9 system*

The term CRISPR was first formulated by Jansen et al. (2002) as a reference to genes associated with DNA repeats in bacteria. According to Barrangou and Horvath (2012), CRISPR is part of a bacterial defense system that plays an essential role in the bacteria's adaptive immunity. Mojica et al. (2005) demonstrated that CRISPR spacers sequences have an extrachromosomal origin derived from plasmids or viruses that could have infected the bacteria, a finding that explained why viruses cannot invade bacteria having DNA spacers corresponding to parts of the virus genomes. In this respect, these

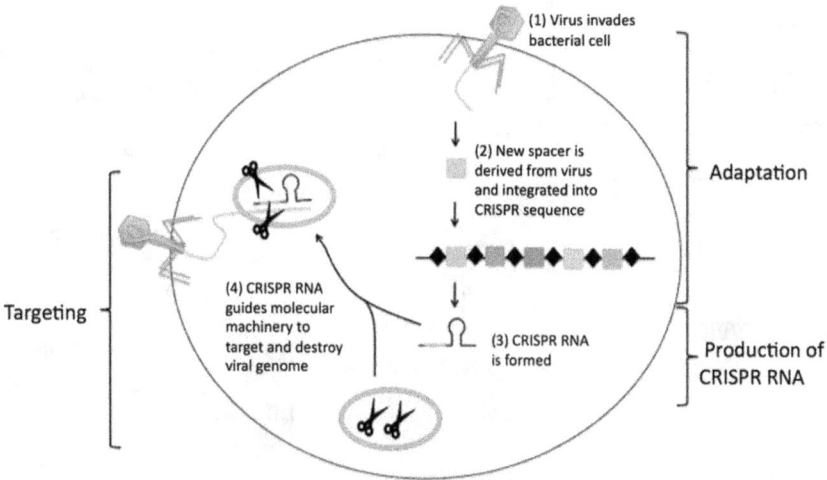

Figure 2.5 The steps of CRISPR-mediated immunity of bacterial cells when invaded by a phage. The CRISPR sequence is transcribed and processed to generate short CRISPR RNA molecules (black diamonds) which guide bacterial molecular machinery to a matching target sequence in the invading page. Figure adapted from Molecular Cell 54, 2014.

spacers serve as a genetic memory of previous infections of bacteria by viruses. CRISPR is associated with Cas genes which encode helicases and nucleases (Barrangou and Marraffini, 2014). The steps of bacterial cell immunity against virus infection mediated by CRISPR are illustrated in Figure 2.5. In brief, the CRISPR sequence is transcribed as short CRISPR RNA (crRNAs) which guides the bacteria molecular machinery to match a sequence in the invading virus and cut up and destroy the invading viral genome.

The Cas9-crRNAs (crRNAs) guide the Cas protein cleavage of target DNA at homologous sequences in three stages (Makarova et al. 2011). The acquisition of a new spacer locus of the CRISPR as new spacers between the clustered palindromic repeated DNA sequences is the first stage. This makes the bacteria immune to infections by the same virus. The second stage involves transcription of the CRISPR locus mediated by a promoter of known lead sequence (L) which is rich in adenine-thymine DNA bases and guides the regeneration of multiple sequences and spacers in long RNA strands that will be divided into small interfering crRNAs. The third stage of crRNAs is the guidance of Cas9 protein cleavage of DNA, known as targeting proteinase, and involves the formation of complexes of mature crRNAs together with Cas proteins which are recognized and destroyed, by DSB, the exogenous DNA of a plasmid, virus, or a transposon which is a mobile DNA in the genome (Makarova et al. 2011). At this stage, crRNAs help the Cas protein to damage invasive DNA invading the bacterial cells

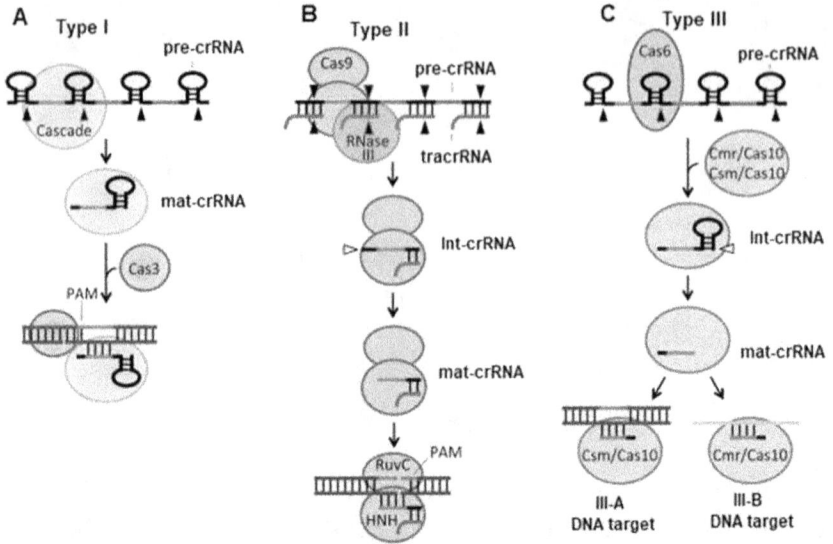

Figure 2.6 Mechanism of crRNA biogenesis and targeting in three types of CRISPR-Cas systems. Figure adapted from Molecular Cell 54, 2014 Elsevier Inc.

(plasmid, viral DNA, or transposon) and thus recall resemblance to the RNA interference which mediates resistance of the nucleic acids to internal and external parasites and pathogens (Westra et al. 2014)

The crRNA production is performed in three CRISPR-Cas systems (see Figure 2.6). In type I (A), pre-crRNA releases mat crRNAs by recruiting the Cas3 nuclease by a cascade to nick the DNA strand complementary to a protospacer adjacent to the motif PAM. The PAM is a 2–6 base pair of DNA sequences following the targeted sequence of Cas9 nuclease (Sinkunas et al. 2013). In Type II (B), primary processing of Cas9 yields mat crRNAs by targeted cleavage of crRNA. In the meantime, each of the RuvC and HNH domains of Cas9 cleaves one DNA strand of the protospacer region, 3 nucleotides upstream of PAM (Shah et al. 2013). The PAM is part of the invading DNA and is not regarded as part of the bacterial CRISPR locus, but the Cas9 cannot cleave the target DNA sequence unless followed by the PAM sequence (Gasiunas et al. 2012; Jinek et al. 2012). In Type III (C), Cas6 cuts the pre-crRNA to generate int-crRNAs incorporated into a Cmr/Cas10 or Csm/Cas10 further maturation is achieved through the trimming of the 3′-end sequences (Hale et al. 2009; Hale et al. 2012).

Doudna and Charpentier (2014) highlighted the remarkable potential of the Cas9 endonuclease in the genome change by their capacity to induce DSB in the DNA three bases upstream of the PAM sequence. These authors promoted CRISPR-Cas9 as a leading genome engineering technology for modifying, regulating, or marking genomic loci in different organisms from

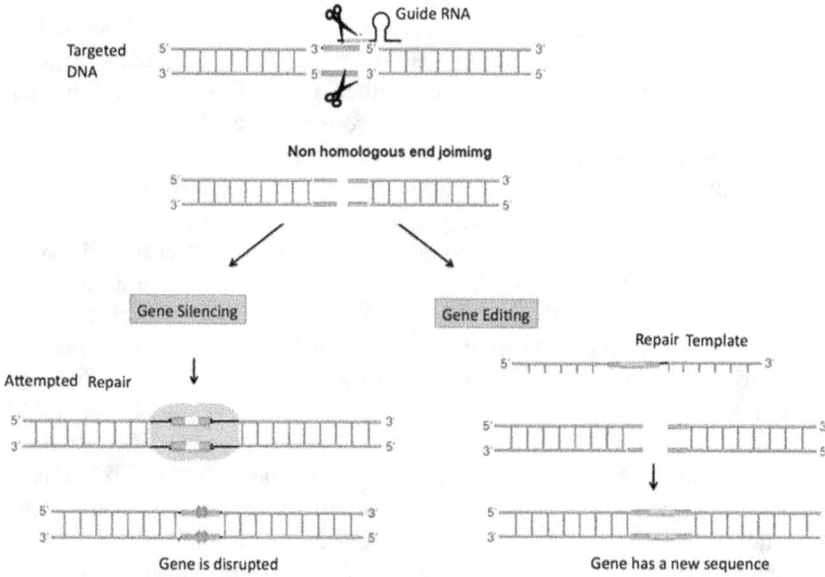

Figure 2.7 Gene silencing and editing with CRISPR. Guide RNA designed to match the DNA region of interest directs molecular machinery to cut both strands of the targeted DNA. During gene silencing, the cell attempts to repair the broken DNA but often does so with errors that disrupt the gene effectively silencing it. For gene editing, a repair template with a specified change in sequence is added to the cell and incorporated into the DNA during the repair process. The targeted DNA is now altered to carry this new sequence.

all kingdoms of life. The DSBs induced by Cas9 are repaired by NHEJ or HR. The NHEJ repair leads to insertion/deletion mutations at the DSB site and is regarded as an error-prone process. In the HR mechanism, more precise mutations are created by exogenous templates, such as gene silencing, HDR, gene editing by DNA-free CRISPR-Cas9, and temporary gene silencing (Chen et al. 2019) as shown in Figure 2.7. CRISPR guides the Cas9 protein to a DNA-matching sequence that can only be used if the target sequence is known. In this way, CRISPR DNA-free gene editing avoids genetic alterations due to random vector integrations. Other CRISPR nucleases have been recently developed by a fusion of Cas1 integrase and Cas6 maturase reverse transcriptase (RT) that produces a specific protein for integration of matched sequence and crRNA production by Wang et al. (2021). Moreover, a cryo-EM structure of a Cas6-RT-Cas1-Cas2 CRISPR integrase complex reveals a hetero-hexamer in which the RT directly contacts the domains of integrase and maturase. The ability of some CRISPR systems to obtain different sequences that mediate CRISPR interference indicating coordination between the functions of all different active sites was highlighted by Wang et al. (2021).

Mikami et al. (2015) used Cas9 and gRNA expression cassettes to assess the occurrence of mutagenesis, separately or sequentially, at the same targeted endogenous sequences in rice callus tissues. More frequent targeted mutagenesis was improved using a combined Cas9/gRNA expression cassette. This approach was optimized by the construction of Cas9/gRNA cassettes and is widely used in various plant genera (Shan et al. 2020). To achieve uniform expression of each gRNA and maintain editing efficiency, a single promoter is the most successful using a small vector fitted to the system. This has been achieved using a polycistronic gene construct with either ribozyme sites (Gao et al. 2016), Csy4 recognition sites (Cˇermák et al. 2017), or transfer RNA sequences (Xie et al. 2015), while gRNA is permeated to produce mature gRNAs for editing in the plant cells. CRISPR-Cas9 was also engineered to expand the scope of gene targeting by recognizing different PAMs (Kleinstiver et al. 2015), by high DNA specificity and broad PAM compatibility (Hu et al. 2018), and expanded targeting space (Nishimasu et al. 2018). In the meantime, the PAM sequences may be at different directions of the targeted sequence to be adapted to different Cas proteins. On the other hand, most of the Cas proteins require PAM at the 30th site of the target DNA while some Cas proteins, such as Cas12, require PAM sequence at the 50th site (Rahman et al. 2022).

2.3.4 CRISPR base editors

Base editing is a major improvement in CRISPR-Cas9 genome engineering. A base editor is mostly adenosine or cytosine deaminase domain fused inactive CRISPR-Cas9 domain to introduce the target region of the genome enabling editing of the targeted point mutations. The cytosine base editors (CBEs) and adenine base editors (ABEs) mediate DNA base transition only, i.e. cytosine to thymine and adenine to guanine, respectively. As reported by Liu et al. (2022), the CBE is comprised of a nickase Cas9 (nCas9) and sgRNA and cleaves a single-stranded sequence using cytosine deaminase and uracil glycosylase inhibitor (UGI). Figure 2.8 illustrates the steps of base editing by the CBE mediating the conversion of cytosine (C) to thymine (T) as shown in Figure 2.8A and the ABE mediating adenine to guanine as shown in Figure 2.8B. The CBE system converts cytosine to uracil in the genomic DNA guided by the sgRNA, while nCas9 cleaves non-target strand, whereas cytosine deaminase modifies the base cytosine (C) instead of the uracil (U). In DNA replication, uracil may be recognized as thymine and pair with adenine. In subsequent DNA replication cycles, adenine pairs with thymine, and thus the conversion of C to T is complete (Komor et al. 2016). In the ABE system, sgRNA guides the nCas9 to cut the non-target strand, while the adenine in the target gene may be converted to inosine (I). In DNA replication, the modified A is recognized as guanine and pairs with cytosine, and thus the conversion of A to G is completed. Esvelt et al. (2014) reported that an ABE can effectively induce the conversion of A to base G in rice and wheat.

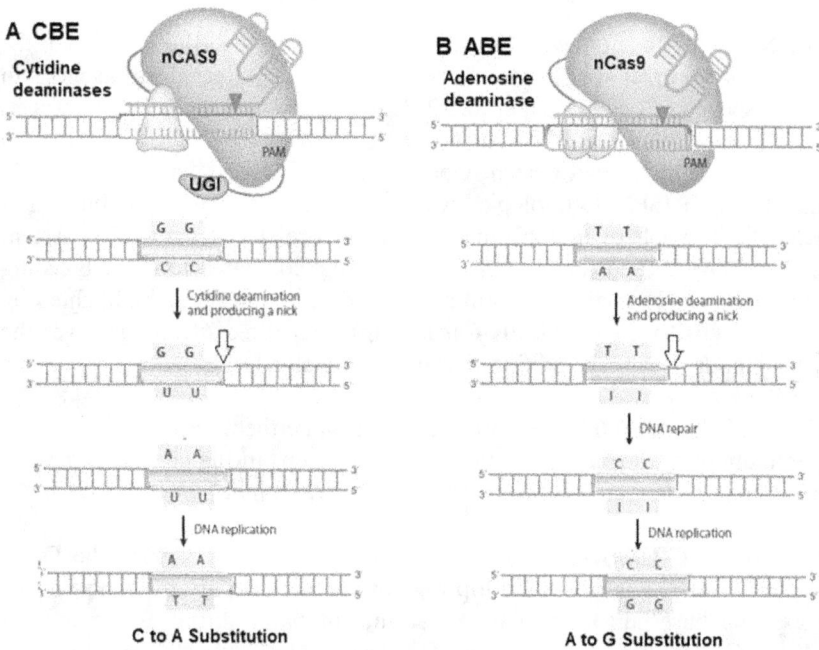

Figure 2.8 Illustrative drawing showing the mechanisms of base editing by the CBE-mediated C-to-T base-editing strategy (A) where the deaminases include rAPOBEC1, hAID, PmCDA1, and hA3A and the ABE-mediated A-to-G base-editing strategy (B) where deaminase is the fusion protein ecTadA-ecTadA*. hAID = human activation-induced cytidine deaminase; nCas9 = a DNA nickase; PAM = protospacer adjacent motif; UGI = uracil glycosylase inhibitor.

The use of ABEs and CBEs made the CRISPR/Cas system widespread tools in genome editing (Gehrke et al. 2018).

A new ABE system was reported by Gao et al. (2017) in which the U-DNA glycosylase replaced the UGI to target the conversion of C-to-G as well as the conversion of either C to T or the conversion of G to A. These systems include methods that can modify bases in DNA without inducing DSBs (Zong et al. 2018). The use of the human APOBEC3A as a plant-based cytidine base editor enhances the efficient conversion of the cytosine(s) to thymine(s) in wheat, rice, and potato, with a 1–17 nucleotide editing window independent of sequence context. The ABE systems of nCas9 achieved A-to-G conversion with high efficiency and have also been optimized for wheat and rice (Li et al. 2018). Double-base editors were used by Gallego-Bartolomé et al. (2018) to convert both C and A to G at target sites guided by the sgRNA. This application has made possible a significant increase and diversity of base mutations and will improve the efficiency of CRISPR/Cas systems as genome-editing tools.

2.4 Applications of gene editing technologies in plants

TALENs were the first easy-to-use genome-editing tool that attracted the attention of scientists worldwide to the application of targeted mutagenesis in multiple species of plants (Gaj et al. 2013; Joung and Sander, 2013). However, the application of TALLENs had limited success due to their low targeting capacity and complicated design. Consequently, TALENs have been largely replaced by CRISPR technologies which are somewhat easier to build and much easier to apply by multiple methods. Therefore, the CRISPR-Cas9 system is a promising method for modern crop breeding compared to other breeding methods such as hybridization, mutations, and gene transfer as highlighted by Chen et al. (2019). The CRISPR-Cas systems offer three advantages over the other methods of genome editing that do not involve DSB.

A Generate far less undesired products that make them more efficient.
B Unlikely to induce large deletions and inversions in the chromosomes.
C Can create nonsense mutations that reduce DSB indels.

However, the CRISPR-Cas systems are not expected to replace the DSB-mediated genome changes but adopting improved Cas9 variants can improve the existing base editors and their variants of base editing as a powerful CRISPR-Cas tool in crop breeding (Mishra et al. 2020). However, further improvement of this technology and more changes in the regulatory system of this approach are needed for more acceptance of genome-edited crop plants in agriculture practice (Hossain et al. 2022).

Most recently, a group led by Friedrich Kragler at the Max Planck Institute of molecular plant physiology in Golm, Germany, presented a new gene editing system that may be used efficiently for the breeding of many important crops which are difficult or impossible to modify with existing methods. Their discovery was recently published in Nature Biotechnology (Yang et al. 2023) as a major step forward that could simplify and speed up the improvement of genetically stable commercial crop varieties by combining a 'mobile' CRISPR tool with the old and well-known methods of grafting that have been used in plant breeding for generations. The research group identified tRNA-like sequences (TLS) that act as signals for the distant movement of RNAs within plants. This discovery with the CRISPR/Cas9 genome-editing system made a breakthrough in simplifying gene editing of plants. The proposed protocol is illustrated in Figure 2.9. When a TLS is added to the CRISPR/Cas9 sequences, plants produce 'mobile' versions of CRISPR/Cas9 RNA which then moves from the root into the shoot, when the root is grafted, and eventually into the flowers that produce the seeds. In the flowers, the CRISPR/Cas9 RNA is converted into the corresponding protein, which is the actual 'genetic scissors' that edit the plant DNA in the flowers. The advantage of this system is that CRISPR/Cas9 system does not remain in the DNA; thus, the seeds developed from these flowers carry only the desired editing. Yang et al. (2013)

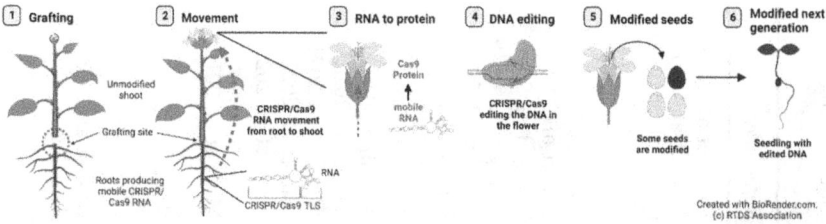

Figure 2.9 Graphical abstract illustrating the protocol of the grafting and mobile CRISPR/Cas9 for genome editing in plants. An animation explaining this protocol is available on the website: https://youtu.be/4gcVBgAr73M.

confirmed that no trace of the CRISPR/Cas9 system is found in the next generation of gene-edited plants and that the protocol works with surprisingly high efficiency.

2.5 CRISPR-Cas technology and production of climate change resilient plants

CRISPR-Cas9 technology in the genome editing of plants is increasingly applied. One important field of their applications is the production of crops resilient to extreme environmental conditions that may prevail in some parts of the world due to the expected climate changes, particularly increased drought, temperature, and salinity. Li et al. (2022) and Rahman et al. (2022) reviewed some examples of genome editing using CRISPR-Cas9 tools and listed examples of improved tolerance to abiotic and biotic stress by using the CRISPR genome-editing system. In this chapter, a brief description of gene editing examples has been presented using the CRISPR-Cas9 system that has been most successful in cereals (rice, wheat, maize, and barley), but also reported in soybean and tomatoes. Previous applications on the use of CRISPR-Cas9, including the examples described here, are listed in Table 2.1.

2.5.1 Rice

Rice is the most engineered crop by the CRISPR-Cas9 gene editing methods to increase crop yield and improve drought and salinity resistance. In this crop, the dense and erect panicle1 (DEP1) gene of nCas9 was deleted by several well-designed CRISPR sgRNAs in the Indica rice line IR58025B (Wang et al. 2017a). Constructs with two sgRNAs were more successful and gave rise to long, dense, and erect panicles with improved yield but plant height was reduced. Miao et al. (2018) edited two groups of PYL genes in rice. These genes associated with the abscisic acid (ABA) receptor (RCAR) are a key role in plant growth and abiotic stress tolerance. Among all the PYL mutants, only two groups showed significant results. Group I mutants showed enhanced growth and improved grain productivity and maintained

Table 2.1 List of crop species engineered for abiotic stress resistance using the CRISPR-Cas9 system and the name of engineered genes and their function, stress type, and the references for the described applications

Crop species	Targeted gene	Function	Stress type	References
Rice	RAV2	Regulation of developmental processes	Salinity	Duan et al. 2016
Rice	MYB30	Negative regulator of cold stress	Cold	Tang et al. 2017
Rice	HSA1	Chloroplast development and protection	Heat	Qiu et al. 2018
Rice	DERF1, PMS3, MSH1, MYB5, SPP	Amino acid synthesis and drought tolerance	Drought	Zhang et al. 2014
Rice	SRL1, SRL2	Regulate leaf rolling, increased yield	Drought	Liao et al. 2019
Rice	SPL10	Regulate trichome development	Salinity	Lan et al. 2019
Rice	RR9, RR10	Negatively regulate cytokinin signaling	Salinity	Wang et al. 2019
Rice	RR22	Involved in cytokinin signaling	Salinity	Zhang et al. 2019
Rice	ERA1	Regulates ABA signaling and dehydration response	Drought	Ogata et al. 2020
Rice	DST	Involved in stomata development	Salinity	Santosh Kumar et al. 2020
Rice	SOS1	Na+/H+ antiporter mediating Na+ transport	Salinity	Lu et al. 2020
Rice	PQT3	E3 ubiquitin ligase	Salinity	Alfatih et al. 2020
Rice	miR535	Involved in Salinity stress regulation	Salinity	Yue et al. 2020
Rice	GI	Circadian clock component	Salinity	Wang et al. 2021
Rice	OsRR22	Involved in cytokinin signaling	Salinity	Han et al. 2022
Rice	bHLH024	Basic helix–loop–helix TF involved in growth and stress responses	Salinity	Alam et al. 2022
Rice	PYL1, PYL4 PYL6	Regulatory component of Abscisic acid	Heat	Miao et al. 2018
Rapeseed	A6.RGA	DELLA protein, negative regulator of gibberellin signaling	Drought	Wu et al. 2020
Wheat	sgRNA(F+E	Drought stress	Drought	Li et al. 2018
Wheat	TaDREB2 TaERF3	Stress-responsive transcription factor genes	Drought	Kim et al. 2018
Wheat	TaSal1	Germination and early growth	Drought	Abdallah et al. 2022

Crop	Gene	Function	Stress	Reference
Barley	HvITPK1	Inositol Trisphosphate 5/6 Kinases 1	Salinity	Vlčko and Ohnoutková, 2020
Maize	ARGOS8	Involved in ethylene response	Drought	Shi et al. 2017
Maize	abh2	Abscisic acid 8'-hydroxylase mediates stomatal opening	Drought	Liu et al. 2020
Maize	STL1	Dirigent protein localized to the Casparian strip	Salinity	Wang et al. 2021
Soybean	GmFT2a,	Photoperiod flowering pathway	Flowering	Cai et al. 2017
Soybean	GmAITR	Regulation of ABA signaling	Salinity	Wang et al. 2021
Tomato	MAPK3	Negative regulator of heat stress	Heat	Yu et al. 2019
Tomato	AGL6	Involved in fruit development	Heat	Zeng et al. 2020
Tomato	GID1a	Gibberellin (GA) receptor	Drought	Illouz-Eliaz et al. 2020
Tomato	LBD40	Involved in jasmonic acid (JA)-mediated stress response	Drought	Liu et al. 2020
Tomato	SlARF4	Auxin Response Factor 4	Salinity	Bouzroud et al. 2020
Tomato	HyPRP1	Negative regulator of salt stress	Salinity	Tran et al. 2021
Tomato	ABIG1	Homeodomain-leucine Zipper (HD-ZIP) TF	Salinity	Ding et al. 2022

seed dormancy. A combination of genes played more important significant roles in seed dormancy, stomatal movement, and growth regulation for group I compared to those in group II.

Homozygous rice mutants edited by CRISPR/Cas9 were developed by Liao et al. (2019) to induce SRL1 and SRL2 gene mutations and showed decreased performance of the drought-responsive traits like transpiration rate, chlorophyll content, and stomatal conductance, as well as reduced vascular bundles and stomatal number but the panicle number was not reduced. CRISPR/Cas9 editing induced curled leaves and improved drought tolerance as indicated by the protein expression patterns and the scavenging of reactive oxygen species. This was associated with higher ABA content, a higher survival rate, and an improved grain filling percentage. The studies of Ogata et al. (2020) indicated that CRISPR/Cas9-induced frameshift mutations in the rice cultivar osera1. Leaf growth of the mutants was similar to control plants but the primary root growth was increased. The osera1 mutant lines also showed an enhanced response to drought stress as indicated by stomatal conductance regulation and an increase of sensitivity to ABA. Usman et al. (2020) used RNA-guided Cas9 nuclease to do precise editing of the OsPYL9 gene which regulates circadian rhythm and abiotic stress-responsive proteins and also confers enhanced drought tolerance and improved grain yield. Most of the DEPs genes related to the above criteria were upregulated in the OsPYL9-edited plants.

Also, in rice, Kumar et al. (2020) generated mutants with different DST alleles; a366 bp deletion in the DNA led to the deletion of 221 amino acids. A homozygous DST mutant with this deletion was selected for phenotypic analysis. This mutant showed improved drought tolerance as indicated by enhanced leaf water retention under drought cognitions and improved abiotic stress-responsive traits, such as broader leaf width and reduced stomatal density. However, the genes associated with stomatal development were downregulated. At the seedling stage, this mutant showed a high salt stress level and moderate osmotic stress tolerance. Kumar et al. (2020) called for engaging their important findings with cloned genes that control important agronomic traits in rice, mostly in Japonica cultivars. The cloned genes may be utilized by genome editing to introduce abiotic stress-responsive alleles and improvement of rice yield.

2.5.2 *Wheat*

Few attempts have been made so far to apply CRISPR/Cas9 genome editing to wheat crop improvement in spite of its global importance as the most widely used crop for human food. Kim et al. (2018) attempted to improve transcription factor genes under abiotic stress by editing the dehydration-responsive element binding protein 2 (TaDREB2) and wheat ethylene-responsive factor 3 (TaERF3) using transient expression of small guide RNA and Cas9 protein. The results showed huge potential for the targeted gene to improve drought resistance by engineering the wheat genes TaDREB2 and TaERF3. Bread wheat mutants harboring the 5'-bisphosphate nucleotides (TaSal1) genes were produced using a sgRNA-CRISPR/Cas9 genome-editing system by Abdallah et al. (2022) in the Egyptian cultivar cv. Giza168. TaSal1 mutations were evident in 41 lines of the M1 progeny, and 5 lines showed full 5 gene knockouts. The Sal1 mutant seedlings germinated and grew better under simulated polyethylene glycol drought than wild-type seedlings, indicating more drought resistance in the Sal1 mutant.

2.5.3 Maize

Svitashev et al. (2016) delivered pre-assembled Cas9-gRNA ribonucleoproteins into maize embryo cells. The results confirmed the success of gene mutagenesis using the Cas9-gRNA technology in maize as indicated by the recovered mutated plants harboring alleles at high frequencies. The CRISPR-Cas9 system was also used to induce allelic variations for producing drought-tolerant maize genotypes by generating novel variants of the ARGOS8 gene (Shi et al. 2017). The native maize GOS2 promoter was used to replace the natural ARGOS8 promoter sequence. Elevated levels of ARGOS8 transcripts matched the molecular evidence and supported a relative increase in the native allele in all the tested tissues, which was the expected result of using the GOS2 promoter. ARGOS8 variants were associated with increased grain yield in the field under flowering stress but yield loss was not evident under well-watered conditions. Wang et al. (2021) reported that an orthologue (ZmESBL) of the salt-tolerant-gene locus 1 (mSTL1) which confers transpiration-dependent salt tolerance (TDST) mediates increased transport of Na+ across the endodermis, leading to salt hypersensitivity.

2.5.4 Soybean

Soybeans were one of the first crops engineered using the CRISPR-Cas9 system. The flowering genes were edited using this method to induce mutations in the GmFT2a gene locus which is an integrator in the photoperiod flowering pathway (Cai et al. 2017). The gene-edited plants showed an increase in plant size and delayed late flowering. This mutation was inherited in the following generations of the gene-edited soybean plants. Wang et al. (2021) successfully generated Cas9-free mutants of the ABA-induced transcription repressors, which confer drought and salinity resistance, and found that BA sensitivity in these mutants was increased, indicating that salinity tolerance in soybean may be improved.

2.5.5 Tomato

The self-pruning 5G gene, which promotes day-neutrality and early yield, was engineered by Soyk et al. (2017) and resulted in enhanced growth of field tomatoes. The result of this study confirmed that yield traits in tomatoes were improved. A mutation in flowering time was identified through mapping of the flowering gene using novel sequencing tools. Further research on the flowering or floral regulator genes was done by Rajendran et al. (2021) through induction of late flowering using the CRISPR/Cas9 system. They suggested a 'yield dynamic model' which deals with the role of genes at the vegetative and reproductive phases in the determination of tomato productivity. These authors proposed a model for floral transition control by genetic regulation. They argued that flowering time variants may provide basic genetic clues to the fine-tuning and optimization of yield in tomatoes and other crops.

Gene-edited tomato plants were also produced by modifying the GID1 to exhibit high leaf water content under drought conditions and reduced transpiration with no effect on growth and productivity (Illouz-Eliaz et al. 2020). High leaf water content under water deficit, normal growth, and better recovery from dehydration was improved in CRISPR-Cas9 gid1 mutants. Liu et al. (2020) explored the function of SlLBD40, a lateral organ boundaries domain harboring genes for lateral organ development by overexpressing the CRISPR/Cas9 mediated SlLBD40 gene. The knockout of this gene resulted in improved tolerance to drought stress as measured by transpiration rate and leaf water potential at mid-day. Engineered tomatoes have reduced xylem vessel proliferation and expansion, indicating that low GA activity was reduced by mechanisms that may include leaf area reduction and promoted stomatal closure. The CRISPR/Cas9 approach was used to manipulate protein domains of tomato hybrid (HyPRP1), which regulate salt stress responses negatively. Elimination of the SlHyPRP1 locus led to more salinity-tolerant tomatoes at the seed germination and seedling stage (Tran et al. 2021).

References

Abdallah NA, Elsharawy H, Abulela HA, Thilmony R, et al. (2022). Multiplex CRISPR/Cas9-mediated genome editing to address drought tolerance in wheat. *GM Crops Food*, 6, 1–17. DOI: 10.1080/21645698.2022.2120313.

Alam MS, Kong J, Tao R, Ahmed T, et al. (2022). CRISPR/Cas9 mediated Knockout of the OsbHLH024 transcription factor improves salt stress resistance in rice (*Oryza sativa* L.). *Plants*, 11, 1184.

Alfatih A, Wu J, Jan SU, Zhang ZS, et al. (2020). Loss of rice PARAQUAT TOLERANCE 3 confers enhanced resistance to abiotic stresses and increases grain yield in the field. *Plant Cell Environ*, 43, 2743–2754.

Barrangou R, Horvath P (2012). CRISPR: New horizons in phage resistance and strain identification. *Annu Rev Food Sci Technol*, 3, 143–162.

Barrangou R, Marraffini LA (2014). CRISPR-Cas systems: Prokaryotes upgrade to adaptive immunity. *Mol Cell*, 54(2), 234–244. DOI: 10.1016/j.molcel.2014.03.011.

Boch J, Scholze H, Schornack S, Landgraf A, et al. (2009). Breaking the code of DNA binding specificity of TAL-type III effectors. *Science*, 326(5959), 1509–1512.

Bouzroud S, Gasparini K, Hu G, Barbosa MA, et al. (2020). Down-regulation and loss of auxin response factor 4 function using CRISPR/Cas9 alter plant growth, and stomatal function and improve tomato tolerance to salinity and osmotic stress. *Genes (Basel)*, 11(3), E272. DOI: 10.3390/genes11030272.

Cai Y, Chen L, Liu X, Guo C, et al. (2017). CRISPR/Cas9-mediated targeted mutagenesis of GmFT2a delays flowering time in soya beans. *Plant Biotech J*. DOI: 10.1111/pbi.12758/.

Carroll D (2011). Genome engineering with zinc-finger nucleases. *Genetics*, 88(4), 773–782. DOI: 10.1534/genetics.111.131433.

Carroll, D (2014). Genome engineering with targetable nucleases. *Annu Rev Biochem*, 83, 409–439. DOI: 10.1146/annurev-biochem-060713-035418.

Chen K, Wang Y, Zhang R, Zhang H, Gao C (2019). CRISPR/Cas genome editing and precision plant breeding in agriculture. *Annu Rev Plant Biol*, 70, 667–697. DOI: 10.1146/annurev-arplant-050718-100049.

Christian M, Cermak T, Doyle EL, Schmidt C, et al. (2010). Targeting DNA double-strand breaks with TAL effector nucleases. *Genetics*, 186, 757–761. DOI: 10.1534/genetics.110.120717.

Cong L, Zhou R, Kuo YC, Cunniff M, Zhang F (2012). Comprehensive interrogation of natural TALE DNA-binding modules and transcriptional repressor domains. *Nat Commun*, 3, 968.

Cong L, Ran FA, Cox D, Lin S, et al. (2013). Multiplex genome engineering using CRISPR/Cas systems. *Science*, 339(6121), 819–823.

Cěrmák T, Curtin SJ, Gil-Humanes J, Cěgan R, et al. (2017). A multipurpose tool kit to enable advanced genome engineering in plants. *Plant Cell*, 29, 1196–1217.

Ding F, Qiang X, Jia Z, Lili L, et al. (2022). Knockout of a novel salt-responsive gene SlABIG1 enhances salinity tolerance in tomatoes. *Environ Exp Bot*, 200, 104903.

Doudna JA, Charpentier E (2014). Genome editing. The new frontier of genome engineering with CRISPR-Cas9. *Science*, 346, 1258096. DOI: 10.1126/science.1258096.

Duan YB, Li J, Qin RY, Xu RF (2016). Identification of a regulatory element responsible for salt induction of rice OsRAV2 through ex-situ and in situ promoter analysis. *Plant Mol Biol*, 90, 49–62.

Esvelt KM, Smidler AL, Catteruccia F, Church GM (2014). Concerning RNA-guided gene drives for the alteration of wild populations. *eLife3*, e03401.

Gaj T, Gersbach CA, Barbas CF (2013). ZFN, TALEN, and CRISPR/Cas-based methods for genome engineering. *Trends Biotech*, 31, 397–405. DOI: 10.1016/j.tibtech.2013.04.004.

Gallego-Bartolomé J, Gardiner J, Liu W, Papikian A, Ghoshal B, et al. (2018). Targeted DNA demethylation of the *Arabidopsis* genome using the human TET1 catalytic domain. *PNAS*, 115, E2125–E2134.

Gao L, Cox DBT, Yan WX, Manteiga JC, Schneider MW, et al. (2017). Engineered Cpf1 variants with altered PAM specificities. *Nat Biotechnol*, 35, 789–792.

Gao X, Chen J, Dai X, Zhang D, Zhao Y (2016). An effective strategy for reliably isolating heritable and Cas9-free *Arabidopsis* mutants generated by CRISPR/Cas9-mediated genome editing. *Plant Physiol*, 171, 1794–1800.

Gasiunas G, Barrangou R, Horvath P, Siksnys V (2012). Cas9-crRNA ribonucleo-protein complex mediates specific DNA cleavage for adaptive immunity in bacteria. *Proc Natl Acad Sci USA*, 109, E2579–E2586.

Gehrke JM, Cervantes O, Clement MK, Wu Y, et al. (2018). An APOBEC3A-Cas9 base editor with minimized bystander and off-target activities. *Nat Biotechnol*, 36, 977–982.

Grohmann L, Keilwagen J, Duensing N, Dagan E, et al. (2019). Detection and identification of genome editing in plants: Challenges and opportunities. *Front Plant Sci*, 10, 236. DOI: 10.3389/fpls.2019.00236.

Gu X, Liu L, Zhang H (2021). Transgene-free genome editing in plants. *Front Genome Ed*, 3, 805317. DOI: 10.3389/fgeed.2021.805317.

Hale, CR, Zhao P, Olson, S, Duff, MO, et al. (2009). RNA-guided RNA cleavage by a CRISPR RNA-Cas protein complex. *Cell*, 139, 945–956.

Hale CR, Majumdar S, Elmore J, Pfister N, et al. (2012). Essential features and rational design of CRISPR RNAs that function with the Cas RAMP module complex to cleave RNAs. *Mol Cell*, 45, 292–302.

Han X, Chen Z, Li P, Xu H, et al. (2022). Development of novel rice germplasm for salt tolerance at seedling stage using CRISPR-Cas9. *Sustainability*, 14, 2621. DOI: 10.3390/su14052621.

Holtzman L, Gersbach CA (2018). Editing the epigenome: Reshaping the genomic landscape. *Annu Rev Genomics Hum Genet*, 19, 43–71. DOI: 10.1146/annurev-genom-083117-021632.

Hossain A, Rahman Md. ME, Ali S, et al. (2022). CRISPR-Cas9-mediated genome editing technology for abiotic stress tolerance in crop plants. In: Plant perspectives to global climate changes*: Developing climate-resilient plants*, Aftab, T., Roychoudhury, A. (Eds). First Edition: Chapter 16. Academic Press (Elsevier). DOI: 10.1016/B978-0-323-85665-2.00008-X.

Hu JH, Miller SM, Geurts MH, Tang W, Chen L, et al. (2018). Evolved Cas9 variants with broad PAM compatibility and high DNA specificity. *Nature*, 556, 57–63.

Illouz-Eliaz N, Nissan I, Nir I, Ramon U, et al. (2020). Mutations in the tomato gibberellin receptors suppress xylem proliferation and reduce water loss under water-deficit conditions. *J Exp Bot*, 71, 3603–3612.

IPCC (2019). *IPCC special report on the ocean and cryosphere in a changing climate*, Pörtner HO, et al. (eds.). Working Group II Technical Support Unit. in press.

IPCC (2013). *Climate change 2013: The physical science basis. Contribution of working group I to the fifth assessment report of the intergovernmental panel on climate change*, Stocker TF, et al. (eds.). Cambridge University Press, Cambridge, UK, and New York, USA, p. 1535.

Jansen R, Embden JD, Gaastra W, Schouls LM (2002). Identification of genes that are associated with DNA repeats in prokaryotes. *Mol Microbiol Mar*, 43(6), 1565–1575.

Jinek, M, Chylinski K, Fonfara I, Hauer M (2012). A programmable dual-RNA-guided DNA endonuclease in adaptive bacterial immunity. *Science*, 337, 816–821. DOI: 10.1126/science.1225829.

Joung JK, Sander JD (2013). TALENs: A widely applicable technology for targeted genome editing. *Nat Rev Mol Cell Biol*, 14, 49–55. DOI: 10.1038/nrm3486.

Jouzel J, Masson-Delmotte V, Cattani O, Dreyfus G, et al. (2007). Orbital and millennial Antarctic climate variability over the past 800,000 years. *Science*, 317, 793–796. DOI: 10.1126/science.1141038.

Kim D, Alptekin B, Budak H (2018). CRISPR/Cas9 genome editing in wheat. *Funct Integr Genomics*, 18(1), 31–41. DOI: 10.1007/s10142-017-0572-x.

Kim, J, Kim, J-S (2016). Bypassing GMO regulations with CRISPR gene editing. *Nat Biotechnol*, 34, 1014–1015. DOI: 10.1038/nbt.3680.

Kleinstiver BP, Prew MS, Tsai SQ, Topkar VV, et al. (2015). Engineered CRISPR-Cas9 nucleases with altered PAM specificities. *Nature*, 523, 481–485.

Komor AC, Kim YB, Packe MS, Zuris JA, Liu DR (2016). Programmable editing of a target base in genomic DNA without double-stranded DNA cleavage. *Nature*, 533, 420–424.

Kumar SVV, Verma RK, Yadav SK, et al. (2020). CRISPR-Cas9 mediated genome editing of drought and salt tolerance (OsDST) gene in indica mega rice cultivar MTU1010. *Physiol Mol Biol Plants*, 26(6), 1099–1110. DOI: 10.1007/s12298-020-00819-w.

Lacis AA, et al. (2010). Atmospheric CO2: Principal control knob governing Earth's temperature. *Science*, 330(6002), 356–359.

Lan T, Zheng Y, Su Z, Yu S, et al. (2019). OsSPL10, a SBP-box gene, plays a dual role in salt tolerance and trichome formation in rice (*Oryza sativa* L.). *G3: Genes Genomes Genet*, 9, 4107–4114.

Li C, Zong Y, Wang Y, Jin S, Zhang D, et al. (2018). Expanded base editing in rice and wheat using a Cas9-adenosine deaminase fusion. *Genome Biol*, 19, 59.

Li T, Huang S, Jiang W, Wright, D, et al. (2011). TAL nucleases (TALNs): Hybrid proteins composed of TAL effectors and FokI DNA cleavage domain. *Nucleic Acids Res*, 39, 359–372. DOI: 10.1093/nar/gkq704.

Li Y, Wu X, Zhang Y, Zhang Q (2022). CRISPR/Cas genome editing improves abiotic and biotic stress tolerance of crops. *Front Genome Ed*, 4, 987817. DOI: 10.3389/fgeed.2022.987817

Li Z, Xiong X, Li J-F (2018). New cytosine base editor for plant genome editing. *Sci China Life Sci*, 61, 1602–1603.

Liao S, Qin X, Luo L, Han Y, Wang X, Usman, B, et al. (2019). CRISPR/Cas9-induced mutagenesis of semi-rolled leaf1, 2 confers curled leaf phenotype and drought tolerance by influencing protein expression patterns and ROS scavenging in rice (*Oryza sativa* L.). *Agronomy*, 9(11), 728. DOI: 10.3390/agronomy9110728

Liu H, Chen W, Li Y, et al. (2022). CRISPR/Cas9 technology and its utility for crop improvement. *Int J Mol Sci*, 23, 10442. DOI: 10.3390/ijms231810442.

Liu L, Zhang J, Xu J, Li Y, et al. (2020). CRISPR/Cas9 targeted mutagenesis of SlLBD40, a lateral organ boundaries domain transcription factor enhances drought tolerance in tomato. *Plant Sci*, 301, 110683.

Liu L, Zhang J, Xu J, Li Y, Guo L, Wang Z, et al. (2020). CRISPR/Cas9 targeted mutagenesis of SlLBD40, a lateral organ boundaries domain transcription factor that enhances drought tolerance in tomatoes. *Plant Sci*, 301, 110683. DOI: 10.1016/j.plantsci.2020.110683.

Liu S, Li C, Wang H, Wang S, Yang S, et al. (2020). Mapping regulatory variants controlling gene expression in drought response and tolerance in maize. *Genome Biol*, 21, 163.

Lu Y, Tian Y, Shen R, Yao Q, et al. (2020). Targeted, efficient sequence insertion and replacement in rice. *Nat Biotechnol*, 38, 1402–1407.

Lüthi D, Le Floch M, Bereiter B, et al. (2008). High-resolution carbon dioxide concentration record 650,000–800,000 years before the present. *Nature*, 453, 379–382. DOI: 10.1038/nature06949.

Magnan A, Colombier M, Billé R, et al. (2016). Implications of the Paris agreement for the ocean. *Nature Clim Change*, 6, 732–735. DOI: 10.1038/nclimate3038.

Makarova KS, Haft DH, Barrangou R, Brouns SJ, Charpentier E, Horvath P, et al. (2011). Evolution and classification of the CRISPR-Cas systems. *Nat Rev Microbiol*, 9(6), 467–47732.

Mali P, Yang L, Esvelt KM, Aach J, et al. (2013). RNA-guided human genome engineering via Cas9. *Science*, 339, 823–826. DOI: 10.1126/science.1232033

Marchane A, Tremblay Y, Hanich L, Ruelland D, Jarlan L, et al. (2017). Climate change impacts on surface water resources in the Rheraya catchment (High-Atlas, Morocco). *Hydrol Sci J*, 62(6), 979–995.

Meier KJS, et al. (2014). The role of ocean acidification in *Emilianiahuxleyi coccolith* thinning in the Mediterranean Sea. *Biogeosciences*, 11, 2857–2869.

Meyssignac B, Calafat FM, Somot S, Rupolo S, et al. (2011). Two-dimensional reconstruction of the Mediterranean Sea level over 1970–2006 from tide gage data and regional ocean circulation model outputs. *Glob Planet Change*, 77(1–2), 49–61.

Miao C, Xiao L, Hua K, Zou C, et al. (2018). Mutations in a subfamily of abscisic acid receptor genes promote rice growth and productivity. *Proc Natl Acad Sci USA*, 115, 6058–6063.

Mikami M, Toki S, Endo M (2015). Comparison of CRISPR/Cas9 expression constructs for efficient targeted mutagenesis in rice. *Plant Mol Biol*, 88, 561–572. DOI: 10.1007/s11103-015-0342-x.

Mishra R, Joshi RK, Zhao K (2020). Base editing in crops: Current advances, limitations, and future implications. *Plant Biotechnol J*, 20–31. DOI: 10.1111/pbi.13225.

Mitchell JFB, Johns TC, Ingram, WJ, Lowe JA (2000). The effect of stabilizing atmospheric carbon dioxide concentrations on global and regional climate change. *Geophys Res Lett*, 27(18), 2977–2980.

Mojica FJ, Díez-Villaseñor C, García-Martínez J, Soria E (2005). Intervening sequences of regularly spaced prokaryotic repeats derive from foreign genetic elements. *J Mol Evol*, 60(2), 174–182.

Morice CP, Kennedy JJ, Rayner NA, Jones PD (2012). Quantifying uncertainties in global and regional temperature change using an ensemble of observational estimates: The HadCRUT4 dataset. *J Geophys Res*, 117, D08101. DOI: 10.1029/2011JD017187.

Moscou MJ, Bogdanove AJ (2009). A simple cipher governs DNA recognition by TAL effectors. *Science*, 326(5959), 1501.

Nishimasu H, Shi X, Ishiguro S, Gao L, Hirano S, et al. (2018). Engineered CRISPR-Cas9 nuclease with expanded targeting space. *Science*, 361, 1259–1262.

Ogata T, Ishizaki T, Fujita M, Fujita Y (2020). CRISPR/Cas9-targeted mutagenesis of OsERA1 confers enhanced responses to abscisic acid and drought stress and increased primary root growth under non-stressed conditions in rice. *PLoS One*, 15(12), e0243376. DOI: 10.1371/journal.pone.0243376.

Qiu Z, Kang S, He L, Zhao J, Zhang S, et al. (2018). The newly identified heat-stress sensitive albino 1 gene affects chloroplast development in rice. *Plant Sci*, 267, 168–179.

Rahman M-U, Zulfiqar S, Raza MA, Ahmad N, Zhang B (2022). Engineering abiotic stress tolerance in crop plants through CRISPR genome editing. *Cells*, 11, 3590. DOI: 10.3390/cells11223590.

Rajendran S, Heo J, Kim YJ, et al. (2021). Optimization of tomato productivity using flowering time variants. *Agronomy*, 11, 285. DOI: 10.3390/agronomy11020285.

Rees HA, Liu, DR (2018). Base editing: Precision chemistry on the genome and transcriptome of living cells. *Nat Rev Genet* 19, 770–788. DOI: 10.1038/s41576-018-0059-1.

Samset BH, Fuglestvedt JS, Lund MT (2020). Delayed emergence of a global temperature response after emission mitigation. *Nature Commun*, 11, 3261. DOI: 10.1038/s41467-020-17001-1.

Schleussner, CF, Tabea KL, Fischer EM, Wohland J, et al. (2016). Differential climate impacts for policy-relevant limits to global warming: The case of 1.5°C and 2°C. *Earth Syst Dynam*, 7, 327–351.

Shah SA, Erdmann S, Mojica FJ, Garrett RA (2013). Protospacer recognition motifs: Mixed identities and functional diversity. *RNA Biol*, 10(5), 891–899. DOI: 10.4161/rna.23764.

Shan S, Soltis PS, Soltis DE, Yang, B (2020). Considerations in adapting CRISPR/Cas9 in nongenetic model plant systems. *Appl Plant Sci*, 8, e11314. DOI: 10.1002/aps3.11314.

Shi J, Gao H, Wang H, Lafitte HR, et al. (2017). ARGOS8 variants generated by CRISPR-Cas9 improve maize grain yield under field drought stress conditions. *Plant Biotechnol J*, 15(2), 207–216. DOI: 10.1111/pbi.12603.

Sinkunas T, Gasiunas G, Waghmare SP, Dickman MJ (2013). In vitro reconstitution of Cascade-mediated CRISPR immunity in *Streptococcus thermophilus*. *EMBO J*, 32, 385–394.

Smith P, Nkem J, Calvin K, et al. (2019). Interlinkages between desertification, land degradation, food security, and greenhouse gas fluxes: Synergies, trade-offs, and integrated response options. In: *Climate change and land: An IPCC special report on climate change, desertification, land degradation, sustainable land management, food security, and greenhouse gas fluxes in terrestrial ecosystems*, PR Shukla, et al. (eds.). DOI: 10.1017/9781009157988.008.

Soyk S, Müller NA, Park SJ, Schmalenbach I, et al. (2017). Variation in the flowering gene SELF PRUNING 5G promotes day-neutrality and early yield in tomatoes. *Nature Genet*, 49, 162–168.

Svitashev S, Schwartz C, Lenderts B, et al. (2016). Genome editing in maize directed by CRISPR–Cas9 ribonucleoprotein complexes. *Nat Commun*, 7, 13274. DOI: 10.1038/ncomms13274.

Tang L, Mao B, Li Y, Lv Q, et al. (2017). Knockout of OsNramp5 using the CRISPR/Cas9 system produces low Cd-accumulating indica rice without compromising yield. *Sci Rep*, 7, 14438.

Toda E, Koiso N, Takebayashi A, Ichikawa M, et al. (2019). An efficient DNA- and selectable-marker-free genome-editing system using zygotes in rice. *Nat Plants*, 5(4), 363–368. DOI: 10.1038/s41477-019-0386-z. Epub Mar 25. PMID: 30911123.

Tran MT, Doan DTH, Kim J, Song YJ, et al. (2021). CRISPR/Cas9-based precise excision of SlHyPRP1 domain(s) to obtain salt stress-tolerant tomato. *Plant Cell Rep*, 40(6), 999–1011. DOI: 10.1007/s00299-020-02622-z.

Turnbull C, Lillemo M, Hvoslef-Eide, TAK (2021). Global regulation of genetically modified crops amid the gene-edited crop boom - A review. *Front Plant Sci*, 12, 630396. DOI: 10.3389/fpls.2021.630396.

Urnov FD, Miller JC, Lee YL, Beausejour CM, et al. (2005). Highly efficient endogenous human gene correction using designed zinc-finger nucleases. *Nature*, 435, 646–651. DOI: 10.1038/nature03556.

Usman B, Nawaz G, Zhao N, Liao S, Liu Y, Li R (2020). Precise editing of the OsPYL9 gene by RNA-guided Cas9 nuclease confers enhanced drought tolerance and grain yield in rice (Oryza sativa L.) by regulating circadian rhythm and abiotic stress-responsive proteins. *Int J Mol Sci*, 21(21), E7854. DOI: 10.3390/ijms21217854.

Vlčko T, Ohnoutková L (2020). Allelic variants of CRISPR/Cas9 induced mutation in an inositol trisphosphate 5/6 kinase gene manifest different phenotypes in barley. *Plants (Basel)*, 9(2), E195. DOI: 10.3390/plants9020195.

Wang M, Mao Y, Lu Y, Tao X, Zhu J-k (2017a). Multiplex gene editing in rice using the CRISPR-Cpf1system. *Mol Plant*, 10, 1011–1013.

Wang T, Xun H, Wang W, Ding X, et al. (2021). Mutation of GmAITR Genes by CRISPR/Cas9 genome editing results in enhanced salinity stress tolerance in soybean. *Front Plant Sci*, 12, 779598.

Wang WC, Lin TC, Kieber J, Tsai YC (2019). Response regulators 9 and 10 negatively regulate salinity tolerance in rice. *Plant Cell Physiol*, 60, 2549–2563.

Wang X, He Y, Wei H, Wang L, et al. (2021). A clock regulatory module is required for salt tolerance and control of the heading date in rice. *Plant Cell Environ*, 44, 3283–3301.

Westra ER, Buckling A, Fineran PC (2014). CRISPR-Cas systems: Beyond adaptive immunity. *Nat Rev Microbiol*, 12, 317–326.

Wu J, Yan, G, Duan Z, Wang Z, et al. (2020). Roles of the Brassica napus DELLA protein BnaA6. RGA, in modulating drought tolerance by interacting with the ABA signaling component BnaA10. ABF2. *Front Plant Sci*, 11, 577.

Xie K, Minkenberg B, Yang Y (2015). Boosting CRISPR/Cas9 multiplex editing capability with the endogenous tRNA-processing system. *PNAS*, 112, 3570–3575.

Yang L, Machin F, Wang S, Saplaoura E, Kragler F (2023). Heritable transgene-free genome editing in plants by grafting wild-type shoots to transgenic donor rootstocks. *Nature Biotechnol*. DOI: 10.1038/s41587-022-01585-8.

Yu W, Wang L, Zhao R, Sheng J, et al. (2019). Knockout of SlMAPK3 enhances tolerance to heat stress involving ROS homeostasis in tomato plants. *BMC Plant Biol*, 19, 354.

Yue E, Cao H, Liu B (2020). OsmiR535, a potential genetic editing target for drought and salinity stress tolerance in *Oryza sativa*. *Plants*, 9, 1337.

Zeng Y, Wen J, Zhao W, Wang, Q, et al. (2020). Rational improvement of rice yield and cold tolerance by editing the three genes OsPIN5b, GS3, and OsMYB30 with the CRISPR–Cas9 system. *Front Plant Sci*, 10, 1663.

Zhang A, Liu Y, Wang F, Li T, et al. (2019). Enhanced rice salinity tolerance via CRISPR/Cas9-targeted mutagenesis of the OsRR22 gene. *Mol Breed*, 39, 47.

Zhang H, Zhang J, Wei P, Zhang B, et al. (2014). The CRISPR/Cas9 system produces specific and homozygous targeted gene editing in rice in one generation. *Plant Bioethanol J*, 12, 797–807.

Zong Y, Song Q, Li C, Jin S, Zhang D, et al. (2018). Efficient C-to-T base editing in plants using a fusion of nCas9 and human APOBEC3A. *Nat Biotechnol*, 36, 950–953.

Abbreviations

ABA	Abscisic acid
ABEs	Adenine base editors
CBEs	Cytosine base editors
CRISPR	Clustered regularly interspaced short palindromic repeats

crRNAs	Ribonucleoprotein complexes preloaded with small, interfering CRISPR RNAs
DSBs	Double-strand breaks
GET	Genome-editing techniques
GHGs	Greenhouse gases
GMOs	Genetically modified organisms
HDR	Homology-directed repairs
HR	Homologous recombination
IPCC	Intergovernmental Panel on Climate Change
NHEJ	Non-homologous end joining
PAM	Protospacer adjacent motif
SDNs	Site-directed nucleases
TALENs	Transcription activator-like effector nucleases
TLS	tRNA-like sequences
UGI	Uracil glycosylase inhibitor
UNFCCC	United Nations Parties on the Convention Framework on Climate Change
ZFNs	Zinc-finger nucleases

3 Genome editing by different site-specific nucleases and their applications in improving horticultural crops

Nicoleta Anca Şuţan and Aurel Popescu

3.1 Introduction

Horticultural crops are important dietary components for human life, providing essential nutrients. The constantly growing world population involves the need to significantly increase fruit and vegetable crop production, while ensuring healthy food products and environmental sustainability (Erpen-Dalla Corte et al. 2019).

Despite the fact that conventional breeding methods have an irrefutable contribution to the development of higher-yielding and better-adapted varieties, new approaches are needed for the added value of horticultural crops. Moreover, crop quality, which is in high demand from consumers and therefore has always been a very important breeding goal, is determined by both intrinsic and extrinsic factors, and consequently, it is difficult to make rapid advances through the crossing and other common breeding strategies (Yang et al. 2022). And not least, the horticultural plant breeders are aware of the fact that the new or improved varieties they will provide to producers must possess not only the most desirable traits, such as high-yield and superior quality of product/fruit, but also resistance or tolerance to specific stresses, as the horticultural crops are exposed to a large variety of biotic (e.g. insects, nematodes, fungi, bacteria, viruses, etc.) and abiotic stresses (e.g. drought, heat, cold, and salinity) that can strongly reduce yield and quality of products.

Traditionally, breeding new cultivars of horticultural crops is based on genetic variation. Amplification of genetic variation is fundamentally important in breeding programs (Holme et al. 2019). Therefore, overcoming the limits of genetic variation underlying quantitative traits is a challenge for plant breeders (Rodríguez-Leal et al. 2017). This obstacle of low genetic diversity is more serious in asexually reproducing species such as potatoes, grapes, and strawberries. However, in these species, this disadvantage is compensated by the easy propagation over many generations of a desirable quantitative genotype, once it has been generated (Xing et al. 2020).

Generally, through the process of genetic recombination, better-performing genotypes of fruit and vegetable crops can be generated, but their

DOI: 10.4324/9781003382102-3

selection is carried out over several generations. Moreover, in perennial fruit crops, the process of recurrent selection of progeny with the desired phenotype takes even more time, due to the prolonged juvenile phase that the seedlings go through.

In recent decades, the application of molecular markers for early selection and genomic technologies, for more efficient identification of quantitative trait loci have allowed the development of new horticultural plant varieties in a much shorter time. Also, genetic engineering has been used in combination with traditional breeding to produce genetically modified (transgenic) plant varieties improved for desired traits, and developed more quickly. Although the value and safety of transgenic horticultural crops continue to be demonstrated, their commercialization has lagged far behind. The deepening of research and the expansion of such genetically modified (GM) crops has slowed down, with only a few genetically engineered fruits and vegetables produced as commercial varieties in the last three decades – virus-resistant papaya (Rainbow and SunUp), virus-resistant squash (CZW3 and ZW20 events), insect-resistant eggplant (BARI Bt Begun), non-browning apple (Arctic™ Golden Delicious Apple; Arctic Granny™; Arctic™ Fuji Apple), and pink-fleshed pineapple (Pinkglow Pineapple).

The development of genome editing technologies, starting ten years ago with the CRISPR/Cas9 system, has provided new tools for the generation of improved varieties in horticultural crops.

Genome editing technologies were developed as a result of the continuous effort of scientists to find novel ways to precisely modify DNA, as well as the result of the search of plant breeders for new ways to increase genetic variation and broaden their repertoire of breeding methods. Although the new breeding techniques that have been developed and used by plant breeders over the last two decades allowed significant advances in plant crop improvement, genome editing is seen now as a revolutionary new technology, whose great potential has yet to be fully explored in various fields, including horticulture.

Through genome editing technology, exogenous sequences can be inserted at the cleavage site in a highly precise manner, dictated by an exogenous repair template whose homology arms flank the cleavage site. The sequence of repair template can be designed for precisely targeted changes, including, deletions, insertions, single nucleotide substitutions, as well as knock-ins and knock-outs. Genome-editing tools could be employed to improve any horticultural crops, including those plant species that contain complex genomes, are highly heterozygous (such as most fruit crops), and do not lend themselves to conventional breeding methods (Kim et al. 2021).

3.2 Genome editing by different site-specific nucleases (SSNs)

It has been accepted that site-specific nucleases (SSNs) are powerful, simple, precise, efficient, and rapid methods, allowing for achieving targeted gene/

genome engineering (Voytas, 2013; Doudna and Charpentier, 2014; Abdallah et al. 2015; Zhang et al. 2018; Ahmar et al. 2020).

There are currently available three powerful classes of SSNs that can be designed to create double-strand breaks at essentially any convenient target: (1) zinc-finger nucleases (ZFNs); (2) transcription activator-like effector nucleases (TALENs); and (3) CRISPR/Cas. Although the CRISPR/Cas system is employed most extensively in research laboratories around the world, the other two are still in use for research and various agricultural applications (Khan et al. 2017). As a result of their ability to make highly specific changes in the DNA sequence, SSNs enabled programmable gene editing, to create new crop varieties with altered functions and desired traits (Belhaj et al. 2013; Mao et al. 2013; Ran et al. 2013; Shan et al. 2013; Chen and Gao, 2014; Belhaj et al. 2015; Xiong et al. 2015; Karkute et al. 2017; Limera et al. 2017; Liu et al. 2017; Songstad et al. 2017; Nadakuduti et al. 2018; Chen et al. 2019; Erpen-Dalla Corte et al. 2019; Nadakuduti et al. 2019; Ahn et al. 2020; He and Zhao, 2020; Monsur et al. 2020; Selvakumar et al. 2020; Xu et al. 2020; Zhou et al. 2020; Ahmad, 2021; Nadakuduti and Enciso-Rodríguez, 2021; Savadi et al. 2021; Wan et al. 2021; Xia et al. 2021; ButiucKeul et al. 2022; Kumar et al. 2022; Liu et al. 2022; Tiwari et al. 2022).

3.2.1 Zinc-finger nucleases

Zinc-finger nucleases (ZFNs) are restriction enzymes specifically designed to recognize and cleave any long stretch of double-stranded DNA sequence (Weinthal et al. 2010). These engineered nucleases are obtained by binding zinc-finger protein domains and a nonspecific DNA cleavage domain from the restriction enzyme FokI (Carroll, 2011; Petolino, 2015; Xu et al. 2019). The zinc-finger protein domain in the constitution of such an artificial nuclease binds specifically to a DNA target sequence, and the FokI nuclease domain cleaves the DNA sequence generating double-strand breaks (DSBs) (Kim et al. 2021).

The cleavage event induced by such nucleases triggers cellular repair processes, resulting in efficient modification of the targeted locus. If the cleavage is removed via non-homologous end joining (NHEJ), small deletions or insertions can be generated, leading to gene knockout. If the break is repaired through a homology-based process, either gene correction or gene addition can be generated.

ZFNs, the first generation of SSNs, enable fast and precise (targeted) genetic and epigenetic modifications. Such SSNs have been designed to change various gene sequences in many crop plants (Davies et al. 2017; Kim et al. 2021), including several horticultural species, such as *Petunia* (Marton et al. 2010), apple, and fig (Peer et al. 2015). Based on the results obtained in apple and fig plants, in which *uidA* gene was mutated, then was repaired and plants were regenerated, Peer et al. (2015) concluded that ZNF may be successfully employed in the knockout of a target gene in apple or fig. They also emphasized that genome editing offers an advantage to non-transgenic breeding programs by being a marker gene-free technology related to antibiotic resistance.

Recent results showed that in a polyploidy crop (wheat), ZFNs can induce simultaneous multiple gene knockouts and modifications of the original gene sequence via NHEJ-directed DNA repair. The results of the same study also demonstrated that ZFNs can generate precise genome changes at both selectable and non-selectable trait loci (Ran et al. 2018). Analysis of the transformed plants confirmed that silenced target genes and edited genes via ZFNs are heritable and adopt the Mendelian inheritance pattern (Ran et al. 2018).

Despite the promising results of ZFNs use for genome editing, their range of application was restricted by several disadvantages, the most important being their imprecision and the risks of side effects, such as cellular toxicity (Mushtaq et al. 2019; Kim et al. 2021).

3.2.2 Transcription activator-like effector nucleases

Similar to ZFNs, transcription activator-like effector nucleases (TALENs) are synthesized by fusion of transcriptional activator-like effector (TALE) repeats and the FokIrestriction endonuclease (Joung and Sander, 2012), being designed and applied to improve plant resistance to biotic and abiotic stress factors and to increase yield and product quality. However, compared to ZFNs, each individual TALE repeat targets a single nucleotide, and consequently, the number of targeted sites is higher (Khan et al. 2017; Xu et al. 2019; Kim et al. 2021). Genome editing by TALENs has been applied to modify genes in some horticultural crops, such as potatoes, tomatoes, and cabbage (Xu et al. 2019; Alvarez et al. 2021). Clasen et al. (2016) reported that knocking out the vacuolar invertase gene (*VInv*) resulted in mutant potato plants (cv. Ranger Russet) characterized by significantly lower levels of reducing sugars in tubers and acrylamide in heat-treated products (chips) and were lightly colored. By targeting *FRIGIDA* gene with TALENs (engineered by method "unit assembly"), Sun et al. (2013) obtained early-flowering mutants of *Brassica oleracea* var. *capitata.* They reported that the DSBs created by TALENs within the target gene (as shown by sequencing data) were rejoined by NHEJ pathway. By using TALEN technology in tomato (*Solanum lycopersicum* L.), purple fruits with significantly higher levels of anthocyanin were obtained (Čermák et al. 2015). The authors increased the number of copies and the efficiency of homologous recombination by incorporating TALENs and donor DNA into geminivirus replicons. In fact, the increased level of anthocyanins was induced by inserting a strong promoter upstream of the gene that controls the synthesis of this secondary metabolite (Čermák et al. 2015). The potential of this state-of-the-art technology in crop plant improvement has been repeatedly demonstrated. However, TALEN genome editing still has obstacles to overcome, such as the increasing of gene-targeting performance and the construction of TALE repeats (Zhang et al. 2018).

3.2.3 Clustered regulatory interspaced short palindromic repeats (CRISPR)/ CRISPR-associated protein (Cas) system

From the recognition of its efficiency in genome editing in 2012 (Jinek et al. 2012), Cas9 from *Streptococcus pyogenes* (*Sp*Cas9) has become the most commonly used site-specific nuclease. Cas9 is an RNA-guided endonuclease able to recognize and cleave a specific DNA sequence in the genome. The specificity of the Cas9 is ensured by designing a programmable guide RNA (gRNA) complex to target any genomic location near the protospacer adjacent motif (PAM). This complex consisting of CRISPR RNA (crRNA) and a trans-activating CRISPR RNA (tracrRNA) leads the Cas9 to the genomic target site that is complementary to the crRNA. To simplify gRNA expression, a synthetic chimeric construct, called a single guide RNA (sgRNA) can be synthesized by fusing tracrRNA and crRNA (Jinek et al. 2012). The PAM required for the targeting of complementary sequences by the Cas9:gRNA complex is positioned downstream of the target DNA sequence (Xu et al. 2019), and the DSB will be created upstream of the PAM. DSB is repaired either by the NHEJ or by the homology-directed repair (HDR) pathways. Although it is the main way to repair DSBs in higher eukaryotic cells, NHEJ is considered an imprecise mechanism due to the simple way of re-joining broken ends, which often leads to disruptive insertions and deletions (indels) in coding exons, and thus determines gene knockouts. In comparison, HDR is a more precise mechanism, requiring a DNA template, but much less efficient than NHEJ (Santa Maria and Llorente, 2018).

More recently, Cas12a (previously known as Cfp1), an endonuclease that offers several desired characteristics not found in Cas9, became an attractive genome editing tool. This nuclease does not require a tracrRNA and cleave dsDNA directed by a crRNA (Zetsche et al. 2015). At the same time, the RuvC domain of Cas12a cut both the target and the non-target DNA sequences (Swarts et al. 2017; Swarts and Jinek, 2018). Both Cas9 and Cas12a endonucleases can induce multiple and simultaneous DSBs, by targeting different genomic loci.

In the past decade, CRISPR-Cas9 technology has recast genomic engineering, enabling targeted gene modification in almost any organism. In addition to multiplex editing capability, the CRISPR-Cas9 system has other important advantages, such as simplicity, greater precision, and reliability.

3.3 Applications of genome editing in the improvement of horticultural crops

Genome editing has received increasing attention in the most recent time and, consequently, the research aiming at the improvement of horticultural crops has been greatly accelerated. Notwithstanding that progress in working with polygenic traits is more problematic in comparison with monogenic traits, especially in polyploid and clonally propagated crops (e.g. potato, strawberry, banana, kiwifruit), remarkable success has been achieved in several horticultural crop plants (such as potato) for some traits.

One of the areas where the CRISPR/Cas9 system is most extensive is gene knockout (gene silencing)with exceptional application in plant functional genomics. In horticultural plant crops, the phytoene desaturase (*PDS*) gene was the most frequent target (Jia and Wang, 2014; Nishitani et al. 2016; Jia et al. 2017; Nakajima et al. 2017; Tian et al. 2017; Zhang et al. 2017; Kaur et al. 2018; Naim et al. 2018; Wang et al. 2018; Xu et al. 2019), its knockout resulting in the generation of albino mutant plantlets (Table 3.1), but many other genes were targeted as well (Čermák et al. 2015; Ito et al. 2015; Nonaka et al. 2017; Andersson et al. 2018; Deng et al. 2018; Li et al. 2018; Nakayasu et al. 2018; Klimek-Chodacka et al. 2018; Wang et al. 2019; Yang et al. 2019; Xu et al. 2019; Gonzalez et al., 2020; Hu et al. 2021). To date, CRISPR/Cas9-mediated genome editing allowed the improvement of at least 26 horticultural species. Among these, tomatoes and potatoes are the species to which the most remarkable results of precise breeding through genome editing have been achieved so far (Table 3.1).

The genes targeted so far by using genome editing tools are genes encoding important traits in horticultural plants, including resistance to pathogens, resistance/tolerance to abiotic stresses (drought, salinity, heat, chilling), fruit quality, fruit color, fruit size, early flowering, delayed flowering time, early ripening, shelf-life, starch quality, shoot branching, plant architecture, resistance/tolerance to herbicides, male sterility, self-incompatibility, day neutrality and early yielding, gynoecious, parthenocarpy/seedlessness (Table 3.1).

3.3.1 Improvement of resistance against fungal, bacterial, and viral pathogens

The achievements in editing genomes of horticultural species for resistance against specific pathogens are already remarkable (Table 3.1) and are so far the best proofs for the great potential of precision breeding using the CRISPR/Cas9 system and, less frequently, TALENs.

Gray mold, caused by *Botrytis cinerea* (*Botryotinia fuckeliana*), is one of the most common and damaging fungal diseases of many vegetable and fruit crops (in greenhouses and fields). Once established it's difficult to bring under control and it may be present all year round (especially in greenhouse crops). In order to perform further functional analysis of *SlMAPK3* in *S. lycopersicum* plant resistance to gray mold, Zhang et al. (2018) used two lines of *slmapk3* mutants and wild-type tomato plants. The authors found that *slmapk3* mutants were more sensitive to *B. cinerea*, and by *SlMAPK3* knockout the antioxidant enzymes activities were diminished and the accumulation of reactive oxygen species (ROS) increased. By detecting the expressions of salicylic acid and jasmonic acid signaling-related genes, they also reported that knockout of *SlMAPK3* enhanced the expressions of *SlPR1*, *SlPAD4*, and *SlEDS1*, whereas reduced the expressions of *SlLoxC, SlPI I,* and *SlPI II* and enhanced the expressions of *SlJAZ1* and *SlMYC2*. Based on

Table 3.1 Applications of genome editing tools in vegetable and fruit crops

Plant	Genome editing tool	Target gene(s)	Trait(s)/phenotype	References
Brassica oleracea	TALEN RNA-guided Cas9 Gibberellin 3-beta-dioxygenase 1	*FRIGIDA* (*BolC.GA4.a*)	Early flowering Plant development	Sun et al. (2013) Lawrenson et al. (2015)
	CRISPR	*BraFLCs*	KO / Early flowering	Jeong et al. (2019)
	CRISPR	Phytoene desaturase(*BoPDS*)	KO / Albino phenotype	Ma et al. (2019a)
	CRISPR	*BoPDS1, BoRK3, BoMS1*	KO / Albino phenotype, male sterility	Ma et al. (2019b)
	CRISPR	Flowering-time regulator *GIGANTEA*	Delayed flowering time	Park et al. (2019)
Brassica oleracea alboglabra	CRISPR	Phytoene desaturase(*BoPDS*)	KO / Albino phenotype	Lee et al. (2020)
	CRISPR	*BaPDS1, BaPDS2 BoaCRTISO*	KO / Albino phenotype Yellow color of Chinese broccoli	Sun et al. (2018)
Capsicum annuum	CRISPR	*CaMLO2*	KO / Susceptibility to mildew	Park et al. (2021)
Cicer arietinum	CRISPR	4-coumarate ligase (*4CL*), Reveille 7 (*RVE7*)	Drought tolerance	Badhan et al. (2021)
Citrulluslanatus	CRISPR	Phytoene desaturase (*ClPDS*)	Carotenoid biosynthesis KO / Albino phenotype	Tian et al. (2017)
	CRISPR	*ClALS*	Herbicide resistance	Tian et al. (2018)
	CRISPR	*Clpsk1*	KO / Resistance to *Fusarium oxysporum* f. sp. *Niveum*	Zhang et al. (2020)
	CRISPR	*ClWIP1*	KO / Gynoecious phenotype	Zhang et al. (2020)
	CRISPR	*ClGRF4-ClGIF1*	Enhanced regeneration from tissue culture	Feng et al. (2021)
Cucumis melo	CRISPR	Constitutive triple response 1 (*CmCTR1*) Repressor of silencing 1 (*CmROS1*)	Fruit ripening (climacteric ripening)	Giordano et al. (2022)
	CRISPR	Eukaryotic translation initiation factor 4E (*eif4e* or *EiF4E*)	Virus resistance, male sterility	Pechar et al. (2022)
	CRISPR	*ERECTA* receptor kinase genes	Plant architecture with shorter internodes	Xin et al. (2022)

Crop	SSN	Gene	Trait / Application	Reference
Cucumis sativus	CRISPR	Eukaryotic translation initiation factor 4E (*EiF4E*)	KO / Resistance to specific viruses (ZYMV and PRSMV)	Chandrasekaran et al. (2016)
	CRISPR	*CmWIP1*	KO / Gynoecious phenotype	Hu et al. (2017)
	CRISPR	*ERECTA* receptor kinase genes	Plant architecture with shorter internodes	Xin et al. (2022)
Cucurbita moschata	CRISPR	*ERECTA* receptor kinase genes	Plant architecture with shorter internodes	Xin et al. (2022)
Daucus carota	CRISPR	*PDS*, *MYB113*-like	KO / Albino phenotype	Xu et al. (2019)
	CRISPR	*F3H*	Altered anthocyanin biosynthesis	Klimek-Chodacka et al. (2018)
Ipomoea batatas	CRISPR	*IbGBSSI*, *IbSBEII*	Starch quality	Wang et al. (2019)
Lactuca sativa	CRISPR	Brassinosteriod insensitive 2 (*BIN2*)	Plant development	Woo et al. (2015)
	CRISPR	*LsNCED4*	Thermo-inhibition of seed germination	Bertier et al.(2018)
	CRISPR	*LsPDS*, *LsGPP2*, *LsBIN2*	KO / Albino phenotype, ascorbic acid biosynthesis	Pan et al. (2022)
Pisum sativum	CRISPR	Phytoene desaturase (*PsPDS*)	Carotenoid biosynthesis, **KO / Albino** phenotype	Li et al. (2022)
Solanum lycopersicum	CRISPR	Argonaute7 (*SlAGO7*)	Leaf development	Brooks et al. (2014)
	TALEN	*PROCERA*	Gibberellin signaling	Lor et al. (2014)
	TALEN, CRISPR	Anthocyanin 1 (*ANTI*)	Anthocyanin biosynthesis	Čermák et al. (2015)
	CRISPR	Ripening inhibitor (*RIN*)	Fruit ripening	Ito et al. (2015)
	CRISPR	Phytoene desaturase (*SlPDS*), Phytochrome interacting factor (*SlPIF4*)	Carotenoid biosynthesis	Pan et al. (2016)
	CRISPR	Phytoene synthase (*PSY1*)	Fruit colour	Filler Hayut et al. (2017)
	CRISPR	S1 agamous LIKE 6 (*S1AGL6*)	Parthenocarpy	Klap et al. (2017)
	CRISPR	Mildew-resistant locus o (*Melo*)	Powdery mildew (*Oidium neolycopersici*) resistance	Nekrasov et al. (2017)
	CRISPR	*SlGAD2*, *SlGAD3*	GABA content in tomato fruits (increased)	Nonaka et al. (2017)

(Continued)

Table 3.1 (Continued)

Plant	Genome editing tool	Target gene(s)	Trait(s)/phenotype	References
	CRISPR	SlCLV3 (promoter)	Fruit size, flower morphology, and locule number	Rodriguez-Leal et al. (2017)
	CRISPR	Solyc12g038510	Jointless mutant, abscission	Roldan et al. (2017)
	CRISPR*	DELLA, ETR1	Herbicide resistance	Shimatani et al. (2017)
	CRISPR	Self-pruning 5G (sp5G), self-pruning (sp)	Plant development	Soyk et al. (2017)
	CRISPR	SlIAA9	Parthenocarpy	Ueta et al. (2017)
	CRISPR	SlMAPK3	Drought stress	Wang et al. (2017)
	CRISPR	ALC	Long shelf-life	Yu et al. (2017)
	CRISPR	Psy1 and CrtR-b2	Carotenoid metabolism	D'Ambrosio et al. (2018)
	CRISPR	SlMYB12	Pink tomato fruit color	Deng et al. (2018)
	CRISPR	Elongation factor-1α (SlEF1α) promoter	Multiple mutations	Hashimoto et al. (2018)
	CRISPR	LeMADS-RIN	Ethylene production (reduced)	Jung et al. (2018)
	CRISPR	SlCBF1	Chilling tolerance	Li et al. (2018)
	CRISPR	SP, SP5G, SlCLV3, SlWUS	de novo-domestication of wild tomato	Li et al. (2018)
	CRISPR	GABA-TP1, GABA-TP2, GABA-TP3, CAT9 and SSADH	Amino-butyric acid synthesis	Li et al. (2018)
	CRISPR	Genes associated with the carotenoid metabolic pathway	Lycopene accumulation (enhanced)	Li et al. (2018)
	CRISPR		Resistance to tomato yellow leaf curl virus	Tashkandi et al. (2018)
	CRISPR	SP, O, FW2.2, CycB, FAS, MULT	Fruit number, size, shape, nutrient content, and plant architecture	Zsögön et al. (2018)
	CRISPR	CCD8	Resistance to the parasitic weed Phelipanche aegyptiaca	Bari et al. (2019)
	CRISPR	SlALS	Herbicide resistance (enhanced)	Danilo et al. (2019)
	CRISPR	SlJAZ2	Resistance to bacterial speck	Ortigosa et al. (2019)
	CRISPR	PROCERA	KO / Derepressed growth	Tomlinson et al. (2019)

Organism	Method	Gene	Trait / function	Reference
	CRISPR*	Acetolactate synthase (ALS)	Resistance to chlorsulfuron herbicide	Veillet et al. (2019a)
	CRISPR	Pectate lyase (PL), polygalacturonase 2a (PG2a), and beta-galactanase (TBG4)	Altered fruit color and firmness	Wang et al. (2019)
	CRISPR	SlMAPK3	KO / Tolerance to heat stress	Yu et al. (2019)
	Cas9*	GAI, ALS2, PDS1		Lu et al. (2021)
	CRISPR	PMR4	Resistance to powdery mildew (*Oidium neolycopersici*)	Santillán Martínez et al. (2020)
	CRISPR	CCD8	Resistance to the parasitic weed *Phelipancheaegyptiaca*	Bari et al. (2021)
	CRISPR	SlINVINH1 and SlVPE5	Soluble sugar content in tomato fruit	Wang et al. (2021)
	CRISPR	KLUH promoter	Fruit weight	Li et al. (2022)
Solanum melongena	CRISPR	SmelPPO	Enzymatic browning	Maioli et al. (2020)
Solanum tuberosum	CRISPR	Acetolactate synthase1 (StALS1)	Herbicide resistance (tolerance)	Butler et al. (2015)
	TALEN	StALS	Herbicide tolerance	Butler et al. (2016)
	TALEN	Vacuolar invertase (VInv)	KO / Accumulation of reducing sugars	Clasen et al. (2016)
	TALEN	Starch branching enzyme (Sbe1, Sbe2)	Starch quality (increased amylose ratio, and elongated amylopectin chains)	Ma et al. (2017)
	CRISPR	StMYB44	KO / Phosphate transport	Zhou et al. (2017)
	CRISPR	Granule bound starch synthase (GBSS)	KO of the GBSS enzyme function	Andersson et al. (2018)
	CRISPR	StALS1, StALS2	Starch biosynthesis	Kusano et al. (2018)
	CRISPR	Steroid 16α-hydroxylase (St16DOX)	α-solanine and α-chaconine biosynthesis	Nakayasu et al. (2018)
	CRISPR	Stilar ribonuclease gene (S-RNase)	Self-incompatibility	Ye et al. (2018)
	CRISPR	Stilar ribonuclease gene (S-RNase)	Self-incompatibility	Enciso-Rodriguez et al. (2019)
	CRISPR*	Acetolactate synthase (ALS)	Resistance to chlorsulfuron herbicide	Veillet et al. (2019a)
	CRISPR	StALS1, StALS2	Herbicide tolerance	Veillet et al. (2019a)
	CRISPR		Virus resistance	Zhan et al. (2019)
	CRISPR	Phytoene desaturase (StPDS)	KO / Albino phenotype	Bánfalvi et al. (2020)

(Continued)

Table 3.1 (Continued)

Plant	Genome editing tool	Target gene(s)	Trait(s)/phenotype	References
	CRISPR	*StALS1, StALS2*	Herbicide tolerance	Chauvin et al. (2021)
	CRISPR	*StDND1, StCHL1,* and *StDMR6-1*	Late blight resistance	Kieu et al. (2021)
	CRISPR	Starch branching enzyme (*Sbe*)	Starch quality (increased amylose ratio, and elongated amylopectin chains)	Zhao et al. (2021)
	CRISPR	Sterol side chain reductase 2 (*StSSR2*)	Concentration of steroidal glycoalkaloids (SGAs are beneficial for providing resistance to bacterial and viral disease, but are toxic to humans and animals)	Zheng et al. (2021)
Actinidia chinensis	CRISPR	Vacuolar invertase (*VInv*)	KO / Cold-induced sweetening	Yasmeen et al. (2022)
	CRISPR PTG/Cas9	Phytoene desaturase (*PDS*)	KO / Albino phenotype	Wang et al. (2018)
Carica papaya	CRISPR	Phytoene desaturase (*PDS*)	KO / Albino phenotype	Brewer and Chambers (2022)
Citrus paradisi	dTALE	*CsLOB1* (promoter)	Susceptibility to *Xanthomonas citri* subsp. *citri* (citrus canker)	Jia et al. (2016)
Citrus sinensis	CRISPR	Phytoene desaturase (*PDS*)	Albino phenotype (KO)	Jia et al. (2017)
	CRISPR Cas9/sgRNA	Phytoene desaturase (*PDS*)	KO / Albino phenotype, decrease of carotenoid content	Jia and Wang (2014)
	CRISPR	Susceptibility gene *CsLOB1* promoter	Citrus canker-resistance	Peng et al. (2017)
	CRISPR	Phytoene desaturase (*CsPDS*)	KO / Albino phenotype	Zhang et al. (2017)
	CRISPR	Phytoene desaturase (*CsPDS*) *CsLOB1* (promoter)	KO / Albino phenotype	Jia et al. (2019)
	CRISPR	*CsWRKY22*	Susceptibility to *Xanthomonas citri* subsp. *citri*	Wang et al. (2019)
	CRISPR HypaCas9/	Phytoene desaturase (*CsPDS*) *CsDMR6*	KO / Albino phenotype	Dutt et al. (2020) Parajuli et al. (2022)

Crop	SSN	Target gene	Application	Reference
			Resistance to bacterial disease (citrus canker)	
Fragaria x ananassa	CRISPR	*FaTM6*	Anther development	Martín-Pizarro et al. (2019)
Fragaria vesca	CRISPR	Phytoene desaturase (*FvPDS*)	KO / Albino phenotype	Xing et al. (2018)
	CRISPR	*TAA1 ARF8*	Auxin biosynthesis and signaling	Zhou et al. (2018)
	BE	*FvebZIPs1.1*	Sugar content	Xing et al. (2020)
Malus domestica	CRISPR	*DIPM-1, DIPM-2, DIPM-4*	Resistance to fire blight disease	Malnoy et al. (2016)
	CRISPR	Phytoene desaturase (*PDS*)	Albino phenotype (KO), decrease of carotenoid content	Nishitani et al. (2016)
	CRISPR	*DIPM-1, DIPM-2, DIPM-4*	Resistance to fire blight disease	Osakabe et al. (2018)
	CRISPR	Phytoene desaturase (*PDS*) Terminal flower (*MdTFL1*)	KO / Albino phenotype Early flowering	Charrier et al. (2019)
	CRISPR	*MdDIPM4*	Resistance to fire blight disease (Reduced susceptibility to *Erwinia amylovora*)	Pompili et al. (2020)
	CRISPR	Phytoene desaturase (*MdPDS*)	KO / Albino phenotype	Malabarba et al. (2021)
	CRISPR	Phytoene desaturase (*MdPDS*)	KO / Albino phenotype	Schröpfer and Flachowsky (2021)
Musa spp.	CRISPR	Phytoene desaturase (*PDS*)	Albino phenotype (KO)	Kaur et al. (2018)
	CRISPR	Phytoene desaturase (*PDS*)	Albino phenotype (KO)	Naim et al. (2018)
	CRISPR	*eBSV*	Control of virus pathogenesis	Tripathi et al. (2019)
	CRISPR	Lycopene epsilon-cyclase (*LCYε*)	β-carotene synthesis	Kaur et al. (2020)
	CRISPR	*DMR6*	Resistance to bacterial disease	Tripathi et al. (2021)
	CRISPR	Susceptibility (*S*) genes	Resistance to bacterial disease	Tripathi et al. (2022)
Pyrus communis	CRISPR	Terminal flower (*PcTFL1*)	KO / Early flowering	Charrier et al. (2019)
	CRISPR	Phytoene desaturase (*PcPDS*)	KO / Albino phenotype	Malabarba et al. (2021)
Theobroma cacao	CRISPR	Non-Expressor of Pathogenesis-Related 3 (*TcNPR3*)	Resistance to cacao pathogen *Phytophthora tropicalis*	Fister et al. (2018)
Vitis vinifera	CRISPR	*MLO-7*	Resistance to powdery mildew	Malnoy et al. (2016)
	CRISPR	*IdnDH*	Biosynthesis of tartaric acid	Ren et al. (2016)
	CRISPR	Phytoene desaturase (*PDS*)		Nakajima et al. (2017)

(Continued)

Table 3.1 (Continued)

Plant	Genome editing tool	Target gene(s)	Trait(s)/phenotype	References
	CRISPR	*VvRKY52*	KO / Albino phenotype, decrease of carotenoid content	Wang et al. (2018)
	CRISPR	Phytoene desaturase (*PDS*)	Biotic stress response	Ren et al. (2019)
	CRISPR	*VvCCD8*	KO / Albino phenotype	Ren et al. (2020)
	CRISPR	*VvMLO3*	Shoot branching	Wan et al. (2020)
			Resistance to powdery mildew (*Erysiphe necator*)	
	CRISPR	*TMT1, TMT2.*	Sugar levels	Ren et al. (2021)

these findings, Zhang et al. (2018) appreciated that *SlMAPK3 has* a significant contribution to tomato plant defense against *B. cinerea* through the regulation of antioxidant response and salicylic acid and jasmonic acid signaling pathways.

By targeting *WRKY52* gene (a member of the grape WRKY transcription factor family, that regulates stress tolerance in plants), Wang et al. (2018) obtained edited grape (*Vitis vinifera* L., cv. Thompson Seedless) transformants resistant to gray mold (*Botrytis cinerea* Fr.), the fungus causing one of the main diseases of grapes (Table 3.1).

In watermelon seedlings (*Citrullus vulgaris* (Thunb.) Schrad), by knocking out the phytosulfokine1(*ClPSK1*) gene, Zhang et al. (2020) induced an enhanced resistance to infection with the filamentous fungus *Fusarium oxysporum*f. sp. *niveum*.

Knowing that *Erwinia amylovora*, bacterium causing fire blight disease in apple (as well in pear, quince, and some other species in the Rosaceae family), triggers its infection through a mechanism involving the susceptibility protein *MdDIPM4*, Pompili et al. (2020) used the CRISPR/Cas9 system to knock out the *MdDIPM4* gene in two apples (*Malus* x *domestica*) susceptible cultivars (Gala and Golden Delicious). An editing efficiency of 75% was calculated for the fifty-seven transgenic lines that were screened (using high-throughput sequencing) to identify CRISPR/Cas9-induced mutations in the *MdDIPM4* target region. Gene-edited apple lines with a loss-of-function mutation showed a highly significant decrease in susceptibility to fireblight (when inoculated with the pathogen), compared to control plants.

CRISPR/Cas technology has also been used successfully to engineer viral resistance (Robertson et al. 2022), as part of the strategy for increasing plant productivity. CRISPR/Cas9 system was used to edit the *elF4E* gene of cucumber (*Cucumis sativus* L.) and improved resistance of the edited plants against three dominant potyviruses infecting the cucurbits, respectively was reported by Chandrasekaran et al. (2016). After three generations of back-crossing, the homozygous non-transgenic plants exhibited effective resistance against zucchini yellow mosaic virus (ZYMV), papaya ringspot mosaic virus-W (PRSV-W) and cucumber vein yellowing virus (CVYV), and, while the heterozygous mutant and non-mutant plants were highly permissive to viral infections. More recently, Tashkandi et al. (2018) used the CRISPR/Cas9 systems to confer durable virus resistance *in planta*. The engineered *S.lycopersicum* lines expressing CRISPR/Cas9 proved to be capable to target and cleave the sequences from tomato yellow leaf curls virus(TYLCV) genome encoding the coat protein or replicase, thus conferring immunity against this virus, as confirmed by low content of TYLCV particles in regenerated tomato plants.

Since banana (*Musa* sp.) production is seriously restricted by many pathogens and pests (especially where they are co-existing), growing disease-resistant banana varieties is desirable to reduce their unfavorable impacts (Tripathi et al. 2020). CRISPR/Cas9-based genome editing can be applied for developing disease-resistant varieties of bananas. Triploid bananas (*M. acuminata* Cavendish) resistant to pathogens are the outcome of hybridizations between diploid bananas *M. balbisiana* and its derivatives, which are characterized by a well-developed root system and abiotic stress endurance, but carriers of endogenous banana streak viruses (BSV). By using the same technology, Tripathi et al. (2019) generated mutations in integrated copies of BSV, and altered transcription and/or translation of functional viral proteins. Later, Tripathi et al. (2022) reported that the same strategy, based on CRISPR/Cas9-mediated gene/genome editing (either by knocking out the disease-causing susceptibility *S* genes or by activating the expression of the plant defense genes), is used to design improved banana plants with low susceptibility to bacterial diseases such as banana *Xanthomonas* wilt (BXW), Moko and blood diseases.

3.3.2 *Improvement of abiotic stress resistance*

Soil salinization has become one of the major environmental and socioeconomic issues globally and this is expected to be exacerbated further because of climate change. Therefore, salt tolerance is an important breeding objective in many important horticultural crops, and the significant contribution of genome editing to the development of plant varieties with increased salinity tolerance was anticipated. Mutation in the *SlARF4* gene, involved in tolerance to salinity and drought in tomatoes, was induced by CRISPR/Cas9. Bouzroud et al. (2020) have shown that down-regulation of *SlARF4*, induced a favorable response of plants to osmotic and saline stress, presenting a well-developed root system, a higher content of soluble sugars and chlorophyll and a lower stomatal conductance, regardless of normal or stress conditions.

Genome editing systems are now being used to unravel the role and understand the function of certain genes presumably involved in drought, temperature, and salinity stresses, as an essential part of strategies aiming at improving tolerance to such abiotic stresses (Li et al. 2022).

Drought tolerance would be an important attribute of certain varieties of horticultural plants, mainly vegetables. For instance, drought stress is one of the most devastating environmental factors with detrimental effects on the growth of tomato plants. Wang et al. (2017) and Li et al. (2019) reported the use of CRISPR/Cas9 system to obtain *SlMAPK3* and respectively *SlNPR1* mutants in *S. lycopersicum*, in order to establish their role in drought tolerance. They found that these genes are involved in drought response, modulating transcription of stress-related genes (such as *SlDREB*, *SlGST*, and *SlLOX*) by either up- or down-regulation of their expression, and having a role in protecting cell membranes from oxidative damage. More recently, Badhan et al. (2021) used CRISPR/Cas9 system to unravel the role of Reveille 7 (*RVE7*) and 4-coumarate ligase (*4CL*) genes in the drought stress mechanism in chickpea (*Cicer arietinum*), by their knockout via targeted mutagenesis. An example of heat tolerance was reported by Klap et al. (2017)

in tomatoes. Improved fruit setting under heat stress was promoted by *SlAGL6* gene knockout.

3.3.3 *Improvement of fruit and vegetable quality*

Improvement of quality in fruits and vegetables is a primary goal for horticultural plant breeders. CRISPR/Cas9 tools have already been used to unravel the role of genes likely to be involved in quality traits (including fruit size, color, and firmness), or even to create varieties with improved quality and nutritional value (Nonaka et al. 2017; Li et al. 2018; Maioli et al. 2020; Wang et al. 2021; Li et al. 2022).

Aiming at the improvement of starch quality, Andersson et al. (2018) used CRISPR-Cas9 to modify the starch contents (quality) in tetraploid potatoes by knocking out the *GBSS* (granule-bound starch synthase) gene in potato (*Solanum tuberosum*). They found an increased amylopectin/amylose ratio in the potato lines in which GBSS enzyme activity was suppressed by the knockout of all *GBSS* four alleles.

More recently, Wang et al. (2019) improved starch quality in sweet potatoes (*Ipomoea batatas*(L.) Lam). By targeting *IbGBSSI* and *IbSBEII* genes, both involved in the starch biosynthetic pathway, they found that the knockout of *IbGBSSI* gene reduced, while the knockout of the *IbSBEII* gene increased the amylose percentage in the allopolyploid sweet potato (Table 3.1).

Kaur et al. (2020) are the first who reported improvement in a nutritional trait by using the CRISPR/Cas9 technology. The genome editing approach allowed them to create β-carotene-enriched lines of the Cavendish banana cultivar. Compared with the unedited plants, selected edited lines showed a significantly increased content of β-carotene in their fruit pulp.

Strawberry (*Fragaria vesca*) is another fruit species in which genome engineering was used to improve a fruit quality trait. Xing et al. (2020) were successful in altering the sugar content of strawberries by editing the gene *FvebZIPs1.1* using a base editor (A3A-PBE). They found that the sugar content in fruits of homozygous T1 mutant lines could be significantly higher than that noted for the wild-type. Also, they produced transgene-free strawberry mutants (35 novel genotypes) whose high sugar content represented a trait that was passed down from one generation to another. Xing et al. (2020) pointed out that genome editing in connection with asexual reproduction may provide great opportunities to improve quantitative traits, that could be easily passed on to the next generation by asexual reproduction.

3.3.4 *Improvement of other traits*

It is well known the fact that the improvement of quantitative traits often relies on the hardworking selection of rare natural and spontaneous mutations that occur by chance in regulatory regions of genes. For example, yield is under the control of quantitative trait loci (QTLs) that affect fruit number, weight, and size. An innovative approach based on genome engineering was reported by Rodríguez-Leal et al. (2017), who proved that the editing of gene promoters via CRISPR/Cas9 tools induces cis-regulatory alleles whose regulatory changes can beneficially impact the quantitative traits in tomatoes. CRISPR/Cas system was also used to knockout genes that prejudice crop yields. Thus, the same authors induced changes in *SlCLV3* promoter and *SlENO* gene and increased tomato fruit size.

Shelf-life is desirable (and sometimes critical) in many horticultural crops and, therefore, there is a real interest of breeders for this trait. Yu et al. (2017) reported that the use of *Agrobacterium tumefaciens*-mediated CRISPR/Cas9 transformation allowed them to obtain tomato plants showing excellent storage performance (shelf-life trait) as a result of the mutation/replacement of *ALC* gene (see abbreviations list).

Parthenocarpy (the ability to develop fruits without pollination) is a useful trait in certain fruit and vegetable crops such as grapes, watermelon, citrus, and banana (especially in those species in which there is a growing market demand for seedless fruits). Klap et al. (2017) demonstrated that parthenocarpy can be achieved by knockout or mutation of the *SlAGL6* and *SlIAA9* genes by using CRISPR/Cas tools.

Gynoecious inbred lines (bearing only female flowers) have great importance in some vegetable crops, such as cucumber and squash, due to their positive impact on yield and cost-effectiveness required for crossing. Hybrid cucumber cultivars are produced by crossing one inbred parents, gynoecious female, and monoecious male, respectively. Targeting the *CsWIP1* gene (encoding a transcription factor) in *Cucumis sativus* via the CRISPR/Cas9 tools, Hu et al. (2017) generated *Cswip1* mutants exhibiting gynoecious phenotype. Zhang et al. (2020) generated gynoecious watermelon (*Citrullus lanatus*) lines by targeting the *ClWIP1* gene using the genome editing approach.

The CRISPR/Cas technology was also used for the identification of genetic factors for male infertility and to the elucidation of the mechanisms underlying this functional inability, and subsequently to generate artificial male-sterile lines. Feng et al. (2021) reported the development of male sterility *C. lanatus* after *eIF4E* knockout through a one-base deletion mediated by CRISPR/Cas9 system. The F2 generation homozygous melon exhibited Moroccan watermelon mosaic virus (MWMV) resistance along with male sterility phenotype.

Herbicide resistance, as a breeding goal, is beneficial primarily for farm management, whereas benefits to consumers are less obvious. However, in many horticultural crops (mainly vegetables) there is a constant interest in introducing this trait. Although the gene transfer technology was proven to be very efficient for creating transgenic plants with resistance/tolerance to various herbicides, such varieties are rarely cultivated, due to the strong public opposition to genetically modified organisms. Genome editing can be an alternative to transgenic engineering with a better chance to gain public acceptance. In watermelon (*C. lanatus*), the CRISPR-Cas9-mediated base editing system was employed to cause single-nucleotide mutation within the acetolactate synthase (*ClALS*) gene (Tian et al. 2018), encoding a catalytic enzyme involved in the synthesis pathway of three important amino acids (valine, leucine, and isoleucine). The edited watermelon plants possessing C to T conversion exhibited resistance to sulfonylurea herbicides without any change in fruit and seed traits. More recently, Veillet et al. (2020) developed *S. lycopersicum* and *S. tuberosum* plants resistant to chlorsulfuron (a selective and systemic herbicide) targeting the acetolactate synthase (*ALS*) gene by a cytidine base editor (CBE) using *Agrobacterium*-mediated transformation.

A really interesting result of precision breeding was described by Zsögön et al. (2018), who enforced CRISPR/Cas9 genome editing to associate relevant agronomic traits with beneficial traits present in wild lines of *Solanum pimpinellifolium*. Targeting 6 quantitative trait loci important for modern tomato cultivars resulted in *de novo* domestication of wild tomatoes. The CRISPR/Cas edited tomato lines have a significant increase in fruit size (threefold) and fruit number (tenfold). Moreover, fruit lycopene accumulation in edited lines was reported to be improved by 500% compared with that in fruits of the cultivated tomato. Having in mind the useful traits of wild species, such as biotic and abiotic stress tolerance, that have been lost in modern varieties of horticultural crops, these results are even more interesting.

3.4 Limitations and perspectives of genome editing in horticultural crops

CRISPR/Cas9 genome editing technology is a powerful and enthralling tool for studying and describing gene functions and the process of improving cultivars/varieties of horticultural crops (mainly vegetables and fruits); however, there are several limitations of the CRISPR/Cas9 editing technology, that are necessary to consider: (i) the requirement for a specific PAM sequence in the target region; (ii) the necessity of gRNA/Cas9 expression optimization for a plant or plant variety; (iii) the Cas9p fidelity: CRISPR/Cas9 predominantly makes indel mutations and is commonly used for gene knockouts (Fizikova et al. 2021). Therefore, more efforts need to be made to enhance the efficiency and specificity of CRISPR-Cas9 techniques.

The use of changed Cas9p partially or completely independent of the PAM sequence seems to be a promising approach. However, PAM limitation remains one of the major inconveniences of SpCas9, and therefore more diversity in CRISPR/Cas toolbox is required (Veillet et al. 2020). Also, we have to consider that many important horticultural traits are multigenic, while CRISPR-Cas9 was used so far, and seems to be suitable to target one or multiple genetic loci (Kim et al. 2021).

Successful genetic transformation of fruit and vegetable crops through gene/genome editing requires the consideration of editing efficiency (Xu et al.

2019; Zhou et al. 2020; Wan et al. 2021). Higher Cas9 expression and effective targets are important factors to increase the efficiency of the CRISPR/Cas system, although designing two or more targets for one gene can improve the probability of obtaining homozygous mutations in the first generation (Fan et al. 2015). This is of particular significance for woody horticultural plants, such as grape (*Vitis vinifera* L.) or apple (*Malus domestica* Borkh.), with long lifespans.

In recent years, alternatives to the wild-type *Streptococcus pyogenes* Cas9 (SpCas9) have been engineered, including high-fidelity variants. The resulting SpCas9-HF1 nuclease displays genome-wide specificity and undetectable off-target effects, all with a single guide RNA.

A very recent report shows that the advances and major improvements needed for high-efficiency gene editing of plants are achievable in a short time. A research team from the University of Texas has designed a new variant of Cas9 (called SuperFi-Cas9), which has similar high rates of DNA cleavage, but acts 4,000 times more precisely compared to Cas9, thus preventing the off-target activity (Bravo et al. 2022).

However, low efficiency in the introduction of edited sequences at target sites was already solved once the prime editing was developed (Anzalone et al. 2019). Cytosine base editors (CBEs) and adenine base editors (ABEs) are now available (Li et al. 2019). CRISPR/Cas9 base editors that have been recently developed can perform precise genome manipulation as they were proven to have broadened editing capabilities and improved editing specificity (Gaudelli et al. 2017). CBEs has been already used in various crops, including tomato and potato (Veillet et al. 2019b), strawberry (Xing et al. 2020), apple and pear (Malabarba et al. 2021), with relatively high efficiency.

It is expected that more efficient gene-/genome-editing systems, with fewer off-target effects, will be available soon and updated continuously (Kim et al. 2021). Therefore, it is not only a hope, but already a certitude, that the new tools provided by genome editing will enable faster production of new horticultural crop varieties, higher-yielding, with improved and better nutritional characteristics, resistant to specific pests and pathogens, better adapted or with greater tolerance to abiotic stresses.

Acknowledgment

N.A.Ş. gratefully acknowledge the financial support of the Romanian Ministry of Research, Innovation and Digitization, CNCS/CCCDI-UEFIS-CDI, through the Project number PN-III-P4-ID-PCE-2020-0620, within PNCDI III.

Authors Contributions The authors equally contributed to writing the manuscript.

The authors declare no competing interests.

References

Abdallah, N.A., Prakash, C.S., & McHughen, A.G. (2015). Genome editing for crop improvement: challenges and opportunities. *GM Crops Food*, 6, 183–205. doi: 10. 1080/21645698.2015.1129937.

Ahmad, I. (2021). Genome Editing in Fruit Tree Crops. Prospects and Challenges. In O.P. Gupta & S.G. Karkute (Eds.), *Genome Editing in Plants. Principles and Applications*. CRC Press

Ahmar, S., Saeed, S., Ullah Khan, M.H., Ullah Khan, S., Mora-Poblete, F., Kamran, M., Faheem, A., Maqsood, A., Rauf, M., Saleem, S., Hong, W.J., & Jung, K.H. (2020). A revolution toward gene-editing technology and its application to crop improvement. *International Journal of Molecular Sciences*, 21, 5665. doi: 10.3390/ijms21165665

Ahn, C.H., Ramya, M., An, H.R., Park, P.M., Kim, Y.J., Lee, S.Y., & Jang, S. (2020). Progress and challenges in the improvement of ornamental plants by genome editing. *Plants*, 9, 687. doi: 10.3390/plants9060687

Alvarez, D., Cerda-Bennasser, P., Stowe, E., Ramirez-Torres, F.,Capell, T., Dhingra, A., & Christou, P. (2021). Fruit crops in the era of genome editing: closing the regulatory gap. *Plant Cell Reports*, 40(6), 915–930. doi: 10.1007/s00299-021-02664-x

Andersson, M., Turesson, H., Olsson, N., Fält, A.-S., Ohlsson, P., Gonzalez, M.N., Samuelsson, M., & Hofvander, P. (2018). Genome editing in potato via CRISPR-Cas9 ribonucleoprotein delivery. *Physiologia Plantarum*, 164(4), 378–384. doi: 10.1111/ppl.12731

Anzalone, A.V., Randolph, P.B., Davis, J.R., Sousa, A.A., Koblan, L.W., Levy, J.M., Chen, P.J., Wilson, C., Newby, G.A., Raguram, A., & Liu, D.R. (2019). Search-and-replace genome editing without double-strand breaks or donor DNA. *Nature*, 576, 149–157. doi: 10.1038/s41586-019-1711-4

Badhan, S., Ball, A., & Mantri, N. (2021). First report of CRISPR/Cas9 mediated DNA-free editing of *4CL* and *RVE7* genes in chickpea protoplasts. *International Journal of Molecular Science*, 22(1), 1–15. doi: 10.3390/ijms22010396

Bánfalvi, Z., Csákvári, E., Villányi, V., & Kondrák, M. (2020). Generation of transgene-free PDS mutants in potato by *Agrobacterium*-mediated transformation. *BMC Biotechnology*, 20, 25. doi: 10.1186/s12896-020-00621-2

Bari, V.K., Nassar, J.A., Kheredin, S.M., Gal-On, A., Ron, M., Britt, A., Steele, D., Yoder, J., & Aly, R. (2019). CRISPR/Cas9-mediated mutagenesis of *carotenoid cleavage dioxygenase 8* in tomato provides resistance against the parasitic weed *Phelipancheaegyptiaca*. *Scientific Reports*, 9, 11438. doi: 10.1038/s41598-019-47893-z

Bari, V.K., Nassar, J.A., & Aly, R. (2021). CRISPR/Cas9 mediated mutagenesis of *more axillary growth 1* in tomato confers resistance to root parasitic weed *Phelipancheaegyptiaca*. *Scientific Reports*, 11, 3905. doi: 10.1038/s41598-021-82897-8

Belhaj, K., Chaparro-Garcia, A., Kamoun, S., & Nekrasov, V. (2013). Plant genome editing made easy: targeted mutagenesis in model and crop plants using the CRISPR/Cas system. *Plant Methods*, 9, 39. doi: 10.1186/1746-4811-9-39

Belhaj, K., Chaparro-Garcia, A., Kamoun, S., Patron, N., & Nekrasov, V. (2015). Editing plant genomes with CRISPR/Cas9. *Current Opinion in Biotechnology*, 32, 76–84. doi: 10.1016/j.copbio.2014.11.007

Bertier, L.D., Ron, M., Huo, H., Bradford, K.J., Britt, A.B., & Michelmore, R.W. (2018). High-resolution analysis of the efficiency, heritability, and editing outcomes of CRISPR/Cas9-induced modifications of NCED4 in lettuce (*Lactuca sativa*). *G3* (Bethesda), *8*, 1513–1521. doi: 10.1534/g3.117.300396

Bouzroud, S., Gasparini, K., Hu, G., Barbosa, M.A.M., Rosa, B.L., Fahr, M., Bendaou, N., Bouzayen, M., Zsögön, A., Smouni, A., & Zouine, M. (2020). Down Regulation and Loss of Auxin Response Factor 4 Function Using CRISPR/Cas9 Alters Plant Growth, Stomatal Function and Improves Tomato Tolerance to Salinity and Osmotic Stress. Genes (Basel), Mar 3; *11*(3), 272. doi: 10.3390/genes11 030272. PMID: 32138192; PMCID: PMC7140898.

Bravo, J.P.K., Liu , M.S., Hibshman, G.N., Dangerfield, T.L., Jung, K., McCool, R.S., Johnson, K.A., & Taylor, D.W. (2022). Structural basis for mismatch surveillance by CRISPR-Cas9. Nature, *603*(7900), 343–347. doi: 10.1038/s41586-022-04470-1. Epub 2022 Mar 2. Erratum in: Nature. 2022 Apr;604(7904):E10. PMID: 35236982; PMCID: PMC8907077.

Brewer, S.E., & Chambers, A.H. (2022). CRISPR/Cas9-mediated genome editing of phytoene desaturase in *Carica papaya* L. *The Journal of Horticultural Science and Biotechnology*, *97*(5), 580–592. doi: 10.1080/14620316.2022.2038699

Brooks, C., Nekrasov, V., Lippman, Z.B., & Van Eck, J. (2014). Efficient gene editing in tomato in the first generation using the clustered regularly interspaced short palindromic repeats/CRISPR-associated 9 system. *Plant Physiology*, *166*(3), 1292–1297. doi: 10.1104/pp.114.247577

ButiucKeul, A., Farkas, A., Carpa, R., Dobrota, C., & Iordache, D. (2022). Development of smart fruit crops by genome editing. *Turkish Journal of Agriculture and Forestry*, *46*(2), 129–140. doi: 10.55730/1300-011X.2965

Butler, N.M., Atkins, P.A., Voytas, D.F., & Douches, D.S. (2015). Generation and inheritance of targeted mutations in potato (*Solanum tuberosum* L.) using the CRISPR/Cas system. *PLoS ONE*, *10*, Article e0144591. doi: 10.1371/journal.pone.0144591

Butler, N.M., Baltes, N.J., Voytas, D.F., & Douches, D.S. (2016). Geminivirus-mediated genome editing in potato (*Solanum tuberosum* L.) using sequence-specific nucleases. *Frontiers in Plant Science*, *7*, 1045. doi: 10.3389/fpls.2016.01045

Carroll, D. (2011). Genome engineering with zinc-finger nucleases. *Genetics*, *188*(4),773–782. doi: 10.1534/genetics.111.131433.

Čermák, T., Baltes, N.J., Čegan, R., Zhang, Y., & Voytas, D.F. (2015). High-frequency, precise modification of the tomato genome. *Genome Biology*, 16, 232. doi: 10.1186/s13059-015-0796-9

Chandrasekaran, J., Brumin, M., Wolf, D., Leibman, D., Klap, C., Pearlsman, M., Sherman, A., Arazi, T., & Gal-On, A. (2016). Development of broad virus resistance in non-transgenic cucumber using CRISPR/Cas9 technology. *Molecular Plant Pathology*, *17*, 1140–1153. doi: 10.1111/mpp.12375.

Charrier, A., Vergne, E., Dousset, N., Richer, A., Petiteau, A., & Chevreau, E. (2019). Efficient targeted mutagenesis in apple and first-time edition of pear using the CRISPR-Cas9 system. *Frontiers in Plant Science*, *10*, 40. doi: 10.3389/fpls.2019.00040

Chauvin, L., Sevestre, F., Lukan, T., Nogué, F., Gallois, J.L., Chauvin, J.E., & Veillet, F. (2021). Gene editing in potato using CRISPR-Cas9 technology. *Methods in Molecular Biology*, *2354*, 331–351. doi: 10.1007/978-1-0716-1609-3_16

Chen, K., & Gao, C. (2014). Targeted genome modification technologies and their applications in crop improvements. *Plant Cell Reports*, *33*, 575–583. doi: 10.1007/s00299-013-1539-6

Chen, K., Wang, Y., Zhang, R., Zhang, H., & Gao, C. (2019). CRISPR/Cas genome editing and precision plant breeding in agriculture. *Annual Review of Plant Biology, 70*, 667–697. doi: 10.1146/annurev-arplant-050718-100049

Clasen, B.M., Stoddard, T.J., Luo, S., Demorest, Z.L., Li, J., Cedrone, F., Tibebu, R., Davison, S., Ray, E.E., Daulhac, A., Coffman, A., Yabandith, A., Retterath, A., Haun, W., Baltes, N.J., Mathis, L., Voytas, D.F., & Zhang, F. (2016). Improving cold storage and processing traits in potato through targeted gene knockout. *Plant Biotechnology Journal, 14*, 169–176. doi: 10.1111/pbi.12370

D'Ambrosio, C., Stigliani, A.L., & Giorio, G. (2018). CRISPR/Cas9 editing of carotenoid genes in tomato. *Transgenic Research, 27*(4), 367–378.doi: 10.1007/s1124 8-018-0079-9

Danilo, B., Perrot, L., Mara, K., Botton, E., Nogué, F., & Mazier, M. (2019). Efficient and transgene-free gene targeting using *Agrobacterium*-mediated delivery of the CRISPR/Cas9 system in tomato. *Plant Cell Reports, 38*, 459–462

Davies, J.P., Kumar, S., & Sastry-Dent, L. (2017). Use of zinc-finger nucleases for crop improvement. *Progress in Molecular Biology and Translational Science, 149*, 47–63. doi: 10.1016/bs.pmbts.2017.03.006.

Deng, L., Wang, H., Sun, C., Li, Q., Jiang, H., Du, M., Li, C.B., & Li, C. (2018). Efficient generation of pink-fruited tomatoes using CRISPR/Cas9 system. *Journal of Genetics and Genomics, 45*, 51–54. doi: 10.1016/j.jgg.2017.10.002

Doudna J.A., & Charpentier E. (2014). The new frontier of genome engineering with CRISPR-Cas9. *Science, 346*, 1258096. doi: 10.1126/science.1258096

Dutt, M., Mou, Z., Zhang, X., Tanwir, S.E., & Grosser, J.W. (2020). Efficient CRISPR/Cas9 genome editing with *Citrus* embryogenic cell cultures. *BMC Biotechnology, 20*, 58. doi: 10.1186/s12896-020-00652-9

Enciso-Rodriguez, F., Manrique-Carpintero, N.C., Nadakuduti, S.S., Buell, C.R., Zarka, D., & Douches, D. (2019). Overcoming self-incompatibility in diploid potato using CRISPR-Cas9. *Frontiers in Plant Science, 10*, 376. doi: 10.3389/fpls.2019.00376

Erpen-Dalla Corte, L., Mahmoud, L.M., Moraes, T.S., Mou, Z., Grosser, J.W., & Dutt, M. (2019). Development of improved fruit, vegetable, and ornamental crops using the CRISPR/Cas9 genome editing technique. *Plants, 8*, 601. doi: 10.3390/plants8120601

Fan, D., Liu, T., Li, C. et al. (2015). Efficient CRISPR/Cas9-mediated Targeted Mutagenesis in Populus in the First Generation. Sci Rep, 5, 12217.

Feng, Q., Xiao, L., He, Y., Liu, M., Wang, J., Tian, S., Zhang, X., & Yuan, L. (2021). Highly efficient, genotype-independent transformation and gene editing in water-melon (*Citrullus lanatus*) using a chimeric *ClGRF4-GIF1* gene. *Journal of Integrative Plant Biology, 63*(12), 2038–2042. doi: 10.1111/jipb.13199

Filler Hayut, S., Melamed Bessudo, C., & Levy, A.A. (2017). Targeted recombination between homologous chromosomes for precise breeding in tomato. *Nature Communications, 8*, 15605. doi: 10.1038/ncomms15605

Fister, A.S., Landherr, L., Maximova, S.N., & Guiltinan, M.J. (2018). Transient expression of CRISPR/ Cas9 machinery targeting TcNPR3 enhances defense response in *Theobroma cacao*. *Frontiers in Plant Science, 9*, 9. doi: 10.3389/fpls.2018.00268

Fizikova, A., Tikhonova, N., Ukhatova, Y., Ivanov, R., & Khlestkina, E. (2021). Applications of CRISPR/Cas9 system in vegetatively propagated fruit and berry crops. *Agronomy, 11*, 1849. doi: 10.3390/agronomy11091849

Gaudelli, N.M., Komor, A.C., Rees, H.A., Packer, M.S., Badran, A.H., Bryson, D.I., & Liu, D.R. (2017). Programmable base editing of A-T to G-C in genomic DNA without DNA cleavage. *Nature, 551*, 464–471. doi: 10.1038/nature24644

Giordano, A., Santo Domingo, M., Quadrana, L., Pujol, M., Martín-Hernández, A.M., & Garcia-Mas, J. (2022). CRISPR/Cas9 gene editing uncovers the roles of CONSTITUTIVE TRIPLE RESPONSE 1 and REPRESSOR OF SILENCING 1 in melon fruit ripening and epigenetic regulation. *Journal of Experimental Botany, 73*(12), 4022–4033. doi: 10.1093/jxb/erac148

González, M.N., Massa, G.A., Andersson, M., Turesson, H., Olsson, N., Fält, A.S., Storani, L., DécimaOneto, C.A., Hofvander, P., & Feingold, S.E. (2020). Reduced enzymatic browning in potato tubers by specific editing of a polyphenol oxidase gene via ribonucleoprotein complexes delivery of the CRISPR/Cas9 system. *Frontiers in Plant Science, 10*, 1649. doi: 10.3389/fpls.2019.01649

Hashimoto, R., Ueta, R., Abe, C., Osakabe, Y., & Osakabe, K. (2018). Efficient multiplex genome editing induces precise, and self-ligated type mutations in tomato plants. *Frontiers in Plant Science, 9*, 916. doi: 10.3389/fpls.2018.00916

He Y., & Zhao Y. (2020). Technological breakthroughs in generating transgene-free and genetically stable CRISPR-edited plants. *aBIOTECH, 1*(1), 88–96. doi: 10. 1007/s42994-019-00013-x

Holme, I.B., Gregersen, P.L., & Brinch-Pedersen, H. (2019). Induced genetic variation in crop plants by random or targeted mutagenesis: convergence and differences. *Frontiers in Plant Science, 10*, 1468. doi: 10.3389/fpls.2019.01468

Hu, B., Li, D., Liu, X., Qi, J., Gao, D., Zhao, S., Huang, S., Sun, J., & Yang, L. (2017). Engineering non-transgenic gynoecious cucumber using an improved transformation protocol and optimized CRISPR/Cas9 system. *Molecular Plant, 10*, 1575–1578. doi: 10.1016/j.molp.2017.09.005

Hu, C., Sheng, O., Deng, G., He, W., Dong, T., Yang, Q., Dou, T., Li, C., Gao, H., Liu, S., Yi, G., & Bi, F. (2021). CRISPR/Cas9-mediated genome editing of MaACO1 (aminocyclopropane-1-carboxylate oxidase 1) promotes the shelf life of banana fruit. *Plant Biotechnology Journal, 19*(4), 654–656. doi: 10.1111/pbi.13534

Ito, Y., Nishizawa-Yokoi, A., Endo, M., Mikami, M., & Toki, S. (2015). CRISPR/ Cas9-mediated mutagenesis of the RIN locus that regulates tomato fruit ripening. *Biochemical and Biophysical Research Communications, 467*, 76–82. doi: 10.1016/ j.bbrc.2015.09.117

Jeong, S.Y., Ahn, H., Ryu, J., Oh, Y., Sivanandhan, G., Won, K.H., Park, Y.D., Kim, J.S., Kim, H., Lim, Y.P., & Kim, S.G. (2019). Generation of early-flowering Chinese cabbage (*Brassica rapa* spp. *pekinensis*) through CRISPR/Cas9-mediated genome editing. *Plant Biotechnology Reports, 13*, 491–499. doi: 10.1007/s11816-019-00566-9

Jia, H., Orbovic, V., Jones, J.B., & Wang, N. (2016). Modification of the PthA4 effector binding elements in type I CsLOB1 promoter using Cas9/sgRNA to produce transgenic Duncan grapefruit alleviating XccDpthA4: DCsLOB1.3 infection. *Plant Biotechnology Journal, 14*, 1291–1301.

Jia, H., Orbovic, V., & Wang, N. (2019). CRISPR-LbCas12a-mediated modification of *Citrus. Plant Biotechnology Journal, 17*, 1928–1937. doi: 10.1111/pbi.12495

Jia, H., & Wang, N. (2014). Targeted genome editing of sweet orange using Cas9/ sgRNA. *PLoS One, 9*(4), e93806. doi: 10.1371/journal.pone.0093806

Jia, H., Zhang, Y., Orbovic, V., Xu, J., White, F.F., Jones, J.B., & Wang, N. (2017). Genome editing of the disease susceptibility gene *CsLOB1* in citrus confers resistance to citrus canker. *Plant Biotechnology Journal, 15*, 817–823. doi: 10.1111/pbi.12677

Jinek, M., Chylinski, K., Fonfara, I., Hauer, M., Doudna, J.A., & Charpentier, E. (2012). A programmable dual-RNA-guided DNA endonuclease in adaptive bacterial immunity. *Science, 337*(6096), 816–821. doi: 10.1126/science.1225829

Joung, J.K., & Sander, J.D. (2012). TALENs: a widely applicable technology for targeted genome editing. *Nature Reviews Molecular Cell Biology, 14*, 49–55. doi: 10.1038/ nrm3486

Jung, Y.J., Lee, G.J., Bae, S., & Kang, K.K. (2018). Reduced ethylene production in tomato fruits upon CRISPR/Cas9-mediated LeMADS-RIN mutagenesis. *Korean Journal of Horticultural Science and Technology, 36*, 396–405

Karkute, S.G., Singh, A.K., Gupta, O.P., Singh, P.M., & Singh, B. (2017). CRISPR/ Cas9 mediated genome engineering for improvement of horticultural crops. *Frontiers in Plant Science, 8*, 1635. doi: 10.3389/fpls.2017.01635

Kaur, N., Alok, A., Shivani, Kaur, N., Pandey, P., Awasthi, P., & Tiwari, S. (2018). CRISPR/Cas9-mediated efficient editing in phytoene desaturase (PDS) demonstrates precise manipulation in banana cv. Rasthali genome. *Functional and Integrative Genomics, 18*(1), 89–99.doi: 10.1007/s10142-017-0577-5

Kaur, N., Alok, A., Shivani, Kumar, P., Kaur, N., Awasthi, P., Chaturvedi, S., Pandey, P., Pandey, A., Pandey, A.K., & Tiwari, S. (2020). CRISPR/Cas9 directed editing of lycopene epsilon-cyclase modulates metabolic flux for β-carotene biosynthesis in banana fruit. *Metabolic Engineering, 59*, 76–86. doi: 10.1016/j.ymben.2020.01.008

Khan, Z., Khan, S.H., Mubarik, M.S., Sadia, B., & Ahmad, A. (2017). Use of TALEs and TALEN technology for genetic improvement of plants. *Plant Molecular Biology Reporter, 35*, 1–19. doi: 10.1007/s11105-016-0997-8

Kieu, N.P., Lenman, M., Wang, E.S., Petersen, B.L., & Andreasson, E. (2021). Mutations introduced in susceptibility genes through CRISPR/Cas9 genome editing confer increased late blight resistance in potatoes. *Scientific Reports, 11*, 4487. doi: 10.1038/s41598-021-83972-w

Kim, Y.C., Kang, Y., Yang, E.Y., Cho, M.C., Schafleitner, R., Lee, J.H., & Seonghoe Jang, S. (2021). Applications and major achievements of genome editing in vegetable crops: a review. *Frontiers in Plant Science, 12*, 688980. doi: 10.3389/fpls.2021.688980

Klap, C., Yeshayahou, E., Bolger, A.M., Arazi, T., Gupta, S.K., Shabtai, S., Usadel, B., Salts, Y., & Barg, R. (2017). Tomato facultative parthenocarpy results from SlAGAMOUS-LIKE 6 loss of function. *Plant Biotechnology Journal, 15*(5), 634–647. doi: 10.1111/pbi.12662

Klimek-Chodacka, M., Oleszkiewicz, T., Lowder, L.G., Qi, Y., & Baranski, R. (2018). Efficient CRISPR/Cas9-based genome editing in carrot cells. *Plant Cell Reports, 37*(4), 575–586. doi: 10.1007/s00299-018-2252-2

Kumar, D., Yadav, A., Ahmad, R., Dwivedi, U.N., & Yadav, K. (2022). CRISPR-based genome editing for nutrient enrichment in crops: a promising approach toward global food security. *Frontiers in Genetics, 13*, 932859. 10.3389/fgene.2022.932859

Kusano, H.,Ohnuma, M., Mutsuro-Aoki, H., Asahi, T., Ichinosawa, D., Onodera, H., Asano, K., Noda, T., Horie, T., Fukumoto, K., Kihira, M., Teramura, H., Yazaki, K., Umemoto, N., Muranaka, T., & Shimada, H. (2018). Establishment of a modified CRISPR/Cas9 system with increased mutagenesis frequency using the translational enhancer dMac3 and multiple guide RNAs in potato. *Scientific Reports, 8*(1), 13753. doi: 10.1038/s41598-018-32049-2

Lawrenson, T., Shorinola, O., Stacey, N., Li, C., Østergaard, L., Patron, N., Uauy, C., & Harwood, W. (2015). Induction of targeted, heritable mutations in barley and *Brassica oleracea* using RNA-guided Cas9 nuclease. *Genome Biology, 16*, 258. doi: 10.1186/s13059-015-0826-7

Lee, M.H., Lee, J., Choi, S.A., Kim, Y.S., Koo, O., Choi, S.H., Ahn, W.S., Jie, E.Y., & Kim, S.W. (2020). Efficient genome editing using CRISPR-Cas9 RNP delivery into cabbage protoplasts via electro-transfection. *Plant Biotechnology Reports, 14*, 695–702.doi: 10.1007/ s11816-020-00645-2

Li, G., Liu, R., Xu, R., Varshney, R.K., Ding, H., Li, M., Yan, X., Huang, S., Li, J., Wang, D., Ji, Y., Wang, C., He, J., Luo, Y., Gao, S., Wei, P., Zong, X., & Yang, T. (2022). Development of an *Agrobacterium*-mediated CRISPR/Cas9 system in pea (*Pisum sativum* L.). *The Crop Journal, (in press)*. doi: 10.1016/j.cj.2022.04.011

Li, Q., Feng, Q., Snouffer, A., Zhang, B., Rodríguez, G.R., & van der Knaap, E. (2022). Increasing fruit weight by editing a *cis*-regulatory element in tomato *KLUH* promoter using CRISPR/Cas9. *Frontiers in Plant Science, 13*, 879642. doi: 10.3389/fpls.2022.879642

Li, R., Liu, C., Zhao, R., Wang, L., Chen, L., Yu, W., Zhang, S., Sheng, J., & Shen, L. (2019). CRISPR/Cas9-mediated *SlNPR1* mutagenesis reduces tomato plant drought tolerance. *BMC Plant Biology, 19*, 38

Li, X., Wang, Y., Chen, S., Tian, H., Fu, D., Zhu, B., Luo, Y., & Zhu, H. (2018). Lycopene is enriched in tomato fruit by CRISPR/Cas9-mediated multiplex genome editing. *Frontiers in Plant Science, 9*, 1–12. doi: 10.3389/fpls.2018.00559

Li, X., Xu, S., Fuhrmann-Aoyagi, M.B., Yuan, S., Iwama, T., Kobayashi, M., & Miura, K. (2022). CRISPR/Cas9 technique for temperature, drought, and salinity stress responses. *Current Issues in Molecular Biology, 44*, 2664–2682. doi: 10.3390/cimb44060182

Limera, C., Sabbadini, S., Sweet, J.B., & Mezzetti, B. (2017). New biotechnological tools for the genetic improvement of major woody fruit species. *Frontiers in Plant Science, 8*, 1418. doi: 10.3389/fpls.2017.01418

Liu, X., Xie, C., Si, H., & Yang, J. (2017). CRISPR/Cas9-mediated genome editing in plants. *Methods, 121–122*, 94–102. doi: 10.1016/j.ymeth.2017.03.009

Liu, Y., Zhang, C., Wang, X., Li, X., & You, C. (2022). CRISPR/Cas9 technology and its application in horticultural crops. *Horticultural Plant Journal, 8*(4), 395–407.doi: 10.1016/j.hpj.2022.04.007

Lor, V.S., Starker, C.G., Voytas, D.F., Weiss, D., & Olszewski, N.E. (2014). Targeted mutagenesis of the tomato *PROCERA* gene using transcription activator-like effector nucleases. *Plant Physiology, 166*, 1288–1291. doi: 10.1104/pp.114.247593

Lu, Y., Tian, Y., Shen, R., Yao, Q., Zhong, D., Zhang, X., & Zhu, J.K. (2021). Precise genome modification in tomato using an improved prime editing system. *Plant Biotechnology Journal, 19*(3), 415–417. doi: 10.1111/pbi.13497

Ma, C., Liu, M., Li, Q., Si, J., Ren, X., & Song, H. (2019a). Efficient BoPDS gene editing in cabbage by the CRISPR/Cas9 system. *Horticultural Plant Journal, 5*(4), 164–169. doi: 10.1016/j.hpj.2019.04.001

Ma, C., Zhu, C., Zheng, M., Liu, M., Zhang, D., Liu, B., Li, Q., Si, J., Ren, X., & Song, H. (2019b). CRISPR/Cas9-mediated multiple gene editing in *Brassica oleracea* var. *capitata* using the endogenous tRNA-processing system. *Horticulture Research*, 6, 20. doi: 10.1038/s41438-018-0107-1

Ma, J., Xiang, H., Donnelly, D.J., Meng, F.R., Xu, H., Durnford, D., & Li, X.Q. (2017). Genome editing in potato plants by *Agrobacterium*-mediated transient expression of transcription activator-like effector nucleases. *Plant Biotechnology Reports*, 11, 249–258.doi: 10.1007/s11816-017-0448-5

Maioli, A., Gianoglio, S., Moglia, A., Acquadro, A., Valentino, D., Milani, A.M., Prohens, J., Orzaez, D., Granell, A., Lanteri, S., & Comino, C. (2020). Simultaneous CRISPR/Cas9 editing of three PPO genes reduces fruit flesh browning in *Solanum melongena* L. *Frontiers in Plant Science, 11*, 607161. doi: 10.33 89/fpls.2020.607161

Malabarba, J., Chevreau, E., Dousset, N., Veillet, F., Moizan, J., & Vergne, E. (2021). New strategies to overcome present CRISPR/Cas9 limitations in apple and pear: efficient dechimerization and base editing. *International Journal of Molecular Sciences, 22*, 319. doi: 10.3390/ijms22010319

Malnoy, M., Viola, R., Jung, M.H., Koo, O.J., Kim, J.S., Velasco, R., & Kanchiswamy, C.N. (2016). DNA-free genetically edited grapevine and apple protoplast using CRISPR/Cas9 ribonucleo-proteins. *Frontiers in Plant Science, 7*, 1904. doi: 10.3389/ fpls.2016.01904

Mao, Y., Zhang, H., Xu, N., Zhang, B., Gou, F., & Zhu, J.K. (2013). Application of the CRISPR-Cas system for efficient genome engineering in plants. *Molecular Plant, 6*, 2008–2011. doi: 10.1093/mp/sst121

Martín-Pizarro, C., Triviño, J.C., & Posé, D. (2019). Functional analysis of the TM6 MADS-box gene in the octoploid strawberry by CRISPR/Cas9-directed mutagenesis. *Journal of Experimental Botany, 70*, 885–895. doi: 10.1093/jxb/ery400

Marton, I., Zuker, A., Shklarman, E., Zeevi, V., Tovkach, A., Roffe, S., Ovadis, M., Tzfira, T., & Alexander Vainstein, A. (2010). Nontransgenic genome modification in plant cells. *Plant Physiology, 154*(3), 1079–1087. doi: 10.1104/pp.110.164806

Monsur, M.B., Shao, G., Lv, Y., Ahmad, S., Wei, X., Hu, P., & Tang, S. (2020). Base editing: the ever expanding clustered regularly interspaced short palindromic repeats (CRISPR) tool kit for precise genome editing in plants. *Genes, 11*, 466. doi: 10.3390/genes11040466

Mushtaq, M., Sakina, A., Wani, S.H., Shikari, A.B., Tripathi, P., Zaid, A., Galla, A., Abdelrahman, M., Sharma, M., Singh, A.K., & Salgotra, R.K. (2019). Harnessing genome editing techniques to engineer disease resistance in plants. *Frontiers in Plant Science, 10*, 550. doi: 10.3389/fpls.2019.00550

Nadakuduti, S.S., & Enciso-Rodríguez, F. (2021). Advances in genome editing with CRISPR systems and transformation technologies for plant DNA manipulation. *Frontiers in Plant Science, 11*, 637159. doi: 10.3389/fpls.2020.637159

Nadakuduti, S.S., Buell, C.R., Voytas, D.F., Starker, C.G., & Douches, D.S. (2018). Genome editing for crop improvement – Applications in clonally propagated polyploids with a focus on potato (*Solanum tuberosum* L.). *Frontiers in Plant Science, 9*, 1607. doi: 10.3389/fpls.2018.01607

Nadakuduti, S.S., Starker, C.G., Voytas, D.F., Buell, C.R., & Douches, D.S. (2019). Genome editing in potato with CRISPR/Cas9. *Methods in Molecular Biology, 1917*, 183–201. doi: 10.1007/ 978-1-4939-8991-114

Naim, F., Dugdale, B., Kleidon, J., Brinin, A., Shand, K., Waterhouse, P., & Dale, J. (2018). Gene editing the phytoene desaturase alleles of Cavendish banana using CRISPR/Cas9. *Transgenic Research*, 27, 451–460. doi: 10.1007/s11248-018-0083-0

Nakajima, I., Ban, Y., Azuma, A., Onoue, N., Moriguchi, T., Yamamoto, T., Toki, S., & Endo, M. (2017). CRISPR/Cas9-mediated targeted mutagenesis in grape. *PLoS One*, 12, e177966. doi: 10.1371/journal.pone.0177966

Nakayasu, M., Akiyama, R., Lee, H.J., Osakabe, K., Osakabe, Y., Watanabe, B., Sugimoto, Y., Umemoto, N., Saito, K., Muranaka, T., & Mizutani, M. (2018). Generation of α-solanine-free hairy roots of potato by CRISPR/Cas9 mediated genome editing of the St16DOX gene. *Plant Physiology and Biochemistry*, 131, 70–77. doi: 10.1016/j.plaphy.2018.04.026

Nekrasov, V., Wang, C., Win, J., Lanz, C., Weigel, D., & Kamoun, S. (2017). Rapid generation of a transgene-free powdery mildew resistant tomato by genome deletion. *Scientific Reports*, 7, 482. doi: 10.1038/s41598-017-00578-x

Nishitani, C., Hirai, N., Komori, S., Wada, M., Okada, K., Osakabe, K., Yamamoto, T., & Osakabe, Y. (2016). Efficient genome editing in apple using a CRISPR/Cas9 system. *Scientific Reports*, 6, 1–8. doi: 10.1038/srep31481

Nonaka, S., Arai, C., Takayama, M., Matsukura, C., & Ezura, H. (2017). Efficient increase of Ɣ-aminobutyric acid (GABA) content in tomato fruits by targeted mutagenesis. *Scientific Reports*, 9(1), 19822. doi: 10.1038/s41598-017-06400-y

Ortigosa, A., Gimenez, S., Leonhardt, N., & Solano, R. (2019). Design of a bacterial speck resistant tomato by CRISPR/Cas9-mediated editing of SlJAZ2. *Plant Biotechnology Journal*, 17(3), 665–673. doi: 10.1111/pbi.13006

Osakabe, Y., Liang, Z., Ren, C., Nishitani, C., Osakabe, K., Wada, M., Komori, S., Malnoy, M., Velasco, R., Poli, M., Jung, M.H., Koo, O.J., Viola, R., & Kanchiswamy, C.N. (2018). CRISPR–Cas9-mediated genome editing in apple and grapevine. *Nature Protocols*, 13, 2844–2863. doi: 10.1038/s41596-018-0067-9

Pan, C., Ye, L., Qin, L., Liu, X., He, Y., Wang, J., Chen, L., & Lu, G. (2016). CRISPR/Cas9-mediated efficient and heritable targeted mutagenesis in tomato plants in the first and later generations. *Scientific Reports*, 6, 24765. doi: 10.1038/srep24765

Pan, W., Liu, X., Li, D., & Zhang, H. (2022). Establishment of an efficient genome editing system in lettuce without sacrificing specificity. *Frontiers in Plant Science*, 13, 930592.doi: 10.3389/ fpls.2022.930592

Parajuli, S., Huo, H., Gmitter, F.G. Jr., Duan, Y., Luo, F., & Deng, Z. (2022). Editing the CsDMR6 gene in citrus results in resistance to the bacterial disease citrus canker. *Horticulture Research*, 9, Article. doi: 10.1093/hr/uhac082

Park, S.C., Park, S., Jeong, Y.J., Lee, S.B., Pyun, J.W., Kim, S., Kim, T.H., Kim, S.W., Jeong, J.C., & Kim, C.Y. (2019). DNA-free mutagenesis of GIGANTEA in *Brassica oleracea* var. *capitata* using CRISPR/Cas9 ribonucleoprotein complexes. *Plant Biotechnology Reports*, 13, 483–489.doi:10.1007/s11816-019-00585-6

Park, S.I., Kim, H.B., Jeon, H.J., & Kim, H. (2021). *Agrobacterium*-mediated *Capsicum annuum* gene editing in two cultivars, hot pepper CM334 and bell pepper Dempsey. *International Journal of Molecular Sciences*, 22(8), 3921. doi: 10.3390/ijms22083921

Pechar, G.S., Donaire, L., Gosalvez, B., García-Almodovar, C., Sánchez-Pina, M.A., Truniger, V., &Aranda, M.A. (2022). Editing melon *eIF4E* associates with virus resistance and male sterility. *Plant Biotechnology Journal*, 20(10), 2006–2022. doi: 10.1111/pbi.13885

Peer, R., Rivlin, G., Golobovitch, S., Lapidot, M., Gal-On, A., Vainstein, A., Tzfira, T., & Flaishman, M.A. (2015). Targeted mutagenesis using zinc-finger nucleases in perennial fruit trees. *Planta, 241*(4), 941–951. doi: 10.1007/s00425-014-2224-x

Peng, A., Chen, S., Lei, T., Xu, L., He, Y., Wu, L., Yao, L., & Zou, X. (2017). Engineering canker-resistant plants through CRISPR/Cas9-targeted editing of the susceptibility gene *CsLOB 1* promoter in citrus. *Plant Biotechnology Journal, 15*(12), 1509–1519. doi: 10.1111/pbi.12733

Petolino, J.F. (2015). Genome editing in plants via designed zinc finger nucleases. *In Vitro Cellular and Developmental Biology – Plant, 51*(1), 1–8.doi: 10.1007/s11627-015-9663-3

Pompili, V., Dalla Costa, L., Piazza, S., Pindo, M., & Malnoy, M. (2020). Reduced fire blight susceptibility in apple cultivars using a high-efficiency CRISPR/Cas9-FLP/FRT-based gene editing system. *Plant Biotechnology Journal, 18*(3), 845–858. doi: 10.1111/pbi.13253

Ran, F.A., Hsu, P.D., Wright, J., Agarwala, V., Scott, D.A., & Zhang, F. (2013). Genome engineering using the CRISPR-Cas9 system. *Nature Protocols, 8*, 2281–2308. doi: 10.1038/nprot.2013.143

Ran, Y., Patron, N., Kay, P., Wong, D., Buchanan, M., Cao, Y.Y., Sawbridge, T., Davies, J.P., Mason, J., Webb, S.R., Spangenberg, G., Ainley, W.M., Walsh, T.A., & Hayden, M.J. (2018). Zinc finger nuclease-mediated precision genome editing of an endogenous gene in hexaploid bread wheat (*Triticum aestivum*) using a DNA repair template. *Plant Biotechnology Journal, 16*(12), 2088–2101. doi: 10.1111/pbi.12941

Ren, C., Guo, Y., Kong, J., Lecourieux, F., Dai, Z., Li, S., & Liang, Z. (2020). Knockout of VvCCD8 gene in grapevine affects shoot branching. *BMC Plant Biology, 20*, 47. doi: 10.1186/s12870-020-2263-3

Ren, C., Liu, X., Zhang, Z., Wang, Y., Duan, W., Li S., & Liang Z. (2016). CRISPR/Cas9-mediated efficient targeted mutagenesis in Chardonnay (*Vitis vinifera* L.). *Scientific Reports, 6*, 32289. doi: 10.1038/srep32289

Ren, C., Liu, Y., Guo, Y., Duan, W., Fan, P., Li, S., & Liang, Z. (2021). Optimizing the CRISPR/Cas9 system for genome editing in grape by using grape promoters. *Horticulture Research, 8*, 52. doi: 10.1038/s41438-021-00489-z

Ren, F., Ren, C., Zhang, Z., Duan, W., Lecourieux, D., Li, S., & Liang, Z. (2019). Efficiency optimization of CRISPR/Cas9-mediated targeted mutagenesis in grape. *Frontiers in Plant Science, 10*, 612. doi:10.3389/fpls.2019.00612

Robertson, G., Burger, J., & Campa, M. (2022). CRISPR/Cas-based tools for the targeted control of plant viruses. *Plant Molecular Pathology, 23*(11), 1701–1718

Rodríguez-Leal, D., Lemmon, Z.H., Man, J., Bartlett, M.E., & Lippman, Z.B. (2017). Engineering quantitative trait variation for crop improvement by genome editing. *Cell, 171*(2), Article 470.e478–480.e478. doi: 10.1016/j.cell.2017.08.030

Roldan, M.V.G., Périlleux, C., Morin, H., Huerga-Fernandez, S., Latrasse, D., Benhamed, M., & Bendahmane, A. (2017). Natural and induced loss of function mutations in *SlMBP21* MADS-box gene led to *jointless-2* phenotype in tomato. *Scientific Reports, 7*, 4402. doi: 10.1038/ s41598-017-04556-1.

Santa Maria, S., & Llorente, B. (2018). DNA recombination, mechanisms of. In: Wells, R.D., Bond, J.S., Klinman, J., & Masters, B.S.S. (Eds) *Molecular Life Sciences.* Springer, New York, NY. doi: 10.1007/978-1-4614-1531-2_74

Santillán Martínez, M.I., Bracuto, V., Koseoglou, E., Appiano, M., Jacobsen, E., Visser, R.G.F., Wolters, A.M.A., & Bai, Y. (2020). CRISPR/Cas9-targeted mutagenesis of the tomato susceptibility gene *PMR4* for resistance against powdery mildew. *BMC Plant Biology*, *20*(1), 284. doi: 10.1186/s12870-020-02497-y

Savadi, S., Mangalassery, S., & Sandesh, M.S. (2021). Advances in genomics and genome editing for breeding next generation of fruit and nut crops. *Genomics*, *113*(6), 3718–3734. doi: 10.1016/j.ygeno.2021.09.001

Selvakumar, R., Praveen Kumar, S., Manjunathagowda, D.C., & Gangadhar, K. (2020). Genome editing for improvement of vegetable crops. *Food Science Reports*, *1*(10), 26–30.

Schröpfer, S., & Flachowsky, H. (2021). Tracing CRISPR/Cas12a mediated genome editing events in apple using high-throughput genotyping by PCR capillary gel electrophoresis. *International Journal of Molecular Sciences*, *22*, 12611. doi: 10.3390/ijms222212611

Shan, Q., Wang, Y., Li, J., Zhang, Y., Chen, K., Liang, Z., Zhang, K., Liu, J., Xi, J.J., Qiu, J.L., & Gao, C. (2013). Targeted genome modification of crop plants using a CRISPR-Cas system. *Nature Biotechnology*, *31*(8), 686–688. doi: 10.1038/nbt.2650

Shimatani, Z., Kashojiya, S., Takayama, M., Terada, R., Arazoe, T., Ishii, H., Teramura, H., Yamamoto, T., Komatsu, H., Miura, K., Ezura, H., Nishida, K., Ariizumi, T., & Kondo A. (2017). Targeted base editing in rice and tomato using a CRISPR-Cas9 cytidine deaminase fusion. *Nature Biotechnology*, *35*(5), 441–443. doi: 10.1038/nbt.3833

Songstad, D.D., Petolino, J.F., Voytas, D.F., & Reichert, N.A. (2017). Genome editing of plants. *Critical Reviews in Plant Sciences*, *36*(1), 1–23. doi: 10.1080/07352689.2017.1281663

Soyk, S., Müller, N.A., Park, S.J., Schmalenbach, I., Jiang, K., Hayama, R., Zhang, L., Van Eck, J., Jiménez-Gómez, J.M., & Lippman, Z.B. (2017). Variation in the flowering gene SELF PRUNING 5G promotes day-neutrality and early yield in tomato. *Nature Genetics*, *49*(1), 162–168. doi: 10.1038/ng.3733

Sun, B., Zheng, A., Jiang, M., Xue, S., Yuan, Q., Jiang, L., Chen, Q., Li, M., Wang, Y., Zhang, Y., Luo, Y., Wang, X., Zhang, F., & Tang, H. (2018). CRISPR/Cas9-mediated mutagenesis of homologous genes in Chinese kale. *Scientific Reports*, *8*(1),16786. doi: 10.1038/s41598-018-34884-9

Sun, Z., Li, N., Huang, G., Xu, J., Pan, Y., Wang, Z., Tang, Q., Song, M., & Wang, X. (2013). Site-specific gene targeting using transcription activator-like effector (TALE)-based nuclease in *Brassica oleracea*. *Journal of Integrative Plant Biology*, *55*(11), 1092–1103. doi: 10.1111/jipb.12091

Swarts, D.C., & Jinek, M. (2018). Cas9 versus Cas12a/Cpf1: structure-function comparisons and implications for genome editing. *WIREs RNA*, *9*(5), e1481. doi: 10.1002/wrna.1481

Swarts, D.C., van der Oost, J., & Jinek, M. (2017). Structural basis for guide RNA processing and seed-dependent DNA targeting by CRISPR-Cas12a. *Molecular Cell*, *66*(2), 221–233.e4. doi: 10.1016/j.molcel.2017.03.016

Tashkandi, M., Ali, Z., Aljedaani, F., Shami, A., & Mahfouz, M.M. (2018). Engineering resistance against Tomato yellow leaf curl virus via the CRISPR/Cas9 system in tomato. *Plant Signaling and Behavior*, *13*(10), Article e1525996. doi: 10.1080/15592324.2018.1525996

Tian, S., Jiang, L., Gao, Q., Zhang, J., Zong, M., Zhang, H., Ren, Y., Guo, S., Gong, G., Liu, F., & Xu, Y. (2017). Efficient CRISPR/Cas9-based gene knockout in watermelon. *Plant Cell Reports, 36*(3), 399–406. doi: 10.1007/s00299-016-2089-5

Tian, S., Jiang, L., Cui, X., Zhang, J., Guo, S., Li, M., Zhang, H., Ren, Y., Gong, G., Zong, M., Liu, F., Chen, Q., & Xu, Y. (2018). Engineering herbicide-resistant watermelon variety through CRISPR/Cas9-mediated base-editing. *Plant Cell Reports, 37*(9), 1353–1356. doi: 10.1007/s00299-018-2299-0

Tiwari, J.K., Buckseth, T., Challam, C., Zinta, R., Bhatia, N., Dalamu, D., Naik, S., Poonia, A.K., Singh, R.K., Luthra, S.K., Kumar, V., & Kumar, M. (2022). CRISPR/Cas genome editing in potato: current status and future perspectives. *Frontiers in Genetics, 13*, 827808. doi: 10.3389/fgene.2022.827808

Tomlinson, L., Yang, Y., Emenecker, R., Smoker, M., Taylor, J., Perkins, S., Smith, J., MacLean, D., Olszewski, N.E., & Jones, J.D.G. (2019). Using CRISPR/Cas9 genome editing in tomato to create a gibberellin-responsive dominant dwarf DELLA allele. *Plant Biotechnology Journal, 17*(1), 132–140. doi: 10.1111/pbi.12952

Tripathi, J.N., Ntui, V.O., Ron, M., Muiruri, S.K., Britt, A., & Tripathi, L. (2019). CRISPR/Cas9 editing of endogenous banana streak virus in the B genome of *Musa* spp. overcomes a major challenge in banana breeding. *Communications Biology, 2*, 46. doi: 10.1038/s42003-019-0288-7

Tripathi, J.N., Ntui, V.O., Shah, T., & Tripathi, L. (2021). CRISPR/Cas9-mediated editing of DMR6 orthologue in banana (*Musa* spp.) confers enhanced resistance to bacterial disease. *Plant Biotechnology Journal, 19*(7), 1291–1293. doi: 10.1111/pbi.13614

Tripathi, L., Ntui, V.O., & Tripathi, J.N. (2020). CRISPR/Cas9-based genome editing of banana for disease resistance. *Current Opinion in Plant Biology, 56*, 118–126. doi: 10.1016/j.pbi.2020.05.003

Tripathi, L., Ntui, V.O., & Tripathi, J.N. (2022). Control of bacterial diseases of banana using CRISPR/Cas-based gene editing. *International Journal of Molecular Sciences, 23*(7), 3619. doi: 10.3390/ijms23073619

Ueta, R., Abe, C., Watanabe, T., Sugano, S.S., Ishihara, R., Ezura, H., Osakabe, Y., & Osakabe, K. (2017). Rapid breeding of parthenocarpic tomato plants using CRISPR/Cas9. *Scientific Reports, 7*(1), 507. doi: 10.1038/s41598-017-00501-4

Veillet, F., Chauvin, L., Kermarrec, M.P., Sevestre, F., Merrer, M., Terret, Z., Szydlowski, N., Devaux, P., Gallois, J.L., & Chauvin, J.E. (2019a). The *Solanum tuberosum* GBSSI gene: A target for assessing gene and base editing in tetraploid potato. *Plant Cell Reports, 38*, 1065–1080. doi: 10.1007/s00299-019-02426-w

Veillet, F., Perrot, L., Chauvin, L., Kermarrec, M.P., Guyon-Debast, A., Chauvin, J.E., Nogué, F., & Mazier, M. (2019b). Transgene-free genome editing in tomato and potato plants using *Agrobacterium*-mediated delivery of a CRISPR/Cas9 cytidine base editor. *International Journal of Molecular Sciences, 20*(2), 402. doi: 10.3390/ijms20020402

Veillet, F., Perrot, L., Guyon-Debast, A., Kermarrec, M.P., Chauvin, L., Chauvin, J.E., Gallois, J.L., Mazier, M., & Nogué, F. (2020). Expanding the CRISPR toolbox in *P. patens* using SpCas9-NG variant and application for gene and base editing in *Solanaceae* crops. *International Journal of Molecular Sciences, 21*(3), 1024. doi: 10.3390/ijms21031024

Voytas, D.F. (2013). Plant genome engineering with sequence-specific nucleases. *Annual Review of Plant Biology, 64*, 327–350. doi: 10.1146/annurev-arplant-042811-105552

Wan, D., Guo, Y., Cheng, Y., Hu, Y., Xiao, S., Wang, Y., & Wen, Y.Q. (2020). CRISPR/Cas9- mediated mutagenesis of VvMLO3 results in enhanced resistance to powdery mildew in grapevine (*Vitis vinifera*). *Horticulture Research*, 7, 116. doi: 10.1038/s41438-020-0339-8

Wan, L., Wang, Z., Tang, M., Hong, D., Sun, Y., Ren, J., Zhang, N., & Zeng, H. (2021). CRISPR-Cas9 gene editing for fruit and vegetable crops: strategies and prospects. *Horticulturae*, 7(7), 193. doi: 10.3390/horticulturae7070193

Wang, B., Li, N., Huang, S., Hu, J., Wang, Q., Tang, Y., Yang, T., Asmutola, P., Wang, J., & Yu, Q. (2021). Enhanced soluble sugar content in tomato fruit using CRISPR/Cas9-mediated SlINVINH1 and SlVPE5 gene editing. *PeerJ*, 9, Article e12478. doi: 10.7717/peerj.12478

Wang, D., Samsulrizal, N., Yan, C., Allcock, N.S., Craigon, J., Blanco-Ulate, B., Ortega-Salazar, I., Marcus, S.E., Moeiniyan Bagheri, H., Perez Fons, L., Fraser, P.D., Foster, T., Fray, R., Knox, J.P., & Seymour, G.B. (2019). Characterization of CRISPR mutants targeting genes modulating pectin degradation in ripening tomato. *Plant Physiology*, 179(2), 544–557. doi: 10.1104/pp.18.01187

Wang, L., Chen, L., Li, R., Zhao, R., Yang, M., Sheng, J., & Shen, L. (2017). Reduced drought tolerance by CRISPR/Cas9-mediated SlMAPK3 mutagenesis in tomato plants. *Journal of Agricultural and Food Chemistry*, 65(39), 8674–8682. doi: 10.1021/acs.jafc.7b02745.

Wang, L., Chen, S., Peng, A., Xie, Z., He, Y., & Zou, X. (2019). CRISPR/Cas9-mediated editing of CsWRKY22 reduces susceptibility to *Xanthomonas citri* subsp. *citri* in Wanjincheng orange (*Citrus sinensis* L.) Osbeck. *Plant Biotechnology Reports*, 13, 501–510.doi: 10.1007/s11816-019-00556-x.

Wang, X., Tu, M., Wang, D., Liu, J., Li, Y., Li, Z., Wang, Y., & Wang, X. (2018). CRISPR/Cas9-mediated efficient targeted mutagenesis in grape in the first generation. *Plant Biotechnology Journal*, 16(4), 844–855. doi: 10.1111/pbi.12832

Wang, X., Tu, M., Wang, Y., Yin, W., Zhang, Y., Wu, H., Gu, Y., Li, Z., Xi, Z., & Wang, X. (2021). Whole-genome sequencing reveals rare off-target mutations in CRISPR/ Cas9-edited grapevine. *Horticulture Research*, 8, 114. doi: 10.1038/s4143 8-021-00549-4

Wang, Z., Wang, S., Li, D., Zhang, Q., Li, L., Zhong, C., Liu, Y., & Huang, H. (2018). Optimized paired-sgRNA/Cas9 cloning and expression cassette triggers high-efficiency multiplex genome editing in kiwifruit. *Plant Biotechnology Journal*, 16(8), 1424–1433. doi: 10.1111/pbi.12884

Weinthal, D., Tovkach, A., Zeevi, V., & Tzfira, T. (2010). Genome editing in plant cells by zinc finger nucleases. *Trends in Plant Science*, 15(6), 308–321. doi: 10.1016/j.tplants.2010.03.001

Woo, J.W., Kim, J., Kwon, S.I., Corvalán, C., Cho, S.W., Kim, H., Kim, S.G., Kim, S.T., Choe, S., & Kim, J.S. (2015). DNA-free genome editing in plants with pre-assembled CRISPR-Cas9 ribonucleoproteins. *Nature Biotechnology*, 33(11), 1162–1174. doi: 10.1038/nbt.3389

Xia, X., Cheng, X., Li, R., Yao, J., Li, Z., & Cheng, Y. (2021). Advances in application of genome editing in tomato and recent development of genome editing technology. *Theoretical and Applied Genetics*, 134(9),2727–2747. doi: 10.1007/s00122-021-03874-3

Xin, T., Tian, H., Ma, Y., Wang, S., Yang, L., Li, X., Zhang, M., Chen, C., Wang, H., Li, H., Xu, J., Huang, S., & Yang, X. (2022). Targeted creation of new mutants with compact plant architecture using CRISPR/Cas9 genome editing by an optimized genetic transformation procedure in cucurbit plants. *Horticulture Research, 9,* Article uhab086. doi: 10.1093/hr/uhab086

Xing, S., Chen, K., Zhu, H., Zhang R., Zhang H., Li B., & Gao C. (2020). Fine-tuning sugar content in strawberry. *Genome Biology, 21,* 230. doi: 10.1186/s13059-020-02146-5

Xing, S., Jia, M., Wei, L., Mao, W., Abbasi, U.A., Zhao, Y., Chen, Y., Cao, M., Zhang, K., Dai, Z., Dou, Z., Jia, W., & Li, B. (2018). CRISPR/Cas9-introduced single and multiple mutagenesis in strawberry. *Journal of Genetics and Genomics, 45*(12), 685–687. doi: 10.1016/j.jgg.2018.04.006

Xiong, J.S., Ding, J., & Li, Y. (2015). Genome-editing technologies and their potential application in horticultural crop breeding. *Horticulture Research, 2*(1), 15019. doi: 10.1038/hortres.2015.19

Xu, J., Hua, K., & Lang, Z. (2019). Genome editing for horticultural crop improvement. *Horticulture Research, 6,* 113–116. doi: 10.1038/s41438-019-0196-5

Xu, X., Yuan, Y., Feng, B., & Deng, W. (2020). CRISPR/Cas9-mediated gene-editing technology in fruit quality improvement. *Food Quality and Safety,4*(4), 159–166. doi: 10.1093/ fqsafe/fyaa028

Xu, Z.S., Feng, K., & Xiong, A.S. (2019). CRISPR/Cas9-mediated multiply targeted mutagenesis in orange and purple carrot plants. *Molecular Biotechnology, 61*(3), 191–199. doi: 10.1007/s12033-018-00150-6

Yang, T., Deng, L., Zhao, W., Zhang, R., Jiang, H., Ye, Z., Li, C.B., & Li, C., 2019. Rapid breeding of pink-fruited tomato hybrids using the CRISPR/Cas9 system. *Journal of Genetics and Genomics, 46*(10), 505–508. doi: 10.1016/j.jgg.2019.10.002

Yang, Y., Xu, C., Shen, Z., & Yan, C. (2022). Crop quality improvement through genome editing strategy. *Frontiers in Genome Editing, 3,* 819687. doi: 10.3389/fgeed.2021.819687

Yasmeen, A., Shakoor, S., Azam, S., Bakhsh, A., Shahid, N., Latif, A., Shahid, A.A., Husnain, T., & Rao, A.Q. (2022). CRISPR/Cas-mediated knockdown of vacuolar invertase gene expression lowers the cold-induced sweetening in potatoes. *Planta, 256*(6), 107. doi: 10.1007/s00425-022-04022-x

Ye, M., Peng, Z., Tang, D., Yang, Z., Li, D., Xu, Y., Zhang, C., & Huang, S. (2018). Generation of self-compatible diploid potato by knockout of S-RNase. *Nature Plants, 4*(9), 651–654. doi: 10.1038/s41477-018-0218-6

Yu, Q.H., Wang, B., Li, N., Tang, Y., Yang, S., Yang, T., Xu, J., Guo, C., Yan, P., Wang, Q., & Asmutola, P. (2017). CRISPR/Cas9-induced targeted mutagenesis and gene replacement to generate long-shelf life tomato lines. *Scientific Reports, 7*(1), 11874. doi: 10.1038/s41598-017-12262-1

Yu, W., Wang, L., Zhao, R., Sheng, J., Zhang, S., Li, R., & Shen, L. (2019). Knockout of SlMAPK3 enhances tolerance to heat stress involving ROS homeostasis in tomato plants. *BMC Plant Biology, 19*(1), 354. doi: 10.1186/s12870-019-1939-z

Zetsche, B., Gootenberg, J.S., Abudayyeh, O.O., Slaymaker, I.M., Makarova, K.S., Essletzbichler, P., Volz, S.E., Joung, J., van der Oost, J., Regev, A., Koonin, E.V., & Zhang, F. (2015). Cpf1 is a single RNA-guided endonuclease of a class 2 CRISPR-Cas system. *Cell, 163*(3), 759–771. doi: 10.1016/j.cell.2015.09.038

Zhan, X., Zhang, F., Zhong, Z., Chen, R., Wang, Y., Chang, L., Bock, R., Nie, B., & Zhang, J. (2019). Generation of virus-resistant potato plants by RNA genome targeting. *Plant Biotechnology Journal, 17*(9), 1814–1822. doi: 10.1111/pbi.13102

Zhang, F., LeBlanc, C., Irish, V.F., & Jacob, Y. (2017). Rapid and efficient CRISPR/ Cas9 gene editing in *Citrus* using the YAO promoter. *Plant Cell Reports, 36*(12), 1883–1887. doi: 10.1007/s00299-017-2202-4

Zhang, M., Liu, Q., Yang, X., Xu, J., Liu, G., Yao, X., Ren, R., Xu, J., & Lou, L. (2020). CRISPR/Cas9-mediated mutagenesis of Clpsk1 in watermelon to confer resistance to *Fusarium oxysporum* f. sp. *niveum. Plant Cell Reports, 39*(5), 589–595. doi: 10.1007/s00299-020-02516-0

Zhang, Y., Massel, K., Godwin, I.D., & Gao, C. (2018). Applications and potential of genome editing in crop improvement. *Genome Biology, 19*, 210. doi: 10.1186/s1305 9-018-1586-y

Zhao, X., Jayarathna, S., Turesson, H., Fält, A.S., Nestor, G., González, M.N., Olsson, N. Beganovic, M., Hofvander, P., Andersson, R., & Andersson, M. (2021). Amylose starch with no detectable branching developed through DNA-free CRISPR-Cas9 mediated mutagenesis of two starch branching enzymes in potato. *Scientific Reports, 11*(1), 4311. doi: 10.1038/s41598-021-83462-z

Zheng, Z., Ye, G., Zhou, Y., Pu, X., Su, W., & Wang, J. (2021). Editing sterol side chain reductase 2 gene (StSSR2) via CRISPR/Cas9 reduces the total steroidal gly-coalkaloids in potato. *All Life, 14*(1), 401–413. doi: 10.1080/26895293.2021.1925358

Zhou, J., Li, D., Wang, G., Wang, F., Kunjal, M., Joldersma, D., & Liu, Z. (2020). Application and future perspective of CRISPR/Cas9 genome editing in fruit crops. *Journal of Integrative Plant Biology, 62*, 269–286. doi: 10.1111/jipb.12793

Zhou, J., Wang, G., & Liu, Z. (2018). Efficient genome editing of wild strawberry genes, vector development and validation. *Plant Biotechnology Journal, 16*(11), 1868–1877. doi: 10.1111/pbi.12922

Zhou, X., Zha, M., Huang, J., Li, L., Imran, M., & Zhang, C. (2017). StMYB44 negatively regulates phosphate transport by suppressing expression of PHOSPH-ATE1 in potato. *Journal of Experimental Botany, 68*, 1265–1281. doi:10.1093/jxb/ erx026

Zsögön, A., Čermák, T., Naves, E.R., Notini, M.M., Edel, K.H., Weinl, S., Freschi, L., Voytas, D.F., Kudla, J., & Peres, L.E.P. (2018). *De novo* domestication of wild tomato using genome editing. *Nature Biotechnology, 36*, 1211–1216. doi: 10.1038/ nbt.4272

Abbreviations list

ABE	adenine base editor
ALS	acetolactate synthase
BSV	banana streak virus
BXW	banana *Xanthomonas* wilt
Cas	*CRISPR-associated protein*
CBE	cytidine base editor
CBE	cytosine base editor
ClALS	*Citrullus lanatus* acetolactate synthase
CRISPR/Cas	clustered regularly interspaced short palindromic repeat/ CRISPR associated proteins
crRNA	CRISPR RNA
CVYV	cucumber vein yellowing virus

DNA	deoxyribonucleic acid
DSB	double-strand break
dsDNA	double-stranded DNA
GBSS	granule bound starch synthase
GM	genetically modified
gRNA	guide RNA
GUS	B-glucuronidase
HDR	homology-directed repair
IbGBSSI	*Ipomoea batatas* granule-bound starch synthase I
IbSBEII	*Ipomoea batatas* starch branching enzyme II
KO	knockout
MWMV	Moroccan watermelon mosaic virus
NHEJ	on-homologous end joining
PAM	protospacer adjacent motif
PCR	polymerase chain reaction
PDS	phytoene desaturase
PRSV-W	papaya ringspot mosaic virus-W
QTLs	quantitative trait loci
RNA	ribonucleic acid
ROS	reactive oxygen species
sgRNA	single guide RNA
***Sp*Cas9**	Cas9 from *Streptococcus pyogenes*
SSN	site-specific nuclease
TALE	transcriptional activator-like effector
TALENs	transcription activator-like effector nucleases
tracrRNA	trans-activating CRISPR RNA
TYLCV	tomato yellow leaf curl virus
ZFNs	zinc-finger nuclease
ZYMV	zucchini yellow mosaic virus

4 Precision genome engineering and designer nucleases for crop improvement

Mohd Hadi Yunus and
Mohd Yunus Khalil Ansari

4.1 Introduction

The Food and Agriculture Organization projects that by 2050, food consumption would rise by more than 60%, requiring 50% more energy and roughly 40% more water to sustain the planet's 10 billion people. Furthermore, climate change may cause a worldwide temperature rise, which may be related to disease outbreaks, affecting agricultural output and reducing the value of the products (Raza et al. 2019). Numerous editing methods and strategies have been developed to solve the issues that arise in plants to adjust for future increases in food demand. Crop improvement in contemporary agriculture was mostly accomplished by cross-mutation or transgenic breeding. Cross-breeding takes several years to introduce beneficial alleles and enhance variety through genetic recombination (Scheben et al. 2017). Breeders were able to generate better varieties of several crops, resulting in enhanced food security, resistance to various abiotic or biotic stress, improved yield, and nutritional content. However, due to uncertain climatic conditions and increased consumer demands, breeders face new problems that must be solved. According to projections, climate change will have a significant effect on agricultural productivity, prompting the development of new kinds of production techniques in various geographic regions (Miladinovic et al. 2021). However, because of their stochastic character, these approaches are limited in their ability to generate and test large numbers of mutants.

Many agriculturally essential features such as shoot branching, tiller number, flowering duration, grain quantity, grain size, nutrient usage efficiency, and resilience to both abiotic and biotic stressors have been found through molecular genetic investigations (Ashikari and Matsuoka, 2006; Hori et al. 2016; Xu et al. 2016). Allelic variations resulting from domestication and subsequent advancements account for significant disparities in crop productivity and other agriculturally relevant features. Crop breeders can spend up to 10 years crossing and backcrossing to introduce only one elite allele into market cultivars. Furthermore, removing any unwanted genes/ agronomic features produced from the parental lines by crossing is extremely difficult if they are strongly connected to the target genes. Thus, transferring

DOI: 10.4324/9781003382102-4

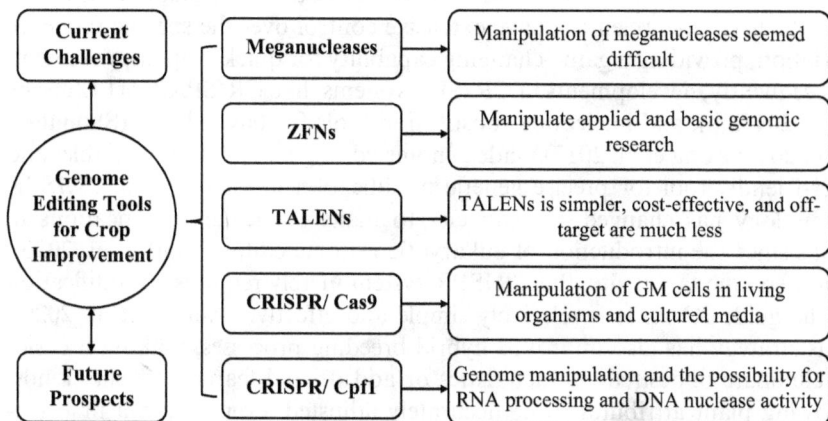

Figure 4.1 An overview of the many features of plant genome editing.

elite alleles from landraces or allied species into marketable crop varieties without introducing undesired genes or DNA fragments can considerably expedite agricultural progress (Li and Xia, 2020).

Scientists have been developing strategies for targeted genome editing in plants ever since 1988, the first gene-targeted experiment in tobacco proto-plasts (Paszkowski et al. 1988), and the discovery of the DNA double-strand breaks (DSBs) in 1993 that it can improve gene-targeting efficacy (Rees et al. 2017). Genome-editing techniques like meganuclease (MNs), zinc-finger nucleases (ZFNs), transcription activator-like effector nucleases (TALENs), and clustered regularly interspaced short palindromic repeats (CRISPR) (Gaj et al. 2013) are salient tools in plant research for future crop modifi-cation (see Figure 4.1).

The ZFNs were developed in 2005 for tobacco crops (Wright et al. 2005) and are the first genuinely targeted protein reagent, revolutionizing the field of genome engineering. They are DNA-binding domains that can identify three base pairs at a precise location (Rai et al. 2019). ZFNs were widely employed in plants like maize, tobacco, and *Arabidopsis* (Shukla et al. 2009; Osakabe et al. 2010; Petolino et al. 2010). However, TALENs are site-driven mutagenesis gene-editing mechanisms. They are based on a similar principle as ZFNs and are added to the toolkit in 2010 (Christian et al. 2010). TALENs were discovered in plant pathogenic bacteria (*Xanthomonas*). They target only one nucleotide at the target location (rather than three), making them precise (Boch et al. 2009). They have been utilized effectively for genome editing in bryophytes and angiosperms (Kopischke et al. 2017). Even though these two techniques have resulted in significant breakthroughs, each has its own set of restrictions, and their usage in plants is far from common.

CRISPR was developed in 2013 by three distinct groups for application in wheat (*Triticum aestivum*), rice (*Oryza sativa*), *Nicotianabenthamiana,* and arabidopsis *thaliana* (Li et al. 2013; Nekrasov et al. 2013; Shan et al. 2013). For the first time, scientists have extensive control over the specific sequence variation, providing a game-changing capability for quick crop improvement. Subsequently, developments in CRISPR systems, like CRISPR/Cpf1 (Zetsche et al. 2015a,b), and nucleotide substitution tools for base editing (Shimatani et al. 2017; Zong et al. 2017) made genome editing a popular, affordable, and user-friendly tool for precise genetic modification in many crops. CRISPR technology has changed our capacity to make particular modifications in crops since the introduction of eukaryotic genome editing (Butt et al. 2020). For a specific target site, the CRISPR system merely requires a modification in the guide RNA. It is relatively simple and effective (Ahmar et al. 2020). This strategy has also improved hybrid-breeding processes, making it easier to eliminate undesired characteristics or add desired features to elite kinds, allowing plant attributes to be accurately adjusted, even within a first generation(Chen et al. 2019). Because of its feasibility and reagent versatility, CRISPR/Cas9 has shown tremendous promise in a variety of crop species (Namo and Belachew, 2021).

4.2 Traditional technologies

Plant breeders employed traditional breeding methods for crop development long before genetic engineering. They first adjusted the underlying genotype of plants by selecting and reproducing favorable and observable phenotypes to modify them for human food, medical demands, and aesthetic objectives (Gaur et al. 2018). Conventional breeding stresses specific qualities that are already present in the species gene pool without adding new genes.

Plant breeders aim to accumulate beneficial alleles in a plant genome that have a substantial influence on nutritional quality, stress tolerance, and other agrarian features. Desirable genes were derived from local germplasm sources like breeding programs and wild species (Varshney et al. 2011). One of the earliest and most popular applications of biotechnology was clonal propagation or micropropagation by tissue culture because it was efficient and produced elite disease-free propagules. Many crops can now produce planting material free of viruses thanks to meristem culture (Krishna et al. 2016). Plant tissue culture has applications that extend well beyond agriculture. By improving accuracy and genotype-based selection, a molecular marker imparts a systematic foundation for the traditional method of breeding. Furthermore, a good knowledge of genetic and genomic regulation of phenotypes acquired by molecular markers can aid in the development of more effective breeding techniques and map-based gene isolation.

4.3 Genome-editing technologies

Genome-editing technology might be thought of as a type of gene splicing in which intracellular DNA is edited in a sequence-specific manner. Genome editing allows scientists to change the DNA of an organism by altering the genetic material at specific sites within the genome (Namo and Belachew, 2021). Synthetic biology necessitates a thorough grasp of the biological processes that must be incorporated into the genome.

Several techniques based on DNA, RNA, and proteins are being established to alter and introduce appropriate agronomical features into the specific plant. Innovative genome editing enables the breaking and rejoining of the DNA molecules at specific points to effectively change cells' genetic material. The initial attempts were to develop means for editing complicated genomes and the development of artificial enzymes in the form of oligonucleotides that would selectively bind to certain regions within the targeted DNA structure and have a chemical group competent for splitting DNA (Namo and Belachew, 2021).

4.4 Sequence-specific nucleases

Three main gene-editing methods are classified according to their mechanism, and the most commonly used method in plants is DNA DSBs using sequence-specific nucleases (Hiom, 2010). The non-homologous end joining process connects DNA ends and is typically seen at junctions where short lengths of nucleotides are inserted or deleted (indel). Different programmable nucleases are used to target DSB induction. The foremost common nucleases for genome editing are ZFNs, TALENs, and CRISPR/Cas9. These chimeric nucleases are produced directly to the present cell and the appropriate vectors encoding nucleases are delivered into the cell. These engineered vectors contain a nuclear localization signal that allows nuclease to enter the nucleus to access genomic DNA (Namo and Belachew, 2021). These different enzymatic mechanisms, structures, and activity of these nucleases lead to differences in their specificity, target selection, and efficiency are presented in (see Table 4.1).

4.5 Yield enhancement by genome engineering

Crop productivity is heavily influenced by intricate interactions among soil and climate conditions, abiotic and biotic stresses, and crop management strategies. These types of interactions may influence plant development and production, either firmly or skeptically (Jovičić et al. 2019). As a result, the combined influence of several abiotic and biotic stresses on physiology and development has received considerable attention (Pandey et al. 2015; Paul et al. 2019). According to research that used disease forecasting models for different regional climatic scenarios, biotic stress accounts for 15% of global losses in food products owing to potential yield loss (Jevtić et al. 2017). So far, several techniques for increasing agricultural yield have been created with all of this knowledge in mind. For more than 50 years, traditional or conventional breeding strategies have resulted in considerable improvements in yield and other critical qualities (Mladenov et al. 2011). However, preventing climate crisis will rely heavily on novel breeding strategies that can create

Table 4.1 Comparison of genome-editing tools

Characteristic	MNs	ZFNs	TALENs	CRISPR	Ref
Origin	Microbial genetic elements	Eukaryotic gene regulators	Bacterium *Xanthomonas*	Adaptive immune system of bacteria and archaea	(Sauer et al. 2016)
Core components	DNA binding and a DNA cleavage domain	Zinc-finger domain with *FokI* nuclease	TALE-DNA binding domain with *FokI* nuclease	Cas9 protein, crRNA	(Sauer et al. 2016)
Dimerization requirement	Dimeric	Dimeric	Dimeric	Monomeric	(Sauer et al. 2016)
Target sequence length	14–40 bp	24–36 bp	24–59 bp	20–22 bp	(Hsu et al. 2013)
Mode of action	Single/chimeric	Paired	Paired	Single	(Sauer et al. 2016)
Catalytic domain	A catalytic domain containing a DNA binding site	Catalytic domain	*FokI* catalytic domain	HNH Nd RUVCcatalytic domain	(Sauer et al. 2016)
Protein engineering criteria	Required	Required	Required	gRNA testing is not complicated	(Sauer et al. 2016)
Cloning criteria	Essential	Essential	Essential	Essential	(Sauer et al. 2016)
gRNA requirement	Not required	Not required	Not required	Required and simple to produce	(Sauer et al. 2016)
Genome altering	Creates DSBs in the target DNA	Creates DSBs in the target DNA	Creates DSBs in the target DNA	Creates single-strand nicks or DSB in the target DNA	(Sauer et al. 2016)
Target recognition efficiency	Low to moderate	Moderate	Moderate	High	(Sauer et al. 2016)
Rate of mutation	High	Medium	Medium	Low	(Sauer et al. 2016)
Multiplexing	Not possible	Difficult	Difficult	Easier	(Sauer et al. 2016)
Methylation sensitivity	High	High	High	Low	(Sauer et al. 2016)
Off-target effects	Low	Low	Least	Low	(Khandagale and Nadaf, 2016)
Cytotoxic effect	Variable to high	Variable to high	Low	Low	(Sauer et al. 2016)
Vector packaging	Easy	Difficult	Difficult	Moderate	(Sauer et al. 2016)
Cost of development	High	High	Higher	Low	(Swarts and Jinek, 2018)

desirable features more accurately and faster than traditional breeding approaches. One of these techniques is CRISPR/Cas9, had already been successfully used in a variety of plant species (Ricroch et al. 2017). Several studies have shown CRISPR/Cas9 as an efficacious tool for increasing agricultural productivity by knocking out genes that negatively influence yield-related variables (Zhang et al. 2018). For trait pyramiding, a multiplexing GE method was used, and the following findings were obtained including increased kernel weight in wheat (Zhang et al. 2016), increased grain size, and grain weight in rice (Xu et al. 2016), and also early heading in rice (Li et al. 2017).

Huang et al. used gene sequencing and the CRISPR/Cas9 method to identify 57 genes affecting yield-related features in 30 different kinds of phenotype known as 'wonder rice', and then developed knockout mutants of those 57 genes. The phenotyping of these mutants led to the finding of numerous genes linked with the regulation of traits in rice (Huang et al. 2018). This novel technique has the potential to be extremely beneficial in evaluating complicated quantitative qualities. Genome editing in crops has also been used to improve stress tolerance and to alter developmental and metabolic processes (Xu et al. 2019). These changes were employed as indirect methods to increase their performance and production in diverse conditions.

4.6 Novel technical breakthroughs in genome engineering

Genome editing has demonstrated a considerable future in agriculture but is quite constrained by its off-target effects, low efficiency, and many other challenges. To address these constraints, new technologies are continuously emerging.

4.6.1 CRISPR/Cpf1

CRISPR-Cpf1 (CRISPR – Prevotella and Francisella 1) is discovered as a novel mechanism for gene editing (Zetsche et al. 2015a). Cpf1 is denoted as a class II type V endonuclease which identifies as a T-rich PAM (5′-TTTN-3′) located at the 5′ end of the target site and cleaves it with a single 44-nucleotide crRNA. Cpf1 creates sticky or cohesive ends, as opposed to SpCas9's blunt ends, which improves the effectiveness of gene insertion at a specific genomic site. Cpf1 enzymes have lower off-target activity rates than nucleases of Cas9 (Kim et al. 2016). These enhanced characteristics make Cpf1 a more appealing tool than SpCas9 (Syed-Shan-e-Ali et al. 2017). Several studies have shown that this novel approach is a powerful DNA-free genome-editing tool (Kim et al. 2017; Xu et al. 2017). Cpf1 ortholog from *Acidaminococcus*sp. BV3L6 (AsCpf1) was also employed in soybean and rice (Hu et al. 2017; Kim et al. 2017; Tang et al. 2017).

CRISPR implementation extends to creating overexpression or mutant phenotypes. CRISPR in general may target any of the protein domains at certain genomic locations, resulting in locus-specific changes. CRISPR-Cpf1 has a wide range of applications including functional screening using gene knockouts, transcriptional repression or activation, epigenome editing, and cell lineage tracking using DNA-barcoding techniques. These innovative applications will boost agricultural productivity and quality, therefore contributing to food security and sustainability (Syed-Shan-e-Ali et al. 2017). Despite being an excellent tool for genome editing, CRISPR systems must be continuously checked for any off-target effects and changes in the efficiency of cleavage (Iqbal et al. 2020). Accessing next-generation sequencing data of numerous plants may aid in overcoming the system's shortcomings.

4.6.2 Base editing

Base editing has various advantages over its modern counterparts. It targets the sequence of interest with precision, predictability, and efficiency. Furthermore, because base editing does not rely on DSBs, random insertions, deletions, and rearrangements may be avoided. Variations in agricultural plants are generally caused by single-base mutations effective strategies for inducing precise point mutations in plants are required. Base editing using CRISPR/Cas9 may change one DNA base to another without the usage of a DNA repair template (Komor et al. 2016). In base-editing technologies, the Cas9 nickase is coupled with an enzyme having base conversion activity. Base editors are formed when a catalytically inactive domain of CRISPR-Cas9 is coupled with a cytosine deaminase domain to convert G-C base pairs to A-T (Nishida et al. 2016). Few articles demonstrate base-editing optimization in crops, successfully demonstrated in *Arabidopsis*, maize, rice, tomato, and wheat using cytidine-deaminase-mediated base editing (CBE) by employing nCas9-PBE, a plant base editor consisting of a Cas9 variant [Cas9-D10A nickase (nCas9)] and rat cytidine-deaminase APOBEC1 (Zong et al. 2017).

4.6.3 DNA-free genome editing

DNA-free genome editing is a game-changing technique that produces genetically modified plants with a lower chance of unwanted off-target effects while still addressing present and future agricultural demand. Both particle bombardment and protoplast-mediated transformation have been used. The avoidance of transgene assimilation and the limitation of Cas9 off-target activity are essential objectives to consider through the escalation of the CRISPR/Cas9 technique. As a result, it aims to make CRISPR/Cas9 ribonucleoprotein (RNP). RNPs capable of producing appropriate and permanent genome editing. In-vitro evaluation of CRISPR/Cas9 RNP revealed robust editing and a significant reduction in off-targets. Using CRISPR/Cas9RNPs, researchers have successfully altered a variety of crops including *A. thaliana*, lettuce, tobacco, rice, and maize (Woo et al. 2015; Svitashev et al. 2016). Korean researchers have introduced CRISPR/Cas9 RNPs to grape and apple protoplast systems, resulting in crops free of transgenes (Malnoy et al. 2016). To boost powdery mildew tolerance in grape varieties, researchers selected a vulnerable gene called MLO-7, and to improve fire blight resistance in apples, they targeted the DIPM-1, DIPM-2, and DIPM-4 genes. Researchers from China have accomplished genome editing using CRISPR/Cas9RNP in bread wheat immature embryo cells (Liang et al. 2017).

4.7 Crop enhancement through the use of genome-editing techniques

Genome engineering has been successfully utilized to target genes in a broad variety of crops to enhance yield, increased nutritional value, better quality, and disease resistance (see Figure 4.2). By targeting features mostly regulated by negatively regulatory genes, genome-editing technologies have shown considerable promise in improving crop tolerance to various abiotic and biotic challenges (Wang et al. 2016).

4.7.1 Improvement in quality of crops

Genome-editing offers far-reaching large-scale implications for overcoming one of modern biotechnology's milestones, specifically the production of novel crops with high yield, tolerance against various abiotic and biotic stress, and high nutritional value. The only option to enhance this vital staple meal and fruit may be through genome editing. CRISPR technology has made significant development. Only a few species of fruit-producing crops, such as citrus, grapes, tomatoes, strawberries, or watermelons, have been identified with features acquired from CRISPR/Cas9 via germline (Zhou et al. 2020). The suppression of ethylene production was discovered to be critical in the fruit-ripening process (Wang et al. 2018). The production of gibberellin has resulted in the development of a dwarf variety of fruit trees (Begemann et al. 2017),

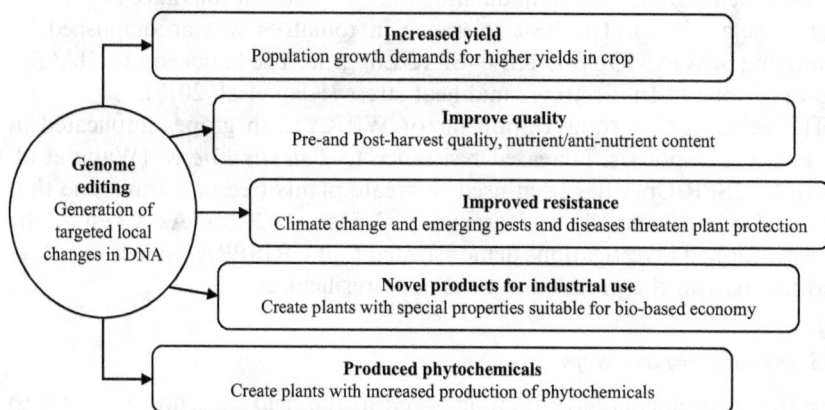

Figure 4.2 Concept of plant genome editing and its target of advanced breeding.

which have the potential for high output through dense planting, reduced water and fertilizer use, and the economic cost of land and labor. Furthermore, its signaling pathways resulted in the production of novel kinds with a longer shelf life (Clasen et al. 2016). Due to the high potency of genome editing, there is no involvement of any foreign DNA, consumers will find it simple to consume genome-edited fruits. Parthenocarpy production regulated by CRISPR/Cas9 is employed in many fruits such as custard apples, citrus, kinnow, grapes, watermelon, and peach. The FAD2-1A and FAD2-1B genes in soybean were modified using CRISPR/Cpf1to increase the oil composition and develop crops with higher yield and oleic acid. Using the CRISPR/Cpf1 technology, researchers effectively improved the quality and yield (Xu et al. 2017). To induce genome editing in higher plants such as soybean, rice, and tobacco, various Cpf1 proteins have been used.

4.7.2 Upgrading of climate-resilient crops

CRISPR technology is widely employed in a crop, such as cotton, corn, rice, tomato, potato, soybeans, and wheat in conjunction with a wide range of abiotic and biotic stresses. Plant breeding initiatives for smart climatic abiotic stress-tolerant crops have been upgraded thanks to the CRISPR technique. High amounts of salt in the soil reduce rice output dramatically. Reactive oxygen species play a crucial function by serving as a signaling molecule for gene expression control, defense against the viral pathogen, and nitrogen fixation symbiotically between soil rhizobia and plants (Wu et al. 2017). CRISPR-Cas9 has permitted the production of novel ARGOS8 variations in maize. The ARGOS8 variation outperformed wild-type alleles in terms of yield during flowering stress circumstances (i.e. five bushels of grain per acre). The results indicate that CRISPR/Cas9 is a reliable method for generating novel allelic variants in crops for plant development, like drought tolerance (Shi et al. 2017). The deletion of two genes, DRB2a and DRB2b, using CRISPR/Cas9 revealed their significance in modulating drought and salt tolerance in soybean (Curtin et al. 2018). The heat tolerance in tomatoes was accomplished by identifying SlAGAMOUS-LIKE 6 (SIAGL6) gene. The knocked-out SIAGL6 improved tomato fruit setting amid heat stress (Klap et al. 2017).

The deletion of a transcription factor WRKY52 in grapes implicated in biotic stress responses, increased resistance to *Botrytis cinerea* (Wang et al. 2018). CRISPR/Cpf1 has been used to create plants free of a transgene that reproduces both sexually and asexually (Vu et al. 2020). As a result, the aforementioned investigations demonstrated that CRISPR/Cas9 editing plays a critical part in the production of climate-resilient crops.

4.7.3 Disease resistance crops

Recent improvements have been achieved in the field of genome editing to develop pathogen-resistant crops. The chosen strategies have been utilized to modify crop immunity at various phases in many crops (Nemudryi et al. 2014). Powdery mildew resistance was developed in wheat genotypes using TALEN and CRISPR/Cas9 methods by manipulation of the mildew resistance locus O (MLO) gene (Wang et al. 2014). Genome editing was also used in the development of bacterial-resistant plant lines such as against the *Xanthomonas oryzae* causing bacterial blight (Li et al. 2016). CRISPR/Cas9 has been tested for its potential to build resistance to geminivirus infection, like in *N. benthamiana* and *Arabidopsis* by introducing sgRNA/Cas9 into the plants (Ji et al. 2015). Resistance against bacterial blight in rice is also enhanced using by modifying the SWEET11, SWEET13, and SWEET14 genes (Oliva et al. 2019). The downy mildew resistance 6 (DMR6) knockouts in tomatoes showed better broad-spectrum resistance to a variety of diseases, including bacteria and oomycetes (de Toledo Thomazella et al. 2016). Such applications showed that CRISPR is a valuable tool for crop improvement against disease resistance.

4.8 Current status in crops improvement

The history of genetic engineering dates back over 70 years, to the discovery of the helix. Since then, researchers have worked to find strategies to modify the genome that balance specificity with time and cost (Namo and Belachew, 2021). This bacterial method has swiftly gained acceptance in eukaryotic host cells for genome editing (Nakayama et al. 2013). Both spontaneous and induced mutations are routinely employed to get sufficient genetic resources with various traits for breeding (Boglioli and Richard, 2015).

Scientists have been encouraged to research strategies to introduce precise mutations in specific areas due to the rarity and unpredictability of those changes (Rocha-Martins et al. 2015). The fast development of gene editing technologies in recent years has sparked renewed interest in how they may be used in agriculture. CRISPR/Cas9 gene editing techniques are less costly and easier to adopt than transgenic technology. Many nations, including the United States, have indicated that they would not regulate gene-edited crops. Because the procedure does not entail the introduction of foreign DNA into the final product. Plant pathogen resistance genes such as *Xanthomonas citri* (Peng et al. 2017) and *Botrytis cinerea* (Wang et al. 2018) are modified in grapes, apples, and citrus. Genes involved in carboxylic acid metabolism are widely targeted in oilseed crops to improve oil quality (Okuzaki et al. 2018). In recent years, genome editing has been utilized to develop crops with enhanced characteristics.

4.9 Prospects and research opportunities

Genome engineering is rapidly being used for plants. The study and development of genome-editing techniques, as well as plant modification through genome editing, is moving at a rapid speed. The design of the CRISPR/Cas is critical to this process, mainly the selection of a target sequence that can efficiently induce DNA DSBs at the target locus. It is extremely useful to predict mutation frequency and spectrum at a certain locus utilizing CRISPR/Cas-mediated targeted mutagenesis and base editing. When it comes to planting breeding, the designing of the genome sequence is just as crucial as genome editing. When the target gene is epigenetically mediated, the genome sequences, DNA and histone modifications, as well as chromatin structure need to be reformed to produce the desired crop (Sukegawa et al. 2021).

4.9.1 Multiplexing and trait stacking

Cellular activities are frequently governed by complicated genetic networks, and agronomic trait modification is reliant on convoluted metabolic pathways, that need a synchronized expression of several genes. One benefit of CRISPR over other technologies is its ability to multiplex or alter many target locations at the same time (Cong et al. 2013). Several laboratories have employed the golden gate cloning or the Gibson assembly approach to merge several sgRNAs into a single Cas9/sgRNA expression vector (Volpi e Silva and Patron, 2017). *Xieet al.*devised a universal technique for generating a large number of sgRNAs from a single polycistronic gene (Xie et al. 2015). They altered the endogenous tRNA processing, resulting in a quite strong foundation to increase CRISPR/Cas9 targeting and multiplex editing capabilities. This tRNA-processing machinery has also been used for multiplex editing by CRISPR/Cpf1 (Ding et al. 2018). In plants with low transformation or editing efficiency, several sgRNAs can be employed to target a single gene.

4.9.2 High-throughput mutant libraries

Whole-genome mutant libraries on a wider scale are extremely valuable for crop improvement and functional genomics. Genomic research will rigorously study the activities of all genes, as the genomes of the majority of crops have already been sequenced. The functions of the majority of them are still unknown, which may help in modifying beneficial agronomic features. Gene knockout is a common and successful technique to find any gene functions. *Lu et al.* created a library of 91,004 mutants by designing 88,541 sgRNAs targeting 34,234 genes. (Lu et al. 2017). *Meng et al.* generated 25,604 sgRNAs for 12,802 genes, yielding almost 14,000 transgenic T0 lines (Meng et al. 2017). Rice was chosen for targeted mutagenesis primarily because of its small genome, abundant genetic resources, and extremely competent transformation method. As the procedures advance, the creation of mutant libraries for other useful plants should not be put off for too long.

4.9.3 Gene regulation

Gene regulation is mostly accomplished by attaching transcriptional repressors or activators to DNA-binding domains, thus targeting the main regions of endogenous genes (Qi et al. 2021). CRISPR/Cas9 may also be utilized to suppress or stimulate plant gene transcription by pairing catalytically inactive dCas9 with particular promoter sequence-targeting sgRNAs (Lowder et al. 2015; Piatek et al. 2015). Both LbCpf1 and AsCpf1 were utilized to repress the transcription in *Arabidopsis*, highlighting Cpf1's significant potential for altering plant transcriptomes (Tang et al. 2017). CRISPR/Cas9 is utilized to boost agricultural yield by controlling the cis-regulation of quantitative trait loci. The mutated SlCLV3 promoters in tomatoes using CRISPR/Cas9 resulted in hundreds of regulatory alterations (Rodríguez-Leal et al. 2017). They may systematically examine the correlation of cis-regulatory areas with phenotypic features in this manner, which should aid in tomato breeding. This concept provides an apparent and efficient way for modifying mRNA translation, which may be used to investigate biological systems and enhance crops. Contrary to applications seeking to change the DNA sequence, the impacts of genome editing on gene regulation at the transcript level related to crop improvement might be utilized to uncover the role of many non-canonical RNAs. Because the majority of non-coding transcripts are still unclear and do not have open reading frames and ideal gene modification that could directly affect the transcription for investigating their function.

4.10 Conclusion

Genome engineering is swiftly evolving into the most popular and versatile approach for functional genomics and crop improvement. It became plausible with the discovery of the very first designer nucleases in the late 1990s. However, the field has grown in prominence since 2012, owing to the invention of CRISPR/Cas9. Genome editing, being a sophisticated molecular biology technology, may make precisely targeted adjustments in any crop. Specific to the availability of several genome-editing techniques for various purposes, it is critical to identify the best system for a given species and goal. For each gene, the guide RNAs or the responsible enzymes should be evaluated to ensure high editing efficiency. To achieve this, considerably more research must be done to develop effective procedures for the regeneration and transformation of many plant species and their variations, particularly in an automated high-throughput way. Direct alteration of meristematic tissues or a germline in a plant would eliminate the tissue culture stages and the species and variety-dependent limitations of the regeneration.

However, more research is needed to examine the permanence and heritability of the epigenetic modifications induced by novel epigenome-editing technologies before they may be utilized to develop new crops. One of the most significant technological challenges involved with genome engineering in crops is the reliance on tissue culture methods, which limit the explants used for delivering genome-editing constructs or components, affecting the effectiveness and determining whether the whole plant can be regenerated. To truly understand the immense potential of genome engineering, the identification of essential genes influencing agronomic features must be researched at the genomic level.

References

Ahmar, S., Gill, R. A., Jung, K.-H., Faheem, A., Qasim, M. U., Mubeen, M., & Zhou, W. (2020). Conventional and molecular techniques from simple breeding to speed breeding in crop plants: Recent advances and future outlook. *International Journal of Molecular Sciences*, 21(7), 2590.

Ashikari, M., & Matsuoka, M. (2006). Identification, isolation and pyramiding of quantitative trait loci for rice breeding. *Trends in Plant Science*, 11(7), 344–350.

Begemann, M. B., Gray, B. N., January, E., Gordon, G. C., He, Y., Liu, H., Wu, X., Brutnell, T. P., Mockler, T. C., & Oufattole, M. (2017). Precise insertion and guided editing of higher plant genomes using Cpf1 CRISPR nucleases. *Scientific Reports*, 7(1), 1–6.

Boch, J., Scholze, H., Schornack, S., Landgraf, A., Hahn, S., Kay, S., Lahaye, T., Nickstadt, A., & Bonas, U. (2009). Breaking the code of DNA binding specificity of TAL-type III effectors. *Science*, 326(5959), 1509–1512.

Boglioli, E., & Richard, M. (2015). *Rewriting the book of life: A new era in precision gene editing.* Boston Consulting Group (BCG).

Butt, H., Zaidi, S. S.-A., Hassan, N., & Mahfouz, M. (2020). CRISPR-based directed evolution for crop improvement. *Trends in Biotechnology*, 38(3), 236–240.

Chen, K., Wang, Y., Zhang, R., Zhang, H., & Gao, C. (2019). CRISPR/Cas genome editing and precision plant breeding in agriculture. *Annual Review of Plant Biology*, 70(1), 667–697. 10.1146/annurev-arplant-050718-100049

Christian, M., Cermak, T., Doyle, E. L., Schmidt, C., Zhang, F., Hummel, A., Bogdanove, A. J., & Voytas, D. F. (2010). Targeting DNA double-strand breaks with TAL effector nucleases. *Genetics*, 186(2), 757–761.

Clasen, B. M., Stoddard, T. J., Luo, S., Demorest, Z. L., Li, J., Cedrone, F., Tibebu, R., Davison, S., Ray, E. E., & Daulhac, A. (2016). Improving cold storage and processing traits in potato through targeted gene knockout. *Plant Biotechnology Journal*, 14(1), 169–176.

Cong, L., Ran, F. A., Cox, D., Lin, S., Barretto, R., Habib, N., Hsu, P. D., Wu, X., Jiang, W., & Marraffini, L. A. (2013). Multiplex genome engineering using CRISPR/Cas systems. *Science*, 339(6121), 819–823.

Curtin, S. J., Xiong, Y., Michno, J.-M., Campbell, B. W., Stec, A. O., Čermák, T., Starker, C., Voytas, D. F., Eamens, A. L., & Stupar, R. M. (2018). Crispr/cas9 and talen s generate heritable mutations for genes involved in small rna processing of glycine max and medicagotruncatula. *Plant Biotechnology Journal*, 16(6), 1125–1137.

de Toledo Thomazella, D. P., Brail, Q., Dahlbeck, D., & Staskawicz, B. (2016). CRISPR-Cas9 mediated mutagenesis of a DMR6 ortholog in tomato confers broad-spectrum disease resistance. *BioRxiv*, 064824.

Ding, D., Chen, K., Chen, Y., Li, H., & Xie, K. (2018). Engineering introns to express RNA guides for Cas9-and Cpf1-mediated multiplex genome editing. *Molecular Plant*, 11(4), 542–552.

Gaj, T., Gersbach, C. A., & Barbas III, C. F. (2013). ZFN, TALEN, and CRISPR/Cas-based methods for genome engineering. *Trends in Biotechnology*, 31(7), 397–405.

Gaur, R. K., Verma, R. K., & Khurana, S. M. P. (2018). Genetic engineering of horticultural crops. In *Genetic engineering of horticultural crops* (pp. 23–46). Elsevier. 10.1016/B978-0-12-810439-2.00002-7

Hiom, K. (2010). Coping with DNA double strand breaks. *DNA Repair*, 9(12), 1256–1263.

Hori, K., Matsubara, K., & Yano, M. (2016). Genetic control of flowering time in rice: Integration of Mendelian genetics and genomics. *Theoretical and Applied Genetics*, 129(12), 2241–2252.

Hsu, P. D., Scott, D. A., Weinstein, J. A., Ran, F., Konermann, S., Agarwala, V., Li, Y., Fine, E. J., Wu, X., & Shalem, O. (2013). DNA targeting specificity of RNA-guided Cas9 nucleases. *Nature Biotechnology*, 31(9), 827–832.

Hu, X., Wang, C., Liu, Q., Fu, Y., & Wang, K. (2017). Targeted mutagenesis in rice using CRISPR-Cpf1 system. *Journal of Genetics and Genomics*, 44(1), 71–73.

Huang, J., Li, J., Zhou, J., Wang, L., Yang, S., Hurst, L. D., Li, W.-H., & Tian, D. (2018). Identifying a large number of high-yield genes in rice by pedigree analysis, whole-genome sequencing, and CRISPR-Cas9 gene knockout. *Proceedings of the National Academy of Sciences*, 115(32), E7559–E7567.

Iqbal, Z., Iqbal, M. S., Ahmad, A., Memon, A. G., & Ansari, M. I. (2020). New prospects on the horizon: Genome editing to engineer plants for desirable traits. *Current Plant Biology*, 24, 100171. 10.1016/j.cpb.2020.100171

Jevtić, R., Župunski, V., Lalošević, M., & Župunski, L. (2017). Predicting potential winter wheat yield losses caused by multiple disease systems and climatic conditions. *Crop Protection*, 99, 17–25.

Ji, X., Zhang, H., Zhang, Y., Wang, Y., & Gao, C. (2015). Establishing a CRISPR–Cas-like immune system conferring DNA virus resistance in plants. *Nature Plants*, 1(10), 1–4.

Jovičić, D., Popović, B. M., Marjanović-Jeromela, A., Nikolić, Z., Ignjatov, M., & Milošević, D. (2019). The interaction between salinity stress and seed ageing during germination of Brassica napus seeds. *Seed Science & Technology*, 47(1), 47–52.

Khandagale, K., & Nadaf, A. (2016). Genome editing for targeted improvement of plants. *Plant Biotechnology Reports*, 10(6), 327–343.

Kim, D., Kim, J., Hur, J. K., Been, K. W., Yoon, S., & Kim, J.-S. (2016). Genome-wide analysis reveals specificities of Cpf1 endonucleases in human cells. *Nature Biotechnology*, 34(8), 863–868.

Kim, H., Kim, S.-T., Ryu, J., Kang, B.-C., Kim, J.-S., & Kim, S.-G. (2017). CRISPR/Cpf1-mediated DNA-free plant genome editing. *Nature Communications*, 8(1), 1–7.

Klap, C., Yeshayahou, E., Bolger, A. M., Arazi, T., Gupta, S. K., Shabtai, S., Usadel, B., Salts, Y., & Barg, R. (2017). Tomato facultative parthenocarpy results from Sl AGAMOUS-LIKE 6 loss of function. *Plant Biotechnology Journal*, 15(5), 634–647.

Komor, A. C., Kim, Y. B., Packer, M. S., Zuris, J. A., & Liu, D. R. (2016). Programmable editing of a target base in genomic DNA without double-stranded DNA cleavage. *Nature*, 533(7603), 420–424.

Kopischke, S., Schüßler, E., Althoff, F., & Zachgo, S. (2017). TALEN-mediated genome-editing approaches in the liverwort *Marchantia polymorpha* yield high efficiencies for targeted mutagenesis. *Plant Methods*, 13(1), 1–11.

Krishna, H., Alizadeh, M., Singh, D., Singh, U., Chauhan, N., Eftekhari, M., & Sadh, R. K. (2016). Somaclonal variations and their applications in horticultural crops improvement. *3 Biotech*, 6(1), 1–18.

Li, C., Unver, T., & Zhang, B. (2017). A high-efficiency CRISPR/Cas9 system for targeted mutagenesis in Cotton (Gossypium hirsutum L.). *Scientific Reports*, 7(1), 1–10.

Li, J.-F., Norville, J. E., Aach, J., McCormack, M., Zhang, D., Bush, J., Church, G. M., & Sheen, J. (2013). Multiplex and homologous recombination–mediated genome editing in Arabidopsis and *Nicotiana benthamiana* using guide RNA and Cas9. *Nature Biotechnology*, 31(8), 688–691.

Li, S., & Xia, L. (2020). Precise gene replacement in plants through CRISPR/Cas genome editing technology: Current status and future perspectives. *Abiotech*, 1(1), 58–73. 10.1007/s42994-019-00009-7

Li, T., Liu, B., Chen, C. Y., & Yang, B. (2016). TALEN-mediated homologous recombination produces site-directed DNA base change and herbicide-resistant rice. *Journal of Genetics and Genomics*, 43(5), 297–305.

Liang, Z., Chen, K., Li, T., Zhang, Y., Wang, Y., Zhao, Q., Liu, J., Zhang, H., Liu, C., & Ran, Y. (2017). Efficient DNA-free genome editing of bread wheat using CRISPR/Cas9 ribonucleoprotein complexes. *Nature Communications*, 8(1), 1–5.

Lowder, L. G., Zhang, D., Baltes, N. J., Paul III, J. W., Tang, X., Zheng, X., Voytas, D. F., Hsieh, T.-F., Zhang, Y., & Qi, Y. (2015). A CRISPR/Cas9 toolbox for multiplexed plant genome editing and transcriptional regulation. *Plant Physiology*, 169(2), 971–985.

Lu, Y., Ye, X., Guo, R., Huang, J., Wang, W., Tang, J., Tan, L., Zhu, J., Chu, C., & Qian, Y. (2017). Genome-wide targeted mutagenesis in rice using the CRISPR/Cas9 system. *Molecular Plant*, 10(9), 1242–1245.

Malnoy, M., Viola, R., Jung, M.-H., Koo, O.-J., Kim, S., Kim, J.-S., Velasco, R., & Nagamangala Kanchiswamy, C. (2016). DNA-free genetically edited grapevine and apple protoplast using CRISPR/Cas9 ribonucleoproteins. *Frontiers in Plant Science*, 7, 1904.

Meng, X., Yu, H., Zhang, Y., Zhuang, F., Song, X., Gao, S., Gao, C., & Li, J. (2017). Construction of a genome-wide mutant library in rice using CRISPR/Cas9. *Molecular Plant*, 10(9), 1238–1241.

Miladinovic, D., Antunes, D., Yildirim, K., Bakhsh, A., Cvejić, S., Kondić-Špika, A., Marjanovic Jeromela, A., Opsahl-Sorteberg, H.-G., Zambounis, A., & Hilioti, Z. (2021). Targeted plant improvement through genome editing: From laboratory to field. *Plant Cell Reports*, 40(6), 935–951. 10.1007/s00299-020-02655-4

Mladenov, N., Hristov, N., Kondic-Spika, A., Djuric, V., Jevtic, R., & Mladenov, V. (2011). Breeding progress in grain yield of winter wheat cultivars grown at different nitrogen levels in semiarid conditions. *Breeding Science*, 61(3), 260–268.

Nakayama, T., Fish, M. B., Fisher, M., Oomen-Hajagos, J., Thomsen, G. H., & Grainger, R. M. (2013). Simple and efficient CRISPR/Cas9-mediated targeted mutagenesis in Xenopus tropicalis. *Genesis*, 51(12), 835–843.

Namo, F. M., & Belachew, G. T. (2021). Genome editing technologies for crop improvement: Current status and future prospective. *Plant Cell Biotechnology and Molecular Biology*, 22, 1–19.

Nekrasov, V., Staskawicz, B., Weigel, D., Jones, J. D., & Kamoun, S. (2013). Targeted mutagenesis in the model plant *Nicotiana benthamiana* using Cas9 RNA-guided endonuclease. *Nature Biotechnology*, 31(8), 691–693.

Nemudryi, A. A., Valetdinova, K. R., Medvedev, S. P., & Zakian, S. M. (2014). TALEN and CRISPR/Cas genome editing systems: Tools of discovery. *Acta Naturae (АнглоязычнаяВерсия)*, 6(3 (22)), 19–40.

Nishida, K., Arazoe, T., Yachie, N., Banno, S., Kakimoto, M., Tabata, M., Mochizuki, M., Miyabe, A., Araki, M., & Hara, K. Y. (2016). Targeted nucleotide editing using hybrid prokaryotic and vertebrate adaptive immune systems. *Science*, 353(6305), aaf8729.

Okuzaki, A., Ogawa, T., Koizuka, C., Kaneko, K., Inaba, M., Imamura, J., & Koizuka, N. (2018). CRISPR/Cas9-mediated genome editing of the fatty acid desaturase 2 gene in Brassica napus. *Plant Physiology and Biochemistry*, 131, 63–69.

Oliva, R., Ji, C., Atienza-Grande, G., Huguet-Tapia, J. C., Perez-Quintero, A., Li, T., Eom, J.-S., Li, C., Nguyen, H., & Liu, B. (2019). Broad-spectrum resistance to bacterial blight in rice using genome editing. *Nature Biotechnology*, 37(11), 1344–1350.

Osakabe, K., Osakabe, Y., & Toki, S. (2010). Site-directed mutagenesis in Arabidopsis using custom-designed zinc finger nucleases. *Proceedings of the National Academy of Sciences*, 107(26), 12034–12039.

Pandey, P., Sinha, R., Mysore, K. S., & Senthil-Kumar, M. (2015). Impact of concurrent drought stress and pathogen infection on plants. In *Combined stresses in plants* (pp. 203–222). Springer.

Paszkowski, J., Baur, M., Bogucki, A., & Potrykus, I. (1988). Gene targeting in plants. *The EMBO Journal*, 7(13), 4021–4026.

Paul, K., Sorrentino, M., Lucini, L., Rouphael, Y., Cardarelli, M., Bonini, P., Reynaud, H., Canaguier, R., Trtílek, M., & Panzarová, K. (2019). Understanding the biostimulant action of vegetal-derived protein hydrolysates by high-throughput plant phenotyping and metabolomics: A case study on tomato. *Frontiers in Plant Science*, 10, 47.

Peng, A., Chen, S., Lei, T., Xu, L., He, Y., Wu, L., Yao, L., & Zou, X. (2017). Engineering canker-resistant plants through CRISPR/Cas9-targeted editing of the susceptibility gene Cs LOB 1 promoter in citrus. *Plant Biotechnology Journal*, 15(12), 1509–1519.

Petolino, J. F., Worden, A., Curlee, K., Connell, J., Strange Moynahan, T. L., Larsen, C., & Russell, S. (2010). Zinc finger nuclease-mediated transgene deletion. *Plant Molecular Biology*, 73(6), 617–628.

Piatek, A., Ali, Z., Baazim, H., Li, L., Abulfaraj, A., Al-Shareef, S., Aouida, M., & Mahfouz, M. M. (2015). RNA-guided transcriptional regulation in planta via synthetic dC as9-based transcription factors. *Plant Biotechnology Journal*, 13(4), 578–589.

Qi, L. S., Larson, M. H., Gilbert, L. A., Doudna, J. A., Weissman, J. S., Arkin, A. P., & Lim, W. A. (2021). Repurposing CRISPR as an RNA-guided platform for sequence-specific control of gene expression. *Cell*, 184(3), 844.

Rai, K. M., Ghose, K., Rai, A., Singh, H., Srivastava, R., & Mendu, V. (2019). Genome engineering tools in plant synthetic biology. In *Current developments in biotechnology and bioengineering* (pp. 47–73). Elsevier.

Raza, A., Razzaq, A., Mehmood, S. S., Zou, X., Zhang, X., Lv, Y., & Xu, J. (2019). Impact of climate change on crops adaptation and strategies to tackle its outcome: A review. *Plants*, 8(2), 34.

Rees, H. A., Komor, A. C., Yeh, W.-H., Caetano-Lopes, J., Warman, M., Edge, A. S., & Liu, D. R. (2017). Improving the DNA specificity and applicability of base editing through protein engineering and protein delivery. *Nature Communications*, 8(1), 1–10.

Ricroch, A., Clairand, P., & Harwood, W. (2017). Use of CRISPR systems in plant genome editing: Toward new opportunities in agriculture. *Emerging Topics in Life Sciences*, 1(2), 169–182.

Rocha-Martins, M., Cavalheiro, G. R., Matos-Rodrigues, G. E., & Martins, R. A. (2015). From gene targeting to genome editing: Transgenic animals applications and beyond. *Anais Da Academia Brasileira de Ciências*, 87, 1323–1348.

Rodríguez-Leal, D., Lemmon, Z. H., Man, J., Bartlett, M. E., & Lippman, Z. B. (2017). Engineering quantitative trait variation for crop improvement by genome editing. *Cell*, 171(2), 470–480.

Sauer, N. J., Mozoruk, J., Miller, R. B., Warburg, Z. J., Walker, K. A., Beetham, P. R., Schöpke, C. R., & Gocal, G. F. (2016). Oligonucleotide-directed mutagenesis for precision gene editing. *Plant Biotechnology Journal*, 14(2), 496–502.

Scheben, A., Wolter, F., Batley, J., Puchta, H., & Edwards, D. (2017). Towards CRISPR/Cas crops–bringing together genomics and genome editing. *New Phytologist*, 216(3), 682–698.

Shan, Q., Wang, Y., Li, J., Zhang, Y., Chen, K., Liang, Z., Zhang, K., Liu, J., Xi, J. J., & Qiu, J.-L. (2013). Targeted genome modification of crop plants using a CRISPR-Cas system. *Nature Biotechnology*, 31(8), 686–688.

Shi, J., Gao, H., Wang, H., Lafitte, H. R., Archibald, R. L., Yang, M., Hakimi, S. M., Mo, H., & Habben, J. E. (2017). ARGOS 8 variants generated by CRISPR-Cas9 improve maize grain yield under field drought stress conditions. *Plant Biotechnology Journal*, 15(2), 207–216.

Shimatani, Z., Kashojiya, S., Takayama, M., Terada, R., Arazoe, T., Ishii, H., Teramura, H., Yamamoto, T., Komatsu, H., & Miura, K. (2017). Targeted base editing in rice and tomato using a CRISPR-Cas9 cytidine deaminase fusion. *Nature Biotechnology*, 35(5), 441–443.

Shukla, V. K., Doyon, Y., Miller, J. C., DeKelver, R. C., Moehle, E. A., Worden, S. E., Mitchell, J. C., Arnold, N. L., Gopalan, S., & Meng, X. (2009). Precise genome modification in the crop species Zea mays using zinc-finger nucleases. *Nature*, 459(7245), 437–441.

Sukegawa, S., Saika, H., & Toki, S. (2021). Plant genome editing: Ever more precise and wide reaching. *The Plant Journal*, 106(5), 1208–1218. 10.1111/tpj.15233

Svitashev, S., Schwartz, C., Lenderts, B., Young, J. K., & Mark Cigan, A. (2016). Genome editing in maize directed by CRISPR–Cas9 ribonucleoprotein complexes. *Nature Communications*, 7(1), 1–7.

Swarts, D. C., & Jinek, M. (2018). Cas9 versus Cas12a/Cpf1: Structure–function comparisons and implications for genome editing. *Wiley Interdisciplinary Reviews: RNA*, 9(5), e1481.

Syed-Shan-e-Ali, Z., Mahfouz, M. M., & Mansoor, S. (2017). CRISPR-Cpf1: A new tool for plant genome editing. *Trends in Plant Science*, 22(7), 550–553.

Tang, X., Lowder, L. G., Zhang, T., Malzahn, A. A., Zheng, X., Voytas, D. F., Zhong, Z., Chen, Y., Ren, Q., & Li, Q. (2017). A CRISPR–Cpf1 system for efficient genome editing and transcriptional repression in plants. *Nature Plants*, 3(3), 1–5.

Varshney, R. K., Bansal, K. C., Aggarwal, P. K., Datta, S. K., & Craufurd, P. Q. (2011). Agricultural biotechnology for crop improvement in a variable climate: Hope or hype? *Trends in Plant Science*, 16(7), 363–371.

Volpi e Silva, N., & Patron, N. J. (2017). CRISPR-based tools for plant genome engineering. *Emerging Topics in Life Sciences*, 1(2), 135–149.

Vu, T. V., Sivankalyani, V., Kim, E.-J., Doan, D. T. H., Tran, M. T., Kim, J., Sung, Y. W., Park, M., Kang, Y. J., & Kim, J.-Y. (2020). Highly efficient homology-directed repair using CRISPR/Cpf1-geminiviral replicon in tomato. *Plant Biotechnology Journal*, 18(10), 2133–2143.

Wang, F., Wang, C., Liu, P., Lei, C., Hao, W., Gao, Y., Liu, Y.-G., & Zhao, K. (2016). Enhanced rice blast resistance by CRISPR/Cas9-targeted mutagenesis of the ERF transcription factor gene OsERF922. *PloS One*, 11(4), e0154027.

Wang, M., Mao, Y., Lu, Y., Wang, Z., Tao, X., & Zhu, J.-K. (2018). Multiplex gene editing in rice with simplified CRISPR-Cpf1 and CRISPR-Cas9 systems. *Journal of Integrative Plant Biology*, 60(8), 626–631.

Wang, X., Tu, M., Wang, D., Liu, J., Li, Y., Li, Z., Wang, Y., & Wang, X. (2018). CRISPR/Cas9-mediated efficient targeted mutagenesis in grape in the first generation. *Plant Biotechnology Journal*, 16(4), 844–855.

Wang, Y., Cheng, X., Shan, Q., Zhang, Y., Liu, J., Gao, C., & Qiu, J.-L. (2014). Simultaneous editing of three homoeoalleles in hexaploid bread wheat confers heritable resistance to powdery mildew. *Nature Biotechnology*, 32(9), 947–951.

Woo, J. W., Kim, J., Kwon, S. I., Corvalán, C., Cho, S. W., Kim, H., Kim, S.-G., Kim, S.-T., Choe, S., & Kim, J.-S. (2015). DNA-free genome editing in plants with preassembled CRISPR-Cas9 ribonucleoproteins. *Nature Biotechnology*, 33(11), 1162–1164.

Wright, D. A., Townsend, J. A., Winfrey Jr, R. J., Irwin, P. A., Rajagopal, J., Lonosky, P. M., Hall, B. D., Jondle, M. D., & Voytas, D. F. (2005). High-frequency homologous recombination in plants mediated by zinc-finger nucleases. *The Plant Journal*, 44(4), 693–705.

Wu, J., Yang, R., Yang, Z., Yao, S., Zhao, S., Wang, Y., Li, P., Song, X., Jin, L., & Zhou, T. (2017). ROS accumulation and antiviral defence control by microRNA528 in rice. Nature *Plants*, 3(1), 1–7.

Xie, K., Minkenberg, B., & Yang, Y. (2015). Boosting CRISPR/Cas9 multiplex editing capability with the endogenous tRNA-processing system. *Proceedings of the National Academy of Sciences*, 112(11), 3570–3575.

Xu, J., Hua, K., & Lang, Z. (2019). Genome editing for horticultural crop improvement. *Horticulture Research*, 6, 113.

Xu, Q., Zhao, M., Wu, K., Fu, X., & Liu, Q. (2016). Emerging insights into heterotrimeric G protein signaling in plants. *Journal of Genetics and Genomics*, 43(8), 495–502.

Xu, R., Qin, R., Li, H., Li, D., Li, L., Wei, P., & Yang, J. (2017). Generation of targeted mutant rice using a CRISPR-Cpf1 system. *Plant Biotechnology Journal*, 15(6), 713–717.

Xu, R., Yang, Y., Qin, R., Li, H., Qiu, C., Li, L., Wei, P., & Yang, J. (2016). Rapid improvement of grain weight via highly efficient CRISPR/Cas9-mediated multiplex genome editing in rice. *Journal of Genetics and Genomics*, 43(8), 529–532.

Zetsche, B., Gootenberg, J. S., Abudayyeh, O. O., Slaymaker, I. M., Makarova, K. S., Essletzbichler, P., Volz, S. E., Joung, J., Van Der Oost, J., & Regev, A. (2015a). Cpf1 is a single RNA-guided endonuclease of a class 2 CRISPR-Cas system. *Cell*, 163(3), 759–771.

Zetsche, B., Volz, S. E., & Zhang, F. (2015b). A split-Cas9 architecture for inducible genome editing and transcription modulation. *Nature Biotechnology*, 33(2), 139–142.

Zhang, J., Zhang, H., Botella, J. R., & Zhu, J.-K. (2018). Generation of new glutinous rice by CRISPR/Cas9-targeted mutagenesis of the Waxy gene in elite rice varieties. *Journal of Integrative Plant Biology*, 60(5), 369–375.

Zhang, Y., Liang, Z., Zong, Y., Wang, Y., Liu, J., Chen, K., Qiu, J.-L., & Gao, C. (2016). Efficient and transgene-free genome editing in wheat through transient expression of CRISPR/Cas9 DNA or RNA. *Nature Communications*, 7(1), 1–8.

Zhou, J., Li, D., Wang, G., Wang, F., Kunjal, M., Joldersma, D., & Liu, Z. (2020). Application and future perspective of CRISPR/Cas9 genome editing in fruit crops. *Journal of Integrative Plant Biology*, 62(3), 269–286.

Zong, Y., Wang, Y., Li, C., Zhang, R., Chen, K., Ran, Y., Qiu, J.-L., Wang, D., & Gao, C. (2017). Precise base editing in rice, wheat and maize with a Cas9-cytidine deaminase fusion. *Nature Biotechnology*, 35(5), 438–440.

Abbreviations

Cas	CRISPR associated protein
CBE	Cytidine-deaminase-mediated base editing
Cpf1	CRISPR – Prevotella and Francisella 1
CRISPR	Clustered regularly interspaced short palindromic repeats
DNA	Deoxyribonucleic acid
DSB	Double-strand breaks
FAO	Food and Agriculture Organization
gRNA	Guide RNA
MN	Meganuclease
NHEJ	Non-homologous end joining
RNA	Ribonucleic acid
RNP	Ribonucleoprotein
ROS	Reactive oxygen species
sgRNAs	Single guide RNA
TALEN	Transcription activator-like effector nucleases
tRNA	Transfer RNA
ZFN	Zinc-finger nuclease

5   CRISPR genome editing to address food security and climate changes

Challenges and opportunities

*Naglaa A. Abdallah, Aladdin Hamwieh,
Khaled Radwan, Nourhan Fouad, and
Michael Baum*

5.1 Introduction

Climate change and population growth have a significant impact on agriculture, biodiversity, and global food security. According to an estimate by the United Nations (UN), The number of hungry humans might approach 840 million by 2030 (*United Nations, Goal 2: Zero Hunger*, 2019). The United Nations established the 17 Sustainable Development Goals (SDGs) in 2015, primarily to eradicate poverty, improve food security, and enhance human health. The UN's Sustainable Development Goal 2 (SDG2) aims to eradicate world hunger. The SDGs are aspirational and ambitious, as the globe must improve food quality and produce 15–20% more food than expected yields (UN World Population Prospects, 2019). To achieve the SDGs in a timely manner, innovation and technology adoption must be exploited in considerable quantities.

Climate change poses a major threat to the environment's future, particularly agriculture, food security, nutrition, and biodiversity. The principal driver of climate change is the rise in greenhouse gas emissions into the atmosphere, which contribute to global warming and numerous other threats. Depending on the number of emissions, it is anticipated that drought, heat, floods, and chronic sea level increase, and global warming will occur more frequently in the near future (Prentice et al. 2001). Moreover, climate change will have devastating effects on species migration and extinction (Nunez et al. 2019). Due to climate change, plants and animals must adjust to novel environments that evolve faster than their capacity for adaptation. New breeding technologies, such as transgenic technology and genome editing, have cleared the path for the enhancement of crop types to address these difficulties. Plants adapt to various environmental situations by modifying the expression of many genes. To use biotechnology strategies for the development of climate-resilient crops, it is essential to have a fundamental knowledge of the genetic structure of plants. Stress tolerance is imparted by genes that are activated in response to biotic or abiotic stimuli (Figure 5.1). Stress-induced genes include both regulatory and functional genes. The stress response genes that are activated in the presence of oxidative injury damage activate a variety of antioxidant response pathways.

DOI: 10.4324/9781003382102-5

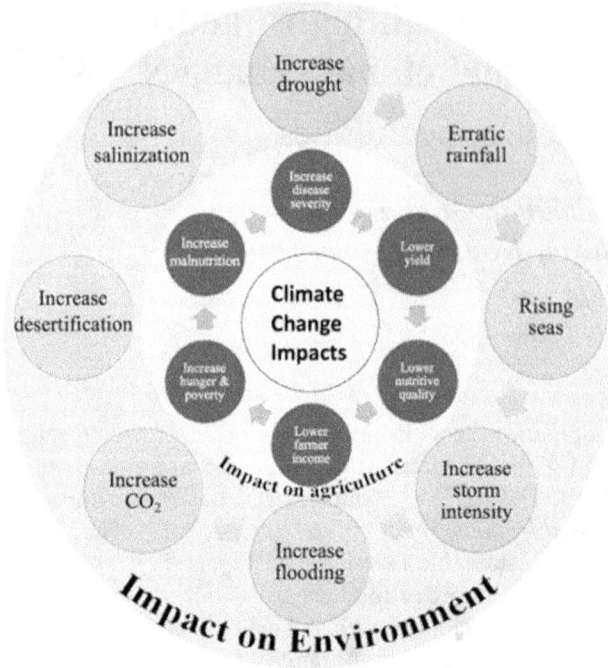

Figure 5.1 Impact of climate change on agriculture and environment.

These pathways involve a series of enzymatic reactions that work to reduce the level of oxidative damage by neutralizing reactive oxygen species and removing them from the cell. The result is a decrease in the amount of damage that is caused by oxidative stress. By keeping the cells protected from this type of damage, the stress response genes play a key role in protecting the overall health of the plants. Abiotic stressors include salinity, drought, heat, cold, heavy metals, radiation, and flooding. In numerous plant species, a variety of stress tolerance genes have been identified, and the overexpression of these genes enhanced plant tolerance. In addition, biotic stressors such as viruses, fungi, and bacteria, insects, directly and indirectly, diminish plant yield through feeding and disease transmission. It is anticipated that climate change would exacerbate the harm caused by plant-infecting diseases and insects. Biotic stressors result in a 20–40% decrease in plant output (Ahmad et al. 2021).

The explosive growth of genome editing has changed the way studying various plants and animals and enabled the direct targeting and exact modification of genomic sequences, as well as the acceleration of crop improvement through precision breeding (Abdallah et al. 2021). Genome editing provides answers to address the wide-ranging consequences of climate change on agriculture while keeping the necessity of its use to preserve biodiversity from climate change's risks. It is imperative and timely to use genome editing for crop development considering the escalating need for answers to the global food problem.

In this chapter, we emphasize the development of gene-editing approaches and research to enhance crop response to climate change, as well as the limitations and opportunities associated with the development of potent applications.

5.2 Overview of genome editing system

Genome editing (GnEd) technologies rely on site-specific endonucleases to make a double-stranded DNA (dsDNA) cut at a specified spot in the host genome. GnEd could be used to generate heritable and predictable mutations in specific parts of the genome precisely as predesigned (Bhattacharya et al. 2021). Using editing, it is possible to induce deletions, insertions, single-nucleotide substitutions, large fragment substitutions, and to modify gene expression. GnEd can result in the knockout of a single gene or numerous genes without the permanent insertion of foreign DNA or knock-in genes from other animals into exact genomic locations. Meganucleases (MeN), transcription activator-like effector nucleases (TALEN), Zinc-Finger Nucleases (ZFN), and the clustered regularly interspaced short palindromic repeats (CRISPR) associated protein (Cas) are GnEd tools (Abdallah et al. 2021; Daboussi et al. 2015; Joung and Sander, 2013; Urnov et al. 2010; Wang et al. 2016). GnEd tools rely on producing a break at a specific place in the host genome, followed by endogenous cell repair through nonhomologous end joining (NHEJ) or homologous direct recombination (HDR). NHEJ repair is utilized to eliminate gene function by generating tiny deletions. While HR requires a template to repair the break and could be used to introduce precise alterations in the DNA sequence by the insertion or substitution of tiny or large DNA fragments in the target region, it could not be used to correct the break. CRISPR-Cas technology is the most commonly used technique for genome editing at present, as it addresses many of the drawbacks of MeNs, ZFNs, and TALENs. In Table 5.1, a comparison of the various GnEd systems is provided.

Novel applications of the CRISPR-Cas system are only limited to dsDNA breaks, but they can also generate other changes. Adapted CRISPR/Cas9 genome editing was created to target various genomic changes for trait improvement in commercially significant crops (Abdallah et al. 2021). Cas9 nickases that cause single-strand breaks can be generated by mutating one of the catalytic residues of the Cas9 nuclease domains (H840A in HNH and D10A in RuvC) (SSBs). Cas nickase was utilized to develop base editing and prime editing and to decrease the CRISPR/Cas system's off-target consequences. Mutating both HNH and RuvC domains will develop a dead Cas (dCas) enzyme, that could be used with epigenome editing. The different CRISPR-Cas approaches are summarized in Table 5.2 and Figure 5.2. First-generation CRISPR approaches use traditional Cas9 and Cas12a enzymes for editing through double strands of DNA cut, while new-generation CRISPR

Table 5.1 Comparative assessment of the various gene-editing approaches

Gene-editing technique	DNA recognition domain	Endonuclease	Advantages	Disadvantages
Meganucleases (MeN)	MeP	MeP	• Large recognition site • high specificity • meganucleaseenabling to cleave unique DNA sequences	Complex design for different target genome sites and off-target effects
Zinc-Finger Nucleases (ZEN)	ZFP	Fok I	• Edit efficiency 10–30 • well-established	Complex design, low manufacturing feasibility, off-target effects, high cytotoxicity and require fusion with endonuclease for dsDNA cut
Transcription activator-like effector nucleases (TALEN)	TALE	Fok I	• Editing efficiency 20–60%, • high specificity • low off-target • low cytotoxicity • easy to design	Difficult to design, lower manufacturing feasibility, low specificity and efficacy, requires extensive sequencing, cost expensive, inability to RNA editing and require fusion with endonuclease for dsDNA cut
Clustered regularly interspaced short palindromic repeats (CRISPR)/Cas	crRNA	Cas enzyme	• Editing efficiency 50–80, • Simply designed, cost-effective, • low off-target; • low cytotoxicity, • reproducible, • high feasibility, • gene knockout is available, • RNA editing, • Enabling quick cycle	Target genome site for Cas9 and Cas12a required to be adjacent to PAM sequence

Table 5.2 Comparisons of genome editing methods assisted by CRISPR-Cas system

Nuclease type		Nuclease modification	Genome modification	Editing types	Reference
Traditional	Cas9 & Cas12a (Cpf1)	Indel with NHEJ Short template for HDR Long template for HDR	No modification	Frameshift mutation Short insertion/substitution Insertion/substitution of gene (s) or long segment	SDN-1 SDN-2 SDN-3	(Feng et al. 2022) (Veley et al. 2021) (Dong et al. 2020)
	Nickaseor dead Cas9	DNA Base editing	Cytidine deaminase fusion	C:G to T:A substation	SDN-1	(Zong et al. 2017) (Li et al. 2018; Yan et al. 2018)
			Adenosine deaminase fusion	A:T to G:C substation	SDN-1	
	Dead Cas13 target ssRNA	RNA Base editing	Adenosine deaminase	A:I Replacement	SDN-1	(Abudayyeh et al. 2019; Cox et al. 2017; Tang et al. 2021)
			Cytosine deaminase	C:U Replacement	SDN-1	
Modified nuclease	Nickase Cas	Prime editing	Reverse transcriptase fusion	Short insertion/substitution with HDR	SDN-2	(Scholefield and Harrison, 2021)
	Dead Cas13	M6A editing Epigenome RNA modification	m6A enzyme fusion	Target m6A editing in the mRNA, writers or erasers	–	(Chennakesavulu et al. 2022)
	Dead Cas	Epigenome DNA modification	TET1, LSD1 or CIB1 fusion	DNA methylation/demethylation	No genome sequence change	(Gallego-Bartolomé et al. 2018) (Ma et al. 2018; Zheng et al. 2019)
			Acetyltransferase	Chromatin acetylation	No genome sequence change	

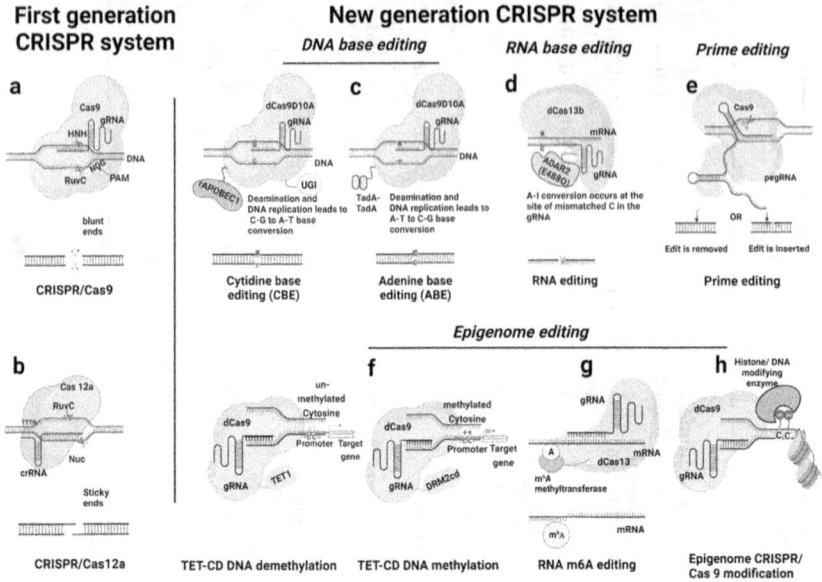

Figure 5.2 Schematic illustration of several genome editing techniques utilizing CRISPR/Cas-based technologies. The first generation of CRISPR-Cas tools (left) generates double-stranded blunt genome cuts; a. Cas 9 system induces bland end cleave at the target sequence adjacent to G-rich PAM sequence and b. By cleaving the target adjacent to a 5′ T-rich PAM region, the Cas12a mechanism generates staggered cuts in the genome. A new generation of CRISPR-Cas tools (right) with modified Cas enzymes. c. DNA base editing for Cytidine base editors using nCas fused to cytidine deaminase that converts G-C to A-T base in the targeted DNA strand and Adenine base editors using nCas or dCas fused to adenosine deaminase that converts T-A to C-G base. d. RNA base editing use nCas9–RT complex dCas13b fused to an adenosine deaminase to convert A into I in the RNA-edited strand. e. Prime editing using nCas9 fused to reverse transcriptase and the prime editing gRNA (pegRNA) which contains a template for reverse transcription. f. Epigenome editing for DNA using dCas9 fused to TET1-CD to demethylate the DNA or to DNMT2-CD to methylate the target DNA. g. epigenome editing of RNA for m6A editing using dCas13 fused to m6A enzyme. h Chromatin methylation using dCas fused to acetyltransferase.

approaches use modified Cas enzymes. For DNA base editing, nickase or dead Cas fused to cytidine deaminase or adenosine deaminase were used to convert C:G to T:A or A:T to G:C, respectively. During RNA base editing, dead Cas13 was fused with cytidine deaminase or adenosine deaminase to convert C to U or A to I, respectively. For prime editing, nickase Cas9 is fused to the reverse transcriptase enzyme and the gRNA contains the template used for replacement. For M6A RNA editing, dCas13 fused to m6A enzyme was used to modify post-transcriptional regulatory mechanisms in eukaryotes. Epigenome modification uses a dead Cas enzyme fused with DNA-methyltransferase-2 catalytic domain (DNMT2-CD) to induce DNA methylation at C, T, or A, or with Ten-Eleven Dioxygenase-1 catalytic domain (TET1-CD) to demethylate the DNA. For chromatin modification, the dead Cas enzyme is fused with acetyltransferase to acetyl the chromatin at the promoter site and cause gene activation (Abdallah et al. 2021).

Edited varieties produced via GnEd could be classified into three site-directed nuclease types (SDN-1, SDN-2, and SDN-3). SDN-1 deletes or adds (indel) a few nucleotides at the target sequence, causing a frameshift mutation. SDN-1 includes CRISPR-Cas9/ CpF1 indel, base editing, RNA editing Cas13, and organelle editing using TALEN approaches. SDN-2 requires a DNA template for homologous direct recombination to add, delete or replace a few nucleotides precisely. SDN-2 includes homologous direct recombination (HDR) with CRISPR-Cas9/CpF1 short template and prime editing. While SDN-3inserts whole gene(s) or replaces large gene segments into the genome, and also employs HDR with the DNA template. The three types of GnEd are different from transgenic techniques, which need to introduce the target gene as well as the selectable marker gene randomly in the genome. Epigenome modification with genome editing will not change the genome sequence but rather changes gene expression. Genome editing is also utilized to create cis-genic plants by introducing a gene-of-interest (GOI) derived from the species or a sexually compatible relative species. Because engineered plants lack undesirable sequences, they resemble conventional breeding. Editing with cis-genic is a form of precision molecular plant breeding, and it is anticipated that plants bred using these techniques would be more widely accepted.

5.3 Omics and CRISPR-Cas9 methods for analyzing gene function

The genomics revolution is the primary factor propelling the development of current agricultural biotechnology. Improving crop tolerance to environmental challenges is necessary to meet global food demand. The integration of multi-omics approaches contributes to a greater knowledge of molecular mechanisms and the production of environmentally adaptable cultivars. Multi-omics approaches, including genomics, transcriptomics, and proteomics, provide large-scale data that has expeditiously contributed to the study of plant variants, resulting in a deeper understanding of the molecular pathways, enzyme activity, and gene networks that regulate stress tolerance in crops. The key data mining methodologies include expanding our knowledge of genes, enabling more accurate result prediction, upgrading traditional breeding programs, and speeding up precision breeding procedures like genetic engineering and genome editing. By using unique and innovative breeding strategies, such as genome editing, data generated from omics techniques can enhance crop productivity and abiotic stress tolerance. CRISPR-Cas9 and Omics are anticipated to collect information about molecular data and provide a snapshot for enhancing an organism's crucial activities.

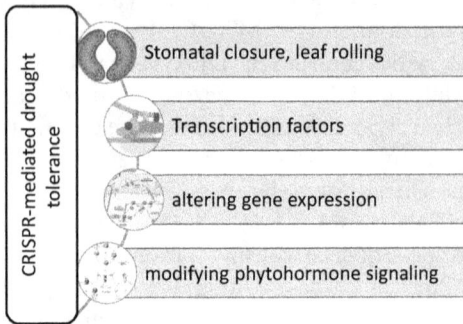

Figure 5.3 Incorporating Omics data sources and biotechnology methods with plant science's creation of stress-resistant cultivars.

Omics data for plant breeding orientation require a huge collection of individuals with genomic, transcriptomic, and phenotypic data, which might be utilized to find causal variations and improve breeding program genomic prediction markers and models. Additional manipulation of causative variables may result in the discovery of genome editing tools for stress tolerance and crop development. Omics data has the potential to be utilized to build a gene function network, which could be characterized to improve data processing and the accuracy of high-throughput experiments. Such a network could potentially provide valuable insight into the regulation of gene functions, allowing for more efficient data analysis, better-informed decisions, and more reliable results. Omics systems, coupled with different databases and bioinformatics generate reproducible information on candidate genes and their biosynthetic pathways, regulators, proteins, and biological networks associated with the plant stress response (Figure 5.3).

Genomic studies participated in identifying important stress-related genes that could be used for improving crop productivity and mitigating the harmful effects of climate change. In the case of salt stress, transcription factors, such as OsTIFY1, OsTIFY6, OsTIFY9, OsTIFY10, and OsTIFY11 have been known as abscisic acid (ABA)-regulated expression of salinity response genes (Ye et al. 2009). Many protein kinases (MAPK, RLK, etc.) are involved in the ABA-dependent pathways of gene expression control. In addition, ABA-independent pathways and transcription factors such as dehydration-responsive element-binding protein (DREB) play a key role, with OsDREB1A, OsDREB1F, and OsDREB2A overexpression resulting in enhanced salt tolerance. *PDH45*, *OsCPK12*, and other genes regulate the buildup of reactive oxygen species (ROS) during salt stress and promote salinity tolerance (Reddy et al. 2017).

DREB transcription factors (TFs) are members of the APETALA2/Ethylene Responsive Element-Binding Factor (AP2/ERF) family with a single AP2 domain that regulate the expression of drought resistance genes (Seki et al. 2001). In *Arabidopsis*, the TFs AtMYC2 and AtMYB2 enhance transgenic plants' tolerance to osmotic stress (Shinozaki and Yamaguchi-Shinozaki, 2007). More than 65% of the genes involved in drought stress responses and communication in plants respond to abscisic acid (ABA) (Lata et al. 2015). ABA-responsive elements (ABRE) regulate *RD29B*, which is also responsible for encoding LEA-like proteins.

Heat shock genes (HSG) protect intracellular proteins by maintaining their structural stability to keep their functions. Heat shock proteins (HSPs) are produced when high temperature alters the expression of a gene during transcription and translation (Khan and Shahwar, 2020). Among all HSPs, small heat shock proteins (sHSPs) are the most diverse (Hasanuzzaman et al. 2013). The genes *Hsp17.7* and *Hsp100* that provide heat tolerance to transgenic plants that overexpress these proteins have been found. Temperature substantially increases NPK1-related transcripts, which confers substantial heat resistance to barley (Hasanuzzaman et al. 2013). Under high-temperature stress, *DREB2A* in its active state affects the level of expression of heat shock-related genes (Sakuma et al. 2006). In addition, HSFA1a and HSFA1b are heat-shock factors (HSFs) that control the early response to heat of a number of genes (Hasanuzzaman et al. 2013).

5.4 Development of climate ready crops through CRISPR-Cas genome editing approach

CRISPR-Cas technologies contribute to the development of climate-ready, stress-tolerant, and high-quality crops. Table 5.3 provides a summary of the most recent research on CRISPR-Cas-mediated genome editing with diverse crops for quality enhancement.

5.4.1 Improvement of abiotic stresses in the context of climate change

Abiotic stressors inhibit plant development and decrease production by 20% to 50%. Abiotic stress-affected land is expected to rise more in the future due to the negative effects of climate change associated with reducing the total area under cultivation and endangering global food security. Better crop plant cultivars can be generated by utilizing current breeding techniques and transgenics, but the process is slow and time-consuming. Genome editing may offer solutions to accelerate the production of climate-smart crops. Drought and salt are key abiotic variables that have a significant negative impact on agricultural productivity. Genome editing holds promise in helping to develop plants with greater tolerance to abiotic stress by introducing targeted changes into the genetic material of crops, as well as introducing useful traits for a variety of other applications, including pest and disease resistance, nutritional content, increased yield, and improved crop quality. By introducing these targeted changes, genome editing has the potential to increase crop resilience and decrease production losses due to adverse environmental conditions. With this in mind, genome editing presents a great opportunity to create climate-smart

Table 5.3 A summary of the most recent gene-editing applications, targeted genes, and modification techniques for enhancing tolerance/resistance to biotic and abiotic challenges, as well as yield and yield quality, across many plant species for climate change mitigation

Trait	Species	Trait targeted	Gene(s)	Edited* method	Transformation method	References
Quality	Rapeseed	Flavonoids and Fatty Acid Composition	*BnTT2*	SDN-1 CRISPR/Cas	*Agrobacterium tumefaciens*	(Xie et al. 2020)
Quality	Rice	Carotenoid accumulation	*Osor*	SDN-1 CRISPR/Cas	*Agrobacterium tumefaciens*	(Endo et al. 2019)
Quality	Rice	Aleurone layer and grain quality	*GRAIN WIDTH & WEIGHT2 (GW2)*	SDN-1 CRISPR/Cas	*Agrobacterium tumefaciens*	(Achary et al. 2021) (Achary and Reddy, 2021)
Quality	Rice	Glutelin content	*BnaFAE1*	SDN-1 CRISPR/Cas	*Agrobacterium tumefaciens*	(Chen et al. 2022)
Quality	Rice	Carotenoid-enriched	*ZmPsy*	SDN-1 CRISPR/Cas	Particle bombardment and protoplast	(Dong et al. 2020)
Quality	Rice and wheat	Grain yield	*TaPGS1*	SDN-1 CRISPR/Cas	*Agrobacterium tumefaciens*	(Liu et al. 2022)
Quality	Rice	Amylose content	*Wx*	SDN-1 CRISPR/Cas	*Agrobacterium tumefaciens*	
Quality	Rice	Lysophospholipid Content, Cooking quality	*OsPLDa1*	SDN-1 CRISPR/Cas	*Agrobacterium tumefaciens*	(Khan et al. 2020)
Quality	Rice	Zinc and Manganese accumulation	*TaCNR5*	SDN-1 CRISPR/Cas	*Agrobacterium tumefaciens*	(Qiao et al. 2019)
Quality	Rice	Sucrose content	*Osgcs1*	SDN-1 CRISPR/Cas	*Agrobacterium tumefaciens*	(Honma et al. 2020)
Quality	Rice	Seed aroma	*OsBADH2*	SDN-1 CRISPR/Cas	*Agrobacterium tumefaciens*	(Tang et al. 2021)

	Crop	Trait	Gene	Method	Delivery	Reference
Quality	Rice	Amylose content	*Wx*	Base Editing	*Agrobacterium tumefaciens*	(Li et al. 2020)
Quality	*Salvia miltiorrhiza*	Phenolic acid biosynthesis	*bZIP2, PAL*	SDN-1 CRISPR/Cas	*Agrobacterium tumefaciens*	(Shi et al. 2021)
Quality	Soybean	Saturated fatty acids	*GmFATB1*	SDN-1 CRISPR/Cas	*Agrobacterium tumefaciens*	(Ma et al. 2021)
Quality	Strawberry	Altered sugar content	*FvebZIPs1.1*	Base editing	*Agrobacterium tumefaciens*	(Xing et al. 2020)
Quality	Tomato	Sugar content	*SlINVINH1 & SlINVINH2*	SDN-1 CRISPR/Cas	*Agrobacterium tumefaciens*	(Kawaguchi et al. 2021)
Quality	Tomato	Increased carotenoid, lycopene, and β-carotene	*SlDDB1, SlDET1 and SlCYC-B*	Base editing	*Agrobacterium tumefaciens*	(Hunziker et al. 2020)
Quality	Watermelons	Seed size and germination	*ClBG1*	SDN-1 CRISPR/Cas	*Agrobacterium tumefaciens*	(Wang et al. 2021)
Quality	Wheat	Amylose content and resistant starch	*TaSBEIIa*	SDN-1 CRISPR/Cas	Particle bombardment	(Li et al. 2021)
Quality	Wheat	Reduced asparagine synthetase	*TaASN2*	SDN-1 CRISPR/Cas	Particle bombardment	(Raffan et al. 2021)
Yield	Banana	Increased shelf life	*MaACO1*		Embryogenic cell suspension	(Hu et al. 2021)
Yield	Barley	Longer seed dormancy period	*Qsd1*	SDN-1 CRISPR/Cas	*Agrobacterium tumefaciens*	(Abe et al. 2019)
Yield	Camelina	Increasing Monounsaturated Fatty Acid Contents	*FAD2*	SDN-1 CRISPR/Cas	*Agrobacterium tumefaciens*	(Lee et al. 2021)
Yield	Ground Cherry	Highly nutritive crop modified for improved agronomic properties	*Ppr -AGO7, -SP, -SP5g*	SDN-1 CRISPR/Cas	*Agrobacterium tumefaciens*	(Lemmon et al. 2018)

(Continued)

Table 5.3 (Continued)

Trait	Species	Trait targeted	Gene(s)	Edited* method	Transformation method	References
Yield	Eggplant	Decreased browning	PP04, PPOS, and PP06	SDN-1 CRISPR/Cas	*Agrobacterium tumefaciens*	(Maioli et al. 2020)
Yield	Peanut	Increased oleic acid content	FAD2A and FAD2B	SDN-1 CRISPR/Cas	*Agrobacterium rhizogenes*	(Yuan et al. 2019)
Yield	Pomegranate	Accumulated gallic acid 3-0- and 4-o-glucosides	*PgUGT84A23* and *PgUGT84A24*	SDN-1 CRISPR/Cas	*Agrobacterium rhizogenes*	(Chang et al. 2019)
Yield	*Populus tremula* × *P. alba*	Reduced-lignin content	CSE1 and CSE2	SDN-1 CRISPR/Cas	*Agrobacterium tumefaciens*	(de Vries et al. 2021)
Yield	Rice	Late flowering time.	*OsLHY,Hd1-* and *Ehd1*	SDN-1 CRISPR/Cas	*Agrobacterium tumefaciens*	(Li et al. 2022)
Yield	Rice	Regulation of flowering time and drought tolerance	*OsFTL4*	SDN-1 CRISPR/Cas	*Agrobacterium tumefaciens*	(Gu et al. 2022)
Yield	Rice	Early heading:	qHD7b	SDN-1 CRISPR/Cas	*Agrobacterium tumefaciens*	(Sohail et al. 2022)
Yield	Banana	Semi-dwarfed	*Ma04g15900, Ma06g27710, Ma08g32850, Ma11g10500, Ma11g17210*	SDN-1 CRISPR/Cas	*Agrobacterium tumefaciens*	(Shao et al. 2020)
Yield	Soybean	Photoperiodic flowering and regional adaptation	GmPRR37	Natural variation and CRISPR/Cas9-mediated mutation	*Agrobacterium tumefaciens*	(Wang et al. 2020)
Yield	*Taraxacum kok-saghyz*	Rubber biosynthesis	(1-FFT)	SDN-1 CRISPR/Cas	*Agrobacterium rhizogenes*	(Iaffaldano et al. 2016)

Yield	Tobacco	Minimal amounts of nicotine	*QPT*	SDN-1 CRISPR/Cas	*Agrobacterium tumefaciens*	(Smith et al. 2022)
Yield	Tomato	Low Cs⁺ / seedless fruits	*slhak5* KO	SDN-1 CRISPR/Cas	*Agrobacterium tumefaciens*	(Nieves-Cordones et al. 2020)
Yield	Tomato	Abscission	*SlHB15A*	SDN-1 CRISPR/Cas	*Agrobacterium tumefaciens*	(Liu et al. 2022)
Biotic	Rice	Rice black-streaked dwarf virus RBSDV	*eIF4G*	SDN1 CRISPR/Cas	*Agrobacterium tumefaciens*	(Wang et al. 2021)
Biotic	Barley	barley mild mosaic virus (BaMMV)	*PDIL5-1*	SDN1 CRISPR/Cas	*Agrobacterium tumefaciens*	(Hoffie et al. 2022)
Biotic	Potato	Potato Virus Y (PVY),	Viral genome	RNA editing Cas13a	Agro-infiltration	(Zhan et al. 2019)
Biotic	Grapevine	Grapevine leafroll-associated virus 3 (GLRaV-3),	Viral genome	RNA editing LshCas13a	*Agrobacterium tumefaciens*	(Jiao et al. 2022)
Biotic	Sweet potato	Sweet potato chlorotic stunt virus (SPCSV) and sweet potato feathery mottle virus (SPFMV)	Viral genome	RNA editing Cas13a	Agro-infiltration	(Yu et al. 2022)
Biotic	Citrus	*Xanthomonas citri*	*CsWRKY22*	SDN1 CRISPR/Cas	*Agrobacterium tumefaciens*	(Long et al. 2021)
Biotic	Rice	*Xanthomonas oryzae*	*EBE$_{AvrXa23}$*	SDN2 CRISPR/Cas	Particle bombardment	(Wei et al. 2021)
Biotic	Brassica	*Sclerotinia sclerotiorum*	*BnMPK3*	SDN2 CRISPR/Cas & transgenic	*Agrobacterium tumefaciens*	(Zhang et al. 2022)
Biotic	Soybean	*Phytophthora sojae*	*GmDRR1*	SDN1 CRISPR/Cas	*Agrobacterium tumefaciens*	(Yu et al. 2022)
Biotic	Wheat	Powdery mildew resistance	*TaTMT3B*	SDN1 CRISPR/Cas	Particle bombardment	(Li et al. 2022)

(Continued)

Table 5.3 (Continued)

Trait	Species	Trait targeted	Gene(s)	Edited* method	Transformation method	References
Biotic	Wheat	Wheat stripe rust	*TaCIPK14*	SDN1 CRISPR/Cas	*Agrobacterium tumefaciens*	(He et al. 2022)
Biotic	Barley	Bipolaris spot blotch and Fusarium root rot	*HvMORC1* and *HvMORC6a*	SDN1 CRISPR/Cas	*Agrobacterium tumefaciens*	(Galli et al. 2022)
Biotic	Soybean	leaf-chewing insects	*Glyma.07g110300*	SDN1 CRISPR/Cas	*Agrobacterium tumefaciens*	(Zhang et al. 2022)
Biotic	Tomato	Plant resistance to herbivorous insects.	*SlZRK1*	SDN1 CRISPR/Cas	*Agrobacterium tumefaciens*	(Sun et al. 2021)
Biotic	Watermelon	Resistance to aphids	*vst1-1*	SDN1 CRISPR/Cas	*Agrobacterium tumefaciens*	(Li et al. 2022)

crops and to reduce the climate change effects on agriculture. As such, it is essential for agriculturalists to recognize the value of genome editing and to develop strategies for its application in order to maximize its potential for agricultural sustainability.

5.4.1.1 Drought tolerance using CRISPR tools

Drought is the most severe growth-limiting factor in agriculture. It is difficult to generate drought-tolerant crops due to the multigenic nature of their drought-tolerance systems. Several CRISPR-based strategies for developing drought-tolerant plants have been examined (Figure 5.4), including drought tolerance through stomatal closure and leaf rolling; TF editing (single or multiple copies); altering the expression of genes (via epigenetic modification, promoter engineering, or miRNA editing); and regulating the level of expression of phytohormone or catabolism pathways (Shelake et al. 2022).

For stomatal closure, CRISPR/SpCas9 was implemented to alter stomatal closure under drought stress by modifying the *OST2* alleles in *Arabidopsis* (Osakabe et al. 2016). In tomato, *npr1* mutants exhibited increased stomatal opening, increased oxidative stress, and decreased antioxidant capacity (Li et al. 2019). In wheat, mutating the 6 alleles of the *TaSal1* gene, which encode the enzyme 3'(2'), 5'-bisphosphate nucleotidase, also altered stomatal closure and caused leaf rolling (Abdallah et al. 2022). In addition, knocking out *SRL1* and *SRL2* (encoding putative glycosylphosphatidylinositol (GPI)-anchored proteins) enhanced rice leaf rolling (Liao et al. 2019). CRISPR

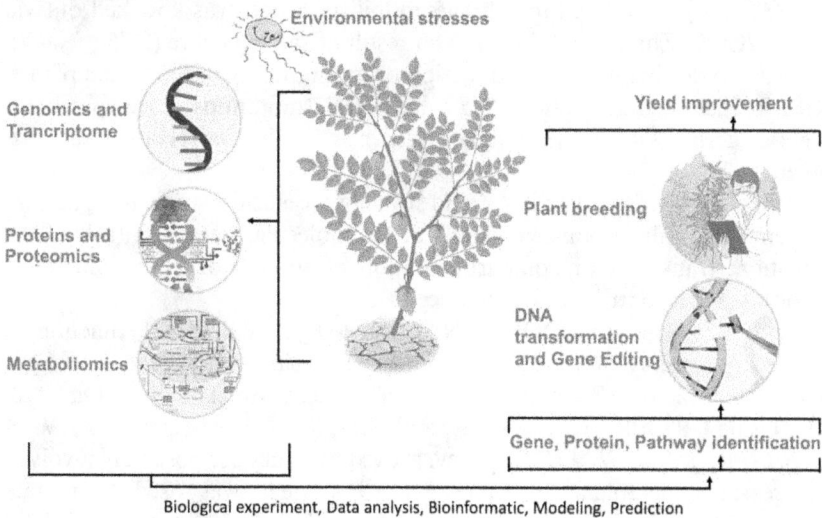

Figure 5.4 Schematic diagram for CRISPR-Cas genome editing tools for developing drought-tolerant crops.

technology was used to target *DREB2* and *ERF3* as key genes for abiotic stress tolerance in wheat (Kim et al. 2018).

Utilizing CRISPR technologies, the drought-responsive transcription factors (TFs) as essential candidates in plant stress tolerance pathways have been identified. The elimination of the *PdNF-YB21* and *GGNC* genes in *Populus implies* indicates positive regulation of drought resistance (Shen et al. 2021; Zhou et al. 2020). In addition, *NAC14* (NAC family) and *ERF83* (AP2/ERF family) are significant drought-stress tolerance regulators(Jung et al. 2021; Liao et al. 2019). In *Arabidopsis* GnEd mutants for the *TRE1* gene, (encoding the trehalase enzyme) enhanced drought tolerance (Nuñez-Muñoz et al. 2021). A mutation produced using CRISPR-edited *ERA1*, increased *Arabidopsis* and rice's drought tolerance. ERA1, a component of the ABA signaling pathway that encodes farnesyltransferase, regulates the dehydration response(Ogata et al. 2020). Epigenetic modification with CRISPR activation (CRISPRa) using dCas9-HAT fusion, was used to upregulate *AREB1* expression and heightens drought tolerance (Roca Paixão et al. 2019). CRISPRa was also used to enhance the expression of *AVP1* to improve performance under drought conditions in *Arabidopsis* (Park et al. 2017).

5.4.1.2 Exploring plant responses to salinity stress using CRISPR tools

Inducing osmotic stress, nutritional imbalance, ionic toxicity, and oxidative stress, salinity stress poses a significant danger to the growth and development of plants (Shrivastava and Kumar, 2015). Some recent accomplishments in plant salinity stress research have been well addressed, but still, several fundamental perennial consequences remain unsolved.

OsRR22, a gene related to salt susceptibility in rice, was knocked out via CRISPR/Cas9 (Zhang et al. 2019). As a result of salt exposure (0.75% NaCl), there was no decline in grain production or grain quality in the edited plants. CRISPR/Cas9 mediated *OsmiR535 and OsRR22* mutations in rice produced novel rice germplasm showing higher tolerance to salt stress (Han et al. 2022; Yue et al. 2020).

CRISPR targeting of TFs is a broad strategy for elucidating their regulatory involvement in salt-response networks at the molecular level. Controlling the absorption, transport, or compartmentation of sodium or potassium, transcription factors contribute to tolerance to salt stress.

Several TFs, including DOF15, NAC041, GTg-2, and PIL14, function as positive salt-stress-responsive factors, and salt-sensitive phenotypes are regulated by these factors (Bo et al. 2019; Liu et al. 2020; Mo et al. 2020; Qin et al. 2019). Using CRISPR to delete kinase and phosphatase genes, such as *FLN2* in rice, *BBS1* in rice, and *ITPK1* in barley, proved that they are positively involved in increasing salt tolerance. CRISPR/Cas9 strategy was used to mutate *OsbHLH024* in rice resulting in overexpression of ion transporter genes. Edited lines exposed to salinity stress indicated a significant increase in the shoot weight, and total chlorophyll content & fluorescence (Alam et al. 2022).

CRISPR tools were also utilized to investigate salt-stress-responsive components, such as *MIR528*, *BG3,* and *NCA1a/OsNCA1b* in rice (Liu et al. 2019; Yin et al. 2020; Zhou et al. 2017), *RBOHD* in pumpkin (Huang et al. 2019), *ACQOS* gene cluster in *Arabidopsis* (Kim et al. 2021), and *HVP10* in barley (Fu et al. 2022). A negative regulator of salt stress in plants, the OsRR22 gene produces a type-B response regulator implicated in cytokinin signaling. CRISPR-Cas9 knocked out the OsRR22 gene, resulting in increased salt tolerance in rice (Zhang et al. 2019).

ABA-induced transcription repressors (AITRs) participate in ABA signaling control. In *Arabidopsis*, silencing of the *AITR* genes increased both drought and salt tolerance. Mutating the *GmAITR* genes (gmaitr36 and gmaitr23456) using CRISPR/Cas9 genome editing in soybean leads to enhance salinity tolerance (Wang et al. 2021).

5.4.2 Improvement of biotic stresses in the context of climate change

Climate change contributes to the severity and widespread of many plant and animal diseases. However, CRISPR tools showed promising development in conferring disease resistance; it can provide a solution for addressing present and growing worldwide risks to plant and animal output caused by intensifying disease (Karavolias et al. 2021). CRISPR technology has the potential to mitigate some of the impacts of climate change on plant and animal health by introducing specific genetic modifications that confer resistance to disease and climate-related stresses. By providing resistance to multiple environmental stresses, CRISPR technology may help reduce the loss of plant and animal productivity due to diseases caused by climate change. Thus, CRISPR technology provides an exciting opportunity to develop crop and livestock varieties that can survive and even thrive in a changing climate. This technology is rapidly advancing and has already been used to create modified plants, animals, and microorganisms that are resistant to disease and able to adapt better in the face of environmental change.

5.4.2.1 Virus resistance using CRISPR tools

CRISPR-mediated genome editing has been reported for DNA and RNA viruses. In potato, CRISPR/Cas9 mediated editing was implemented to modify the potato coilin gene at the C-terminal domain and the edited potato showed an increase in virus resistance (Makhotenko et al. 2019).

CRISPR/Cas9 was used to modify the elongation factor eIF4e gene to reduce symptoms and viral accumulation of cucumber vein yellowing virus (CVYV), zucchini yellow mosaic virus (ZYMV), and papaya ringspot virus-W (PRV-W) in cucumber (Chandrasekaran et al. 2016). Knocking out the *eIF4e* gene in cassava enhanced *Potyviridae* virus resistance by developing plants with lower disease severity and viral titer (Gomez et al. 2019). In rice, rice tungro virus resistant plants were developed by editing *eIF4G* gene.

Edited plants showed no detectable viral proteins and exhibited higher yields relative to the wild type (Macovei et al. 2018).

Furthermore, in tomato, gene-editing techniques employing CRISPR/Cas9-mediated resistance have enhanced the plant's resistance to tomato yellow leaf curl virus (Tashkandi et al. 2018). Multiple guides aimed at multiple sequences in the TYCV genome were introduced into genetically engineered plants resulting in reducing viral accumulation. This resistance was passed down over many generations.

The integrated dsDNA of the banana streak virus (BSV) was removed from the banana genome using CRISPR/Cas9 (Tripathi et al. 2019). BSV integrates into the B sub-genome of the banana genome and remains dormant until the plant is exposed to a stressful condition, such as drought. An integrated plant virus was targeted with the CRISPR technology in a plant genome to develop disease resistance.

5.4.2.2 Bacterial resistance using CRISPR tools

By inhibiting pathogen-mediated stomatal opening, the editing of *JAZ2* in tomato with CRISPR/Cas9 decreased infection by *Pseudomonas syringae*. Edited plants sustained considerably lower *Pseudomonas syringae* pv. tomato levels than wild-type plants (Ortigosa et al. 2019). Climate change causes leaf blight bacteria to cause serious damage in certain places of the planet. Using CRISPR/Cas9 to target the promoter region of various *OsSWEET* genes in rice, resistance lines to numerous *Xanthomonas* infections were generated (Zeng et al. 2020). Although TALENs was not a CRISPR technology, it was utilized to damage the *EBE* site in the promoter region of the *OsSWEET14* (also known as *Os11N3*) gene (Blanvillain Baufumé et al. 2017; Li et al. 2012). Transgene-free bacterial blight-resistant rice was generated by altering the *Xa13* promoter with the CRISPR/Cas9 technology (Li et al. 2020). The elimination of the *Os8N3* gene increased resistance to *Xanthomonas oryzae* pv. *Oryzae* (Oliva et al. 2019; Xu et al. 2019). To combat banana *Xanthomonas* wilt, CRISPR/Cas9-edited bananas with specific mutations in *MusaDMR6* orthologues were created (Tripathi et al. 2021). Using CRISPR/Cas9 to eliminate the *SlDMR6-1* gene resulted in tomato plants with higher resistance to *P. capsici*, *P. syringae*, and *Xanthomonas* spp., than the control (Thomazella et al. 2016).

5.4.2.3 Fungal resistance using CRISPR tools

Rice worldwide can face severe yield losses due to blast disease. The *OsSEC3A* gene was exploited to develop CRISPR/Cas9-mediated resistance to rice blast disease (Ma et al. 2018). In addition, altering the *OsERF922* gene in rice considerably increased blast resistance relative to wild-type rice (Liu et al. 2012). CRISPR/Cas9 was used to eliminate the ethylene gene *OsERF922*in order to implicate blast resistance. Edited rice dramatically decreased lesion diameters without altering agronomic characteristics (Wang et al. 2016).

Edited *hvmorc1*-Knockout in barely showed increased disease resistance to biotrophic *Blumeria graminis* f. sp. *hordei* and necrotrophic *Fusarium graminearum*. Edited plants enhanced the expression of *PR* genes, de-repressed transposable elements, and increased the expression of *HvMORC2* (Kumar et al. 2021).

With the changing climate, wheat productivity is severely impacted by powdery mildew (Tang et al. 2017). CRISPR/Cas9 editing for the *EDR1* gene mechanism in wheat was used for enhancing powdery mildew resistance (Zhang et al. 2017).

In tomato, powdery mildew was controlled by modifying the gene *SlMLO1* using CRISPR/Cas9 tool (Nekrasov et al. 2017). Resistant varieties showed an increase in the accumulation of hydrogen peroxide in the infected plants. Also, in grapevine, editing the gene *VvMLO3* with CRISPR/Cas9 increased powdery mildew resistance with a two-fold reduction of sporulation in edited plants (Wan et al. 2020).

Cotton canker is a severe disease caused by the fungus *Verticillium dahlia*. CRISPR/Cas editing of the *Gh14-3-3d* yielded resistant cotton plants to *Verticillium* showing fewer disease symptoms and pathogenesis (Zhang et al. 2018). *Phytopthora tropicalis* was controlled in cacao by CRISPR/Cas9 knockout-based editing of the *TcNPR3* gene. Developed plants exhibited smaller lesions (Fister et al. 2018).

5.4.2.4 Insect resistance using CRISPR tools

The impact of plant pests on the phenotypic, yield, and spread of crops can result in the starving of millions of people. For successful pest control (without affecting non-target insects) and to assure agricultural improvement and food safety, it is necessary to create innovative ways for integrated pest management.

As it targets their reproductive fitness, CRISPR/Cas9-based pest management solutions, such as integrating the Sterile Insect Technique (SIT) with the CRISPR/Cas9 system, have excellent pest control capabilities. Similarly, the CRISPR/Cas9 technology has been utilized for transcription regulation; however, the transformed organism will be considered genetically modified because it will have a transgene for dead Cas9 (dCas9) and a gRNA expression cassette. Recently, the ovary transportation of the injected Cas9 protein and sgRNA (RNP complex) complex assisted by the ovary targeting peptide ligand was used as a delivery system for CRISPR/Cas complex (Singh et al. 2022).

The SIT using CRISPR-Cas is utilized to create ecologically benign SIT strains by raising, sterilizing, and releasing sterilized males of the target insect species (Kalajdzic and Schetelig, 2017). CRISPR/Cas9-mediated mutagenesis was used to eliminate the white and Sex lethal loci (*Sxl*) genes in *Drosophila Suzuki* in order to reduce the population growth of this destructive pest (Li and Scott, 2016). Sex-specific sterility in *Hyphantria cunea* was obtained using CRISPR/Cas9 Knocked-out *Hcdsx* gene as a successful method for pest control (Li and Scott, 2016). Heat-sensitive *Drosophila melanogaster* is developed by CRISPR-Cas knockout of the *LUBEL* gene to reduce survival in response to heat (Asaoka et al. 2016). Egg production and viability for *Cydiapomonella* were modified by knockout Cas9 of the *CpomOR1* gene (Garczynski et al. 2017).

5.4.3 Genome editing in the context of developing biofortified crops

The CRISPR-Cas approach can increase the value of food by compensating for deficient vitamins or minerals in major crops and eliminating nutrient deficiency. Genome editing for developing biofortified crops has been the focus of many research studies, as this technology has the potential to solve global health and nutrition issues in the long term (Nagamine et al. 2022).

5.4.3.1 Vitamin enriched crops

5.4.3.1.1 VITAMIN D – ENRICHED CROPS

Poor vitamin D status is a global health concern; one billion individuals worldwide are vitamin D deficient. Vitamin D deficiencies may increase the risk of cancer, neurocognitive impairment, and mortality. Food sources of provitamin D3 (7-DHC) are quite limited. 7-DHC was only discovered in immature green fruit and was absent in ripening and mature fruit. Using CRISPR-Cas9 gene editing, the Sl7-DR2 enzyme, which typically converts 7-DHC, was silenced. Increased amounts of 7-DHC in tomato plants because of gene deletion are transformed to vitamin D3 upon exposure to UVB rays (sunshine vitamin). In general, suppressing Sl7-DR2 activity has little effect on tomato plant growth, development, or yield (Li et al. 2022).

5.4.3.1.2 VITAMIN A – ENRICHED CROPS

Vitamin A deficiency is associated with xerophthalmia, night and juvenile blindness, and an increased risk of morbidity and death, especially in children (Reddy et al. 2021; Sommer, 2008). Golden crops (carotenoid-enriched crops) could be produced using CRISPR knockout or knock-in technologies. The b-carotene-enriched Cavendish banana was created using a CRISPR/Cas9-based knockout technique. The absence or extreme reduction of lutein and -carotene, as well as an increase in β-carotene, were seen in genetically modified banana, with no discernible influence on the plant phenotype. A CRISPR-Cas9-based method for the targeted insertion (knock-in) of a 5.2 kb carotenoid biosynthesis cassette including the two carotenoid biosynthetic genes *SSU-crtI* and *ZmPsy* driven by the endosperm-specific glutelin promoter (Dong et al. 2020).

Utilizing the fifth exon of the lycopene epsilon-cyclase (*LCY*) gene, β-carotene-enriched bananas were created using a CRISPR/Cas9-based method. Differential expression of carotenoid pathway genes was detected between altered and unmodified plant lines (Kaur et al. 2020). Using CRISPR-Cas9, the STAYGREEN (*SGR*) negative regulator gene of carotenoid bio-synthesis was silenced in tomato. The lycopene content of modified tomato fruit has been enhanced by approximately 5.1-fold. Lycopene is essential for the treatment of chronic diseases and the prevention of cancer and cardiovascular disease (Li et al. 2018).

5.4.3.1.3 VITAMIN E – ENRICHED CROPS

Vitamin E (also referred to as Tocopherols) is an essential component of the human diet since it helps in maintaining healthy skin and eyes and boosts the body's natural resistance to sickness and infection. Vitamin E deficiency has been linked to cardiovascular diseases and other human illnesses (Rizvi et al. 2014). Tocopherols and tocotrienols are the two classifications of vitamin E. *Hordeum vulgare* homogentisate phytyltransferase (*HvHPT*) and *Hordeum vulgare* homogentisate geranylgeranyl transferase (*HvHGGT*) can be used to increase the vitamin E content of other crops. The α-tocopherol of the eight vitamin E forms is an effective antioxidant for protecting lipids against pho-tooxidation and is considered as a nutraceutical to reduce the risk of human diseases, including cardiovascular diseases, aging, and cancer. Targeted over-expression of *HvHPT* and *HvHGGT* using CRISPR-Cas9 technology in barley led to a considerable increase in tocopherol and tocotrienol content (Zeng et al. 2020). *HvHGGT* deletion resulted in undetectable tocotrienol levels in barley grains without influencing tocopherol levels.

5.4.3.2 Nutrient enriched crops

Iron is required for plants to produce chlorophyll, which is essential for growth and oxygen production. Iron is required by all living species for the synthesis of DNA. Iron is required for the development of chlorophyll in plants, hemoglobin in animals, and the electron transport chain in humans. Plants require iron for oxygen transport and the electron transport chain. Zn is essential for a number of metabolic and immunological processes, particularly protein synthesis, wound healing, DNA synthesis, cell division, and catalytic activity. Phytic acid (PA) is an antinutrient that inhibits the absorption of iron and zinc, resulting in mal-nutrition. The important gene for PA biosynthesis is the inositol pentakispho-sphate 2- kinase 1 (*IPK1*) gene. The disruption of the *TaIPK1* gene in wheat by CRISPR/Cas9 decreases phytic acid and increases iron and zinc accumulation (Ibrahim et al. 2022). Due to the drop in PA content, an analysis of the altered lines revealed a considerable increase in iron and zinc accumulation in grains compared to the control plants. In wheat, the CRISPR-Cas system damages the Inositol Pentakisphosphate 2-kinase 1 (*TaIPK1*) gene, which decreases phytic acid and improves zinc accumulation in grain (Ibrahim et al. 2022).

Downregulated nitrogen absorption genes in plants have a negative effect on the plants' nutrient use efficiency. Using CRISPR-based systems, these genes could be mutated to increase crop yield by boosting nutrient transport, nutrient usage, growth, and yield. The rice nitrate transporter gene *NRT1.1B* was modified using the CRISPR-based cytosine base editing (CBE), which caused a C to T substitution and converted the Thr amino acid at position 327 to Met. In addition, the indel mutation of *NRT1.1B I* resulted in plants with greater nutrient usage efficiency than wild-type plants (Lu et al. 2017).

5.4.4 Genome editing in the context of quality improvement crops

CRISPR-based technologies have been used with amazing accuracy and efficacy to improve the quality of grain, oilseed, and horticultural crops (Ku and Ha, 2020).

5.4.4.1 Photosynthesis improvement using CRISPR-Cas

Photosynthesis increases flag leaf area, which increases CO_2 absorption capacity, decreases photorespiration, and increases flux through the Calvin cycle (Flexas and Carriquí, 2020). knocking out the photosynthesis 1 (*NRP1*) gene (a negative regulator of photosynthesis 1) using CRISPR-based editing, has resulted in an increased photosynthetic efficiency, grain yield, and biomass output in rice(Chen et al. 2021).

Ribulose-1,5-bisphosphate carboxylase/oxygenase (Rubisco) is a crucial CO_2-fixing enzyme that controls the rate of photosynthesis in plants. Three CRISPR-Cas9-mediated mutagenesis of the rubisco small (*rbcS*) subunit family (*rbcS S1a, rbcS S1b, and rbcS T1*) were simultaneously knocked out in tobacco, and the mutant plants exhibited a higher photosynthetic rate than wild-type plants (Donovan et al. 2020).

5.4.4.2 Grain yield and plant architecture improvement using CRISPR-Cas

Grain yield traits in plants are complicated; among the quantitative trait loci that control grain yield, *OsGS3* and *OsGL3*.1 impact grain size and grain yield negatively in rice. Rice grain size, 1,000-grain weight, and overall yield per plant were all improved using CRISPR-based gene editing of two genes (Yuyu et al. 2020). Additionally, altering the three genes *GS3*, *GW2*, and *Gn1a* increased grain production in altered African rice relative to the wild type (Lacchini et al. 2020). The cytokinin-activating enzyme, *OsLOGL5*, inhibits root development, tiller number, and rice yield. Using CRISPR-based editing to mutate the *OsLOGL5* gene increases rice grain production under a variety of field settings(Wang et al. 2020).

Pod shattering in grain crops is one of the primary challenges that pose major hazards to crop productivity. *qSH1* was altered using CRISPR-Cas9 to develop rice lines resistant to shattering. CRISPR gene editing demonstrated greater crop yields than their wild-type plants (Sheng et al. 2020).

5.4.4.3 Grain quality improvement using CRISPR-Cas

Genome editing can be utilized to successfully increase plant nutrients by controlling the pathways involved in their production and metabolism. Amylose production is regulated by the enzyme granule-bound starch synthase 1 (GBSS1), which is encoded by the waxy gene. By disrupting GBSS1 with the CRISPR-Cas9 system, rice and waxy maize lines with low amylose content have been produced (Gao et al. 2020; W. Wang et al. 2020). CRISPR-Cas editing of isoamylase 1 (*ISA1*) produced rice lines with a low amylose content, whereas editing Starch Branching Enzyme I (SBEI) and SBEIIb produced rice lines with a high amylose content (Y. Sun et al. 2017).

The non-protein amino acid gamma-aminobutyric acid (GABA) functions as an inhibitory neurotransmitter. Using CRISPR-Cas multiplex editing to target five genes involved in GABA metabolism, including three GABA-T genes (*GABA-TP1, GABA-TP2,* and *GABA-TP3*), *CAT9*, and *SSADH*, a tomato with a high concentration of GABA was created. Increased GABA levels were seen in modified plants (Li et al. 2018). Additionally, CRISPR/Cas9-induced mutations in the two tomatoes GAD genes *SlGAD2* and *SlGAD3* and edited lines boosted GABA buildup in tomato fruit by up to 15-fold (Nonaka et al. 2017).

Asparagine, a precursor to acrylamide, is produced during the high-temperature baking of wheat. Using four gRNA polycistronic genes, the wheat asparagine synthetase gene, *TaASN2*, was deleted using CRISPR/Cas9. The modified lines yielded seeds with a lower asparagine concentration than the wild type. While the levels of free aspartate, glutamate, and glutamine were greater than those of the wild type (Raffan et al. 2021).

5.5 Limitations and opportunities of GnEd applications

A summary of using genome editing for improving biotic and abiotic stresses, yield, and quality is presented in Figure 5.5. Among the 648 edited crops, rice has the largest number of genes modified via genome editing (215) followed by tomato (92), Maize (47), soybean (36), wheat (35), canola (29), and potato (25). Among the 50 countries that conducted research on genome editing, China was the top country (43.3%) followed by the USA (17.7%), then Japan (4.3%).

The main limitation of genome editing is the construction of CRISPR/Cas vectors. The high frequency of off-target mutations is the greatest disadvantage of CRISPR technology. Although numerous online editing systems have been created to identify and anticipate off-target cleavages, it will be impossible to build gRNA that avoids off-target effects if the genome sequence of the target organism is missing or unassembled.

Cas-nucleases need a PAM sequence, which confines genome editing to specific regions and reduces the number of viable gRNA candidates. Nevertheless, a number of Cas enzymes, such as SpCas9 and Cas12a, are currently accessible and diminishing PAM restriction. New techniques in genome

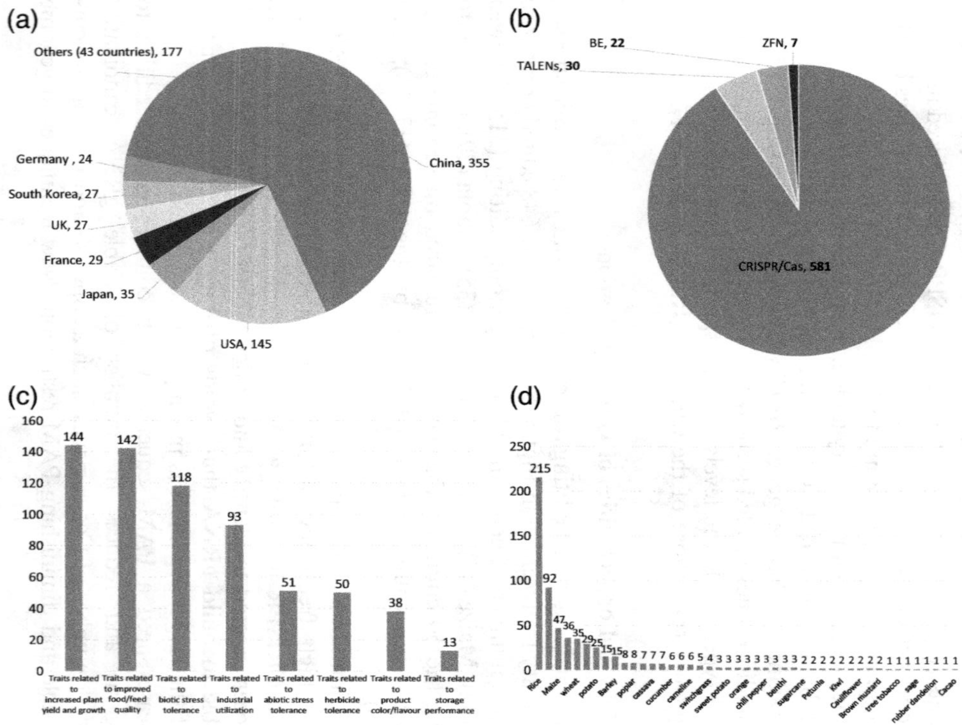

Figure 5.5 A review of the use of gene editing breakthroughs to plants and plant species over time (2016–2022) for climate change mitigation. a) studies on biotic and abiotic stressors, yield, and yield quality. b) on plant research studies. c) studies employing different modification strategies.

editing make it possible to design point mutation without the need for DNA template, double-strand DNA break, or homologous direct recombination.

Base editing, although a precise modification, is the restriction of specific nucleotide changes. Therefore, prime editing is better for precise small modification as it allows to find, replace, and edit the target sequence with high efficiency, facilitating the swap from any one base to any other base. However, prime editing has low editing efficiencies. Also, the use of HDR for target insertion could overcome the base editing limitation but it also has low frequency as it requires donor template DNA.

Maximizing the delivery mechanisms for the target gene-editing components and developing efficient regeneration techniques for the target crops are the obstacles to genome editing. Strategies for delivering CRISPR/Cas are split into direct and indirect methods. Direct techniques include polyethylene glycol (PEG)-mediated delivery and bombardment-mediated delivery, whereas indirect ways include floral dip and *Agrobacterium tumefaciens*-mediated delivery. Nonetheless, somaclonal variation may emerge through classical transformation techniques or *in vitro* cultivation effects. A transient protoplasts system employing PEG-mediated delivery of CRISPR reagents is very effective for gRNA validation, but sustained plant regeneration from protoplasts is required for editing (Woo et al. 2015). The *Agrobacterium* and biolistic bombardment tools for the delivery of the CRISPR constructs are the conventional methods for plant transformation-based systems. However, they have several disadvantages. *Agrobacterium*-mediated transformation is a time-consuming, expensive, and labor-intensive process that must be optimized for each genotype, and biolistic gene transfer is influenced by a few variables, including size, particle type (tungsten or gold), quantity, acceleration, DNA concentration, particle coating, tissue type, and pre-treatment (Jung and Altpeter, 2016). In addition, plasmid viral and non-viral vectors are available for the delivery of CRISPR components. However, utilizing viral vectors like yellow dwarf virus (YDV), tobacco mosaic virus (TMV), potato virus X (PVX), and cowpea mosaic virus (CMV) restrict the applicability of large fragment sequences (Varanda et al. 2021). Several materials, including inorganic nanoparticles, carbon nanotubes, liposomes, protein- and peptide-based nanoparticles, and nanoscale polymeric materials, have the potential to be employed as non-viral vectors for GnEd applications.

Legislation and regulatory frameworks for gene-edited crops are still in the process of development, which is the primary constraint of GnEd applications. Depending on the kind of site-directed nuclease, some nations have developed biosafety rules for genome-edited crops. It is also vital to increase consumer acceptance of genome-edited foods by demonstrating their real benefits.

5.6 Conclusion

Climate change and population increase are the most significant limits on global agriculture yield. It becomes impossible to supply the rising demand for staple foods due to their threads. The development of novel cultivars, yield enhancement, and stress tolerance/adaptations necessitate the application of innovative instruments and techniques. Abiotic stresses are very complicated processes; consequently, the combination of omics and bioinformatics allowed us to decipher the many pathways, mechanisms, and functional regulators of environmental stresses. Several strategies for changing the genetic composition of crops to bestow high salinity with minimal yield loss have been investigated. There are clear differences between transgenic technology (GM) and genome editing technology, although they are closely related. Transgenic technology includes introducing foreign DNA (non-existing) into the crop genome, resulting in random integration in the genome to tailor the crop with new traits. GnEd editing includes manipulation of the genome's specific location by making knockout, replacement, or even changing the gene expression of an existing target gene in a predetermined, precise manner to improve desired traits. Therefore, GnEd is faster, more accurate, and mostly does not need to introduce exogenous DNA for crop improvement. Therefore, it is expected to get people's acceptance and more favorable perceptions of agriculture than genetically modified crops. GnEd tool has enormous potential to improve crop productivity, eliminate nutrient deficiency, provide food security, and develop crop-ready crops.

The authors declare no competing interests.

References

Abdallah, N. A., Elsharawy, H., Abulela, H. A., Thilmony, R., Abdelhadi, A. A., & Elarabi, N. I. (2022). Multiplex CRISPR/Cas9-mediated genome editing to address drought tolerance in wheat. *GM Crops & Food*, 1–17.

Abdallah, N. A., Hamwieh, A., Radwan, K., Fouad, N., & Prakash, C. (2021). Genome editing techniques in plants: A comprehensive review and future prospects toward zero hunger. *GM Crops & Food*, 12(2), 601–615.

Abe, F., Haque, E., Hisano, H., Tanaka, T., Kamiya, Y., Mikami, M., Kawaura, K., Endo, M., Onishi, K., & Hayashi, T. (2019). Genome-edited triple-recessive mutation alters seed dormancy in wheat. *Cell Reports*, 28(5), 1362–1369.

Abudayyeh, O. O., Gootenberg, J. S., Franklin, B., Koob, J., Kellner, M. J., Ladha, A., Joung, J., Kirchgatterer, P., Cox, D. B. T., & Zhang, F. (2019). A cytosine deaminase for programmable single-base RNA editing. *Science*, 365(6451), 382–386.

Achary, V., & Reddy, M. K. (2021). CRISPR-Cas9 mediated mutation in GRAIN WIDTH and WEIGHT2 (GW2) locus improves aleurone layer and grain nutritional quality in rice. *Scientific Reports*, 11(1), 1–13.

Ahmad, S., Tang, L., Shahzad, R., Mawia, A. M., Rao, G. S., Jamil, S., Wei, C., Sheng, Z., Shao, G., & Wei, X. (2021). CRISPR-based crop improvements: A way forward to achieve zero hunger. *Journal of Agricultural and Food Chemistry*, 69(30), 8307–8323.

Alam, M. S., Kong, J., Tao, R., Ahmed, T., Alamin, M., Alotaibi, S. S., Abdelsalam, N. R., & Xu, J.-H. (2022). CRISPR/Cas9 mediated knockout of the OsbHLH024 Transcription factor improves salt stress resistance in rice (*Oryza sativa* L.). *Plants*, 11(9), 1184.

Asaoka, T., Almagro, J., Ehrhardt, C., Tsai, I., Schleiffer, A., Deszcz, L., Junttila, S., Ringrose, L., Mechtler, K., &Kavirayani, A. (2016). Linear ubiquitination by LUBEL has a role in Drosophila heat stress response. *EMBO Reports*, 17(11), 1624–1640.

Bhattacharya, A., Parkhi, V., & Char, B. (2021). Genome editing for crop improvement: A perspective from India. *In Vitro Cellular & Developmental Biology-Plant*, 57(4), 565–573.

Blanvillain-Baufumé, S., Reschke, M., Solé, M., Auguy, F., Doucoure, H., Szurek, B., Meynard, D., Portefaix, M., Cunnac, S., & Guiderdoni, E. (2017). Targeted promoter editing for rice resistance to Xanthomonas oryzaepv. oryzae reveals differential activities for SWEET 14-inducing TAL effectors. *Plant Biotechnology Journal*, 15(3), 306–317.

Bo, W., Zhaohui, Z., Huanhuan, Z., Xia, W., Binglin, L. I. U., Lijia, Y., Xiangyan, H. A. N., Deshui, Y., Xuelian, Z., & Chunguo, W. (2019). Targeted mutagenesis of NAC transcription factor gene, OsNAC041, leading to salt sensitivity in rice. *Rice Science*, 26(2), 98–108.

Chandrasekaran, J., Brumin, M., Wolf, D., Leibman, D., Klap, C., Pearlsman, M., Sherman, A., Arazi, T., & Gal-On, A. (2016). Development of broad virus resistance in non-transgenic cucumber using CRISPR/Cas9 technology. *Molecular Plant Pathology*, 17(7), 1140–1153.

Chang, L., Wu, S., & Tian, L. (2019). Effective genome editing and identification of a regiospecific gallic acid 4-O-glycosyltransferase in pomegranate (Punica granatum L.). *Horticulture Research*, 6.

Chen, F., Zheng, G., Qu, M., Wang, Y., Lyu, M.-J. A., & Zhu, X.-G. (2021). Knocking out NEGATIVE REGULATOR OF PHOTOSYNTHESIS 1 increases rice leaf photosynthesis and biomass production in the field. *Journal of Experimental Botany*, 72(5), 1836–1849.

Chen, Z., Du, H., Tao, Y., Xu, Y., Wang, F., Li, B., Zhu, Q.-H., Niu, H., & Yang, J. (2022). Efficient breeding of low glutelin content rice germplasm by simultaneous editing multiple glutelin genes via CRISPR/Cas9. *Plant Science*, 324, 111449.

Chennakesavulu, K., Singh, H., Trivedi, P. K., Jain, M., & Yadav, S. R. (2022). State-of-the-Art in CRISPR technology and engineering drought, salinity, and thermotolerant crop plants. *Plant Cell Reports*, 41(3), 815–831.

Cox, D. B. T., Gootenberg, J. S., Abudayyeh, O. O., Franklin, B., Kellner, M. J., Joung, J., & Zhang, F. (2017). RNA editing with CRISPR-Cas13. *Science*, 358(6366), 1019–1027.

Daboussi, F., Stoddard, T. J., & Zhang, F. (2015). Engineering meganuclease for precise plant genome modification. In *Advances in new technology for targeted modification of plant genomes* (pp. 21–38). Springer.

de Vries, L., Brouckaert, M., Chanoca, A., Kim, H., Regner, M. R., Timokhin, V. I., Sun, Y., de Meester, B., van Doorsselaere, J., & Goeminne, G. (2021). CRISPR-Cas9 editing of CAFFEOYL SHIKIMATE ESTERASE 1 and 2 shows their importance and partial redundancy in lignification in Populus tremula× P. alba. *Plant Biotechnology Journal*, 19(11), 2221–2234.

Dong, O. X., Yu, S., Jain, R., Zhang, N., Duong, P. Q., Butler, C., Li, Y., Lipzen, A., Martin, J. A., & Barry, K. W. (2020). Marker-free carotenoid-enriched rice generated through targeted gene insertion using CRISPR-Cas9. *Nature Communications*, 11(1), 1–10.

Donovan, S., Mao, Y., Orr, D. J., Carmo-Silva, E., & McCormick, A. J. (2020). CRISPR-Cas9-mediated mutagenesis of the Rubisco small subunit family in Nicotiana tabacum. *Frontiers in Genome Editing*, 28.

Endo, A., Saika, H., Takemura, M., Misawa, N., & Toki, S. (2019). A novel approach to carotenoid accumulation in rice callus by mimicking the cauliflower orange mutation via genome editing. *Rice*, 12(1), 1–5.

Feng, X., Xiong, J., Zhang, W., Guan, H., Zheng, D., Xiong, H., Jia, L., Hu, Y., Zhou, H., & Wen, Y. (2022). ZmLBD5, a class-II LBD gene, negatively regulates drought tolerance by impairing abscisic acid synthesis. *The Plant Journal*.

Fister, A. S., Landherr, L., Maximova, S. N., & Guiltinan, M. J. (2018). Transient expression of CRISPR/Cas9 machinery targeting TcNPR3 enhances defense response in Theobroma cacao. *Frontiers in Plant Science*, 9, 268.

Flexas, J., & Carriquí, M. (2020). Photosynthesis and photosynthetic efficiencies along the terrestrial plant's phylogeny: Lessons for improving crop photosynthesis. *The Plant Journal*, 101(4), 964–978.

Fu, L., Wu, D., Zhang, X., Xu, Y., Kuang, L., Cai, S., Zhang, G., & Shen, Q. (2022). Vacuolar H+-pyrophosphatase HVP10 enhances salt tolerance via promoting Na+ translocation into root vacuoles. *Plant Physiology*, 188(2), 1248–1263.

Gallego-Bartolomé, J., Gardiner, J., Liu, W., Papikian, A., Ghoshal, B., Kuo, H. Y., Zhao, J. M.-C., Segal, D. J., & Jacobsen, S. E. (2018). Targeted DNA demethylation of the Arabidopsis genome using the human TET1 catalytic domain. *Proceedings of the National Academy of Sciences*, 115(9), E2125–E2134.

Galli, M., Hochstein, S., Iqbal, D., Claar, M., Imani, J., & Kogel, K.-H. (2022). CRISPR/Sp Cas9-mediated double knockout of barley Microrchidia MORC1 and MORC6a reveals their strong involvement in plant immunity, transcriptional gene silencing and plant growth. Plant Biotechnology Journal, 20(1), 89-102.. *Plant Biotechnology Journal*, 29(4), 89–102.

Gao, H., Gadlage, M. J., Lafitte, H. R., Lenderts, B., Yang, M., Schroder, M., Farrell, J., Snopek, K., Peterson, D., & Feigenbutz, L. (2020). Superior field performance of waxy corn engineered using CRISPR–Cas9. *Nature Biotechnology*, 38(5), 579–581.

Garczynski, S. F., Martin, J. A., Griset, M., Willett, L. S., Cooper, W. R., Swisher, K. D., & Unruh, T. R. (2017). CRISPR/Cas9 editing of the codling moth (Lepidoptera: Tortricidae) CpomOR1 gene affects egg production and viability. *Journal of Economic Entomology*, 110(4), 1847–1855.

Gomez, M. A., Lin, Z. D., Moll, T., Chauhan, R. D., Hayden, L., Renninger, K., Beyene, G., Taylor, N. J., Carrington, J. C., & Staskawicz, B. J. (2019). Simultaneous CRISPR/Cas9-mediated editing of cassava eIF 4E isoforms nCBP-1 and nCBP-2 reduces cassava brown streak disease symptom severity and incidence. *Plant Biotechnology Journal*, 17(2), 421–434.

Gu, H., Zhang, K., Chen, J., Gull, S., Chen, C., Hou, Y., Li, X., Miao, J., Zhou, Y., & Liang, G. (2022). OsFTL4, an FT-like gene, regulates flowering time and drought tolerance in rice (*Oryza sativa* L.). *Rice*, 15(1), 1–15.

Han, X., Chen, Z., Li, P., Xu, H., Liu, K., Zha, W., Li, S., Chen, J., Yang, G., & Huang, J. (2022). Development of novel rice germplasm for salt-tolerance at seedling stage using CRISPR-Cas9. *Sustainability*, 14(5), 2621.

Hasanuzzaman, M., Nahar, K., Md, M., & Alam, R. (2013). Roy Chowdhury and Fujita, M. *International Journal of Molecular Sciences*, 14(5), 9643–9684.

He, F., Wang, C., Sun, H., Tian, S., Zhao, G., Liu, C., Wan, C., Guo, J., Huang, X., & Zhan, G. (2022). Simultaneous editing of three homoeologs of TaCIPK14 confers broad-spectrum resistance to stripe rust in wheat. *Plant Biotechnology Journal*.

Hoffie, R. E., Perovic, D., Habekuß, A., Ordon, F., & Kumlehn, J. (2022). Novel resistance to the BymovirusBaMMV established by targeted mutagenesis of the PDIL5-1 susceptibility gene in barley. *Plant Biotechnology Journal*.

Honma, Y., Adhikari, P. B., Kuwata, K., Kagenishi, T., Yokawa, K., Notaguchi, M., Kurotani, K., Toda, E., Bessho-Uehara, K., & Liu, X. (2020). High-quality sugar production by osgcs1 rice. *Communications Biology*, 3(1), 1–8.

Hu, C., Sheng, O., Deng, G., He, W., Dong, T., Yang, Q., Dou, T., Li, C., Gao, H., & Liu, S. (2021). CRISPR/Cas9-mediated genome editing of MaACO1 (aminocyclopropane-1-carboxylate oxidase 1) promotes the shelf life of banana fruit. *Plant Biotechnology Journal*, 19(4), 654.

Huang, Y., Cao, H., Yang, L., Chen, C., Shabala, L., Xiong, M., Niu, M., Liu, J., Zheng, Z., & Zhou, L. (2019). Tissue-specific respiratory burst oxidase homolog-dependent H2O2 signaling to the plasma membrane H+-ATPase confers potassium uptake and salinity tolerance in Cucurbitaceae. *Journal of Experimental Botany*, 70(20), 5879–5893.

Hunziker, J., Nishida, K., Kondo, A., Kishimoto, S., Ariizumi, T., & Ezura, H. (2020). Multiple gene substitution by target-AID base-editing technology in tomato. *Scientific Reports*, 10(1), 1–12.

Iaffaldano, B., Zhang, Y., & Cornish, K. (2016). CRISPR/Cas9 genome editing of rubber producing dandelion Taraxacum kok-saghyz using Agrobacterium rhizogenes without selection. *Industrial Crops and Products*, 89, 356–362.

Ibrahim, S., Saleem, B., Rehman, N., Zafar, S. A., Naeem, M. K., & Khan, M. R. (2022). CRISPR/Cas9 mediated disruption of Inositol Pentakisphosphate 2-Kinase 1 (TaIPK1) reduces phytic acid and improves iron and zinc accumulation in wheat grains. *Journal of Advanced Research*, 37, 33–41.

Jiao, B., Hao, X., Liu, Z., Liu, M., Wang, J., Liu, L., Liu, N., Song, R., Zhang, J., & Fang, Y. (2022). Engineering CRISPR immune systems conferring GLRaV-3 resistance in grapevine. *Horticulture Research*, 9.

Joung, J. K., & Sander, J. D. (2013). TALENs: A widely applicable technology for targeted genome editing. *Nature Reviews Molecular Cell Biology*, 14(1), 49–55.

Jung, J. H., & Altpeter, F. (2016). TALEN mediated targeted mutagenesis of the caffeic acid O-methyltransferase in highly polyploid sugarcane improves cell wall composition for production of bioethanol. *Plant Molecular Biology*, 92(1), 131–142.

Jung, S. E., Bang, S. W., Kim, S. H., Seo, J. S., Yoon, H.-B., Kim, Y. S., & Kim, J.-K. (2021). Overexpression of OsERF83, a vascular tissue-specific transcription factor gene, confers drought tolerance in rice. *International Journal of Molecular Sciences*, 22(14), 7656.

Kalajdzic, P., & Schetelig, M. F. (2017). CRISPR/Cas-mediated gene editing using purified protein in D rosophilasuzukii. *EntomologiaExperimentalis et Applicata*, 164(3), 350–362.

Karavolias, N. G., Horner, W., Abugu, M. N., & Evanega, S. N. (2021). Application of gene editing for climate change in agriculture. *Frontiers in Sustainable Food Systems*, 5, 685801.

Kaur, N., Alok, A., Kumar, P., Kaur, N., Awasthi, P., Chaturvedi, S., Pandey, P., Pandey, A., Pandey, A. K., & Tiwari, S. (2020). CRISPR/Cas9 directed editing of lycopene epsilon-cyclase modulates metabolic flux for β-carotene biosynthesis in banana fruit. *Metabolic Engineering*, 59, 76–86.

Kawaguchi, K., Takei-Hoshi, R., Yoshikawa, I., Nishida, K., Kobayashi, M., Kusano, M., Lu, Y., Ariizumi, T., Ezura, H., & Otagaki, S. (2021). Functional disruption of cell wall invertase inhibitor by genome editing increases sugar content of tomato fruit without decrease fruit weight. *Scientific Reports*, 11(1), 1–12.

Khan, M. S. S., Basnet, R., Ahmed, S., Bao, J., & Shu, Q. (2020). Mutations of OsPLDa1 increase lysophospholipid content and enhance cooking and eating quality in rice. *Plants*, 9(3), 390.

Khan, Z., & Shahwar, D. (2020). Role of heat shock proteins (HSPs) and heat stress tolerance in crop plants. *Sustainable agriculture in the era of climate change*. 211–234.

Kim, D., Alptekin, B., & Budak, H. (2018). CRISPR/Cas9 genome editing in wheat. *Functional & Integrative Genomics*, 18(1), 31–41.

Kim, S.-T., Choi, M., Bae, S.-J., & Kim, J.-S. (2021). The functional association of ACQOS/VICTR with salt stress resistance in *Arabidopsis thaliana* was confirmed by CRISPR-Mediated mutagenesis. *International Journal of Molecular Sciences*, 22(21), 11389.

Ku, H.-K., & Ha, S.-H. (2020). Improving nutritional and functional quality by genome editing of crops: Status and perspectives. *Frontiers in Plant Science*, 11, 577313.

Kumar, N., Galli, M., Dempsey, D., Imani, J., Moebus, A., & Kogel, K. (2021). NPR1 is required for root colonization and the establishment of a mutualistic symbiosis between the beneficial bacterium Rhizobium radiobacter and barley. *Environmental Microbiology*, 23(4), 2102–2115.

Lacchini, E., Kiegle, E., Castellani, M., Adam, H., Jouannic, S., Gregis, V., & Kater, M. M. (2020). CRISPR-mediated accelerated domestication of African rice land-races. *PloS One*, 15(3), e0229782.

Lata, C., Muthamilarasan, M., & Prasad, M. (2015). Drought stress responses and signal transduction in plants. In *Elucidation of abiotic stress signaling in plants* (pp. 195–225). Springer.

Lee, K.-R., Jeon, I., Yu, H., Kim, S.-G., Kim, H.-S., Ahn, S.-J., Lee, J., Lee, S.-K., & Kim, H. U. (2021). Increasing monounsaturated fatty acid contents in hexaploid Camelina sativa seed oil by FAD2 gene knockout using CRISPR-Cas9. *Frontiers in Plant Science*, 12.

Lemmon, Z. H., Reem, N. T., Dalrymple, J., Soyk, S., Swartwood, K. E., Rodriguez-Leal, D., van Eck, J., & Lippman, Z. B. (2018). Rapid improvement of domestication traits in an orphan crop by genome editing. *Nature Plants*, 4(10), 766–770.

Li, C., Li, W., Zhou, Z., Chen, H., Xie, C., & Lin, Y. (2020). A new rice breeding method: CRISPR/Cas9 system editing of the Xa13 promoter to cultivate transgene-free bacterial blight-resistant rice. *Plant Biotechnology Journal*, 18(2), 313.

Li, C., Liu, X.-J., Yan, Y., Alam, M. S., Liu, Z., Yang, Z.-K., Tao, R.-F., Yue, E., Duan, M.-H., & Xu, J.-H. (2022). OsLHY is involved in regulating flowering through the Hd1-and Ehd1-mediated pathways in rice (*Oryza sativa* L.). *Plant Science*, 315, 111145.

Li, F., & Scott, M. J. (2016). CRISPR/Cas9-mediated mutagenesis of the white and Sex lethal loci in the invasive pest, Drosophila suzukii. *Biochemical and Biophysical Research Communications*, 469(4), 911–916.

Li, H., Li, X., Xu, Y., Liu, H., He, M., Tian, X., Wang, Z., Wu, X., Bu, Q., & Yang, J. (2020). High-efficiency reduction of rice amylose content via CRISPR/Cas9-mediated base editing. *Rice Science*, 27(6), 445–448.

Li, J., Jiao, G., Sun, Y., Chen, J., Zhong, Y., Yan, L., Jiang, D., Ma, Y., & Xia, L. (2021). Modification of starch composition, structure and properties through editing of TaSBEIIa in both winter and spring wheat varieties by CRISPR/Cas9. *Plant Biotechnology Journal*, 19(5), 937–951.

Li, J., Scarano, A., Gonzalez, N. M., D'Orso, F., Yue, Y., Nemeth, K., Saalbach, G., Hill, L., de Oliveira Martins, C., & Moran, R. (2022). Biofortified tomatoes provide a new route to vitamin D sufficiency. *Nature Plants*, 1–6.

Li, M., Guo, S., Zhang, J., Sun, H., Tian, S., Wang, J., Zuo, Y., Yu, Y., Gong, G., & Zhang, H. (2022). Sugar transporter VST1 knockout reduced aphid damage in watermelon. *Plant Cell Reports*, 41(1), 277–279.

Li, R., Liu, C., Zhao, R., Wang, L., Chen, L., Yu, W., Zhang, S., Sheng, J., & Shen, L. (2019). CRISPR/Cas9-Mediated SlNPR1 mutagenesis reduces tomato plant drought tolerance. *BMC Plant Biology*, 19(1), 1–13.

Li, S., Lin, D., Zhang, Y., Deng, M., Chen, Y., Lv, B., Li, B., Lei, Y., Wang, Y., & Zhao, L. (2022). Genome-edited powdery mildew resistance in wheat without growth penalties. *Nature*, 602(7897), 455–460.

Li, T., Liu, B., Spalding, M. H., Weeks, D. P., & Yang, B. (2012). High-efficiency TALEN-based gene editing produces disease-resistant rice. *Nature Biotechnology*, 30(5), 390–392.

Li, X., Wang, Y., Chen, S., Tian, H., Fu, D., Zhu, B., Luo, Y., & Zhu, H. (2018). Lycopene is enriched in tomato fruit by CRISPR/Cas9-mediated multiplex genome editing. *Frontiers in Plant Science*, 9, 559.

Li, X., Wang, Y., Liu, Y., Yang, B., Wang, X., Wei, J., Lu, Z., Zhang, Y., Wu, J., & Huang, X. (2018). Base editing with a Cpf1-cytidine deaminase fusion. *Nature Biotechnology*, 36(4), 324–327.

Liao, S., Qin, X., Luo, L., Han, Y., Wang, X., Usman, B., Nawaz, G., Zhao, N., Liu, Y., & Li, R. (2019). CRISPR/Cas9-induced mutagenesis of semi-rolled leaf1, 2 confers curled leaf phenotype and drought tolerance by influencing protein expression patterns and ROS scavenging in rice (*Oryza sativa* L.). *Agronomy*, 9(11), 728.

Liu, D., Chen, X., Liu, J., Ye, J., & Guo, Z. (2012). The rice ERF transcription factor OsERF922 negatively regulates resistance to Magnaportheoryzae and salt tolerance. *Journal of Experimental Botany*, 63(10), 3899–3911.

Liu, J., Cui, L., Xie, Z., Zhang, Z., Liu, E., & Peng, X. (2019). Two NCA1 isoforms interact with catalase in a mutually exclusive manner to redundantly regulate its activity in rice. *BMC Plant Biology*, 19(1), 1–10.

Liu, X., Cheng, L., Li, R., Cai, Y., Wang, X., Fu, X., Dong, X., Qi, M., Jiang, C.-Z., & Xu, T. (2022). The HD-Zip transcription factor SlHB15A regulates abscission by modulating jasmonoyl-isoleucine biosynthesis. *Plant Physiology*.

Liu, X., Ding, Q., Wang, W., Pan, Y., Tan, C., Qiu, Y., Chen, Y., Li, H., Li, Y., & Ye, N. (2022). Targeted deletion of the first intron of the Wxb allele via CRISPR/Cas9 significantly increases grain amylose content in rice. *Rice*, 15(1), 1–12.

Liu, X., Wu, D., Shan, T., Xu, S., Qin, R., Li, H., Negm, M., Wu, D., & Li, J. (2020). The trihelix transcription factor OsGTγ-2 is involved adaption to salt stress in rice. *Plant Molecular Biology*, 103(4), 545–560.

Long, Q., Du, M., Long, J., Xie, Y., Zhang, J., Xu, L., He, Y., Li, Q., Chen, S., & Zou, X. (2021). Transcription factor WRKY22 regulates canker susceptibility in sweet orange (Citrus sinensis Osbeck) by enhancing cell enlargement and CsLOB1 expression. *Horticulture Research*, 8.

Lu, H., Liu, S., Xu, S., Chen, W., Zhou, X., Tan, Y., Huang, J., & Shu, Q. (2017). CRISPR-S: An active interference element for a rapid and inexpensive selection of genome-edited, transgene-free rice plants. *Plant Biotechnology Journal*, 15(11), 1371.

Ma, J., Chen, J., Wang, M., Ren, Y., Wang, S., Lei, C., & Cheng, Z. (2018). Disruption of OsSEC3A increases the content of salicylic acid and induces plant defense responses in rice. *Journal of Experimental Botany*, 69(5), 1051–1064.

Ma, J., Sun, S., Whelan, J., & Shou, H. (2021). CRISPR/Cas9-mediated knockout of GmFATB1 significantly reduced the amount of saturated fatty acids in soybean seeds. *International Journal of Molecular Sciences*, 22(8), 3877.

Macovei, A., Sevilla, N. R., Cantos, C., Jonson, G. B., Slamet-Loedin, I., Čermák, T., Voytas, D. F., Choi, I., & Chadha-Mohanty, P. (2018). Novel alleles of rice eIF4G generated by CRISPR/Cas9-targeted mutagenesis confer resistance to Rice tungro spherical virus. *Plant Biotechnology Journal*, 16(11), 1918–1927.

Maioli, A., Gianoglio, S., Moglia, A., Acquadro, A., Valentino, D., Milani, A. M., Prohens, J., Orzaez, D., Granell, A., & Lanteri, S. (2020). Simultaneous CRISPR/ Cas9 editing of three PPO genes reduces fruit flesh browning in *Solanum melongena* L. *Frontiers in Plant Science*, 11, 607161.

Makhotenko, A. v, Khromov, A. v, Snigir, E. A., Makarova, S. S., Makarov, V. v, Suprunova, T. P., Kalinina, N. O., & Taliansky, M. E. (2019). Functional analysis of coilin in virus resistance and stress tolerance of potato *Solanum tuberosum* using CRISPR-Cas9 editing. *Doklady Biochemistry and Biophysics*, 484(1), 88–91.

Mo, W., Tang, W., Du, Y., Jing, Y., Bu, Q., & Lin, R. (2020). PHYTOCHROME-INTERACTING FACTOR-LIKE14 and SLENDER RICE1 interaction controls seedling growth under salt stress. *Plant Physiology*, 184(1), 506–517.

Nagamine, A., Takayama, M., & Ezura, H. (2022). Genetic improvement of tomato using gene editing technologies. *The Journal of Horticultural Science and Biotechnology*, 1–9.

Nekrasov, V., Wang, C., Win, J., Lanz, C., Weigel, D., & Kamoun, S. (2017). Rapid generation of a transgene-free powdery mildew resistant tomato by genome deletion. *Scientific Reports*, 7(1), 1–6.

Nieves-Cordones, M., Lara, A., Silva, M., Amo, J., Rodriguez-Sepulveda, P., Rivero, R. M., Martínez, V., Botella, M. A., & Rubio, F. (2020). Root high-affinity K+ and Cs+ uptake and plant fertility in tomato plants are dependent on the activity of the high-affinity K+ transporter SlHAK5. *Plant, Cell & Environment*, 43(7), 1707–1721.

Nonaka, S., Arai, C., Takayama, M., Matsukura, C., & Ezura, H. (2017). Efficient increase of γ-aminobutyric acid (GABA) content in tomato fruits by targeted mutagenesis. *Scientific Reports*, 7(1), 1–14.

Nunez, S., Arets, E., Alkemade, R., Verwer, C., & Leemans, R. (2019). Assessing the impacts of climate change on biodiversity: Is below 2° C enough? *Climatic Change*, 154(3), 351–365.

Nuñez-Muñoz, L., Vargas-Hernández, B., Hinojosa-Moya, J., Ruiz-Medrano, R., & Xoconostle-Cázares, B. (2021). Plant drought tolerance provided through genome editing of the trehalase gene. *Plant Signaling & Behavior*, 16(4), 1877005.

Ogata, T., Ishizaki, T., Fujita, M., & Fujita, Y. (2020). CRISPR/Cas9-targeted mutagenesis of OsERA1 confers enhanced responses to abscisic acid and drought stress and increased primary root growth under nonstressed conditions in rice. *PloS One*, 15(12), e0243376.

Oliva, R., Ji, C., Atienza-Grande, G., Huguet-Tapia, J. C., Perez-Quintero, A., Li, T., Eom, J.-S., Li, C., Nguyen, H., & Liu, B. (2019). Broad-spectrum resistance to bacterial blight in rice using genome editing. *Nature Biotechnology*, 37(11), 1344–1350.

Ortigosa, A., Gimenez-Ibanez, S., Leonhardt, N., & Solano, R. (2019). Design of a bacterial speck resistant tomato by CRISPR/Cas9-mediated editing of Sl JAZ 2. *Plant Biotechnology Journal*, 17(3), 665–673.

Osakabe, Y., Watanabe, T., Sugano, S. S., Ueta, R., Ishihara, R., Shinozaki, K., & Osakabe, K. (2016). Optimization of CRISPR/Cas9 genome editing to modify abiotic stress responses in plants. *Scientific Reports*, 6(1), 1–10.

Park, J.-J., Dempewolf, E., Zhang, W., & Wang, Z.-Y. (2017). RNA-guided transcriptional activation via CRISPR/dCas9 mimics overexpression phenotypes in Arabidopsis. *PLoS One*, 12(6), e0179410.

Prentice, I. C., Farquhar, G. D., Fasham, M. J. R., Goulden, M. L., Heimann, M., Jaramillo, V. J., Kheshgi, H. S., le Quéré, C., Scholes, R. J., & Wallace, D. W. R. (2001). The carbon cycle and atmospheric carbon dioxide.

Qiao, K., Wang, F., Liang, S., Wang, H., Hu, Z., & Chai, T. (2019). New biofortification tool: Wheat TaCNR5 enhances zinc and manganese tolerance and increases zinc and manganese accumulation in rice grains. *Journal of Agricultural and Food Chemistry*, 67(35), 9877–9884.

Qin, H., Wang, J., Chen, X., Wang, F., Peng, P., Zhou, Y., Miao, Y., Zhang, Y., Gao, Y., & Qi, Y. (2019). Rice Os DOF 15 contributes to ethylene-inhibited primary root elongation under salt stress. *New Phytologist*, 223(2), 798–813.

Raffan, S., Sparks, C., Huttly, A., Hyde, L., Martignago, D., Mead, A., Hanley, S. J., Wilkinson, P. A., Barker, G., & Edwards, K. J. (2021). Wheat with greatly reduced accumulation of free asparagine in the grain, produced by CRISPR/Cas9 editing of asparagine synthetase gene TaASN2. *Plant Biotechnology Journal*, 19(8), 1602–1613.

Reddy, G. B., Pullakhandam, R., Ghosh, S., Boiroju, N. K., Tattari, S., Laxmaiah, A., Hemalatha, R., Kapil, U., Sachdev, H. S., & Kurpad, A. v. (2021). Vitamin A deficiency among children younger than 5 y in India: An analysis of national data sets to reflect on the need for vitamin A supplementation. *The American Journal of Clinical Nutrition*, 113(4), 939–947.

Reddy, I. N. B. L., Kim, B.-K., Yoon, I.-S., Kim, K.-H., & Kwon, T.-R. (2017). Salt tolerance in rice: Focus on mechanisms and approaches. *Rice Science*, 24(3), 123–144.

Rizvi, S., Raza, S. T., Ahmed, F., Ahmad, A., Abbas, S., & Mahdi, F. (2014). The role of vitamin E in human health and some diseases. *Sultan Qaboos University Medical Journal*, 14(2), e157.

Roca Paixão, J. F., Gillet, F.-X., Ribeiro, T. P., Bournaud, C., Lourenço-Tessutti, I. T., Noriega, D. D., Melo, B. P. de, de Almeida-Engler, J., & Grossi-de-Sa, M. F. (2019). Improved drought stress tolerance in Arabidopsis by CRISPR/dCas9 fusion with a Histone AcetylTransferase. *Scientific Reports*, 9(1), 1–9.

Sakuma, Y., Maruyama, K., Osakabe, Y., Qin, F., Seki, M., Shinozaki, K., & Yamaguchi-Shinozaki, K. (2006). Functional analysis of an Arabidopsis transcription factor, DREB2A, involved in drought-responsive gene expression. *The Plant Cell*, 18(5), 1292–1309.

Scholefield, J., & Harrison, P. T. (2021). Prime editing–an update on the field. *Gene Therapy*, 28(7), 396–401.

Seki, M., Narusaka, M., Abe, H., Kasuga, M., Yamaguchi-Shinozaki, K., Carninci, P., Hayashizaki, Y., & Shinozaki, K. (2001). Monitoring the expression pattern of 1300 Arabidopsis genes under drought and cold stresses by using a full-length cDNA microarray. *The Plant Cell*, 13(1), 61–72.

Shao, X., Wu, S., Dou, T., Zhu, H., Hu, C., Huo, H., He, W., Deng, G., Sheng, O., Bi, F., Gao, H., Dong, T., Li, C., Yang, Q., & Yi, G. (2020). Using CRISPR/Cas9 genome editing system to create *MaGA20ox2* gene-modified semi-dwarf banana. *Plant Biotechnology Journal*, 18(1), 17–19. doi: 10.1111/pbi.13216. Epub 2019 Aug 20.

Shelake, R. M., Kadam, U. S., Kumar, R., Pramanik, D., Singh, A. K., & Kim, J.-Y. (2022). Engineering drought and salinity tolerance traits in crops through CRISPR-mediated genome editing: Targets, tools, challenges, and perspectives. *Plant Communications*, 3, 100417.

Shen, C., Zhang, Y., Li, Q., Liu, S., He, F., An, Y., Zhou, Y., Liu, C., Yin, W., & Xia, X. (2021). PdGNC confers drought tolerance by mediating stomatal closure resulting from NO and H2O2 production via the direct regulation of PdHXK1 expression in Populus. *New Phytologist*, 230(5), 1868–1882.

Sheng, X., Sun, Z., Wang, X., Tan, Y., Yu, D., Yuan, G., Yuan, D., & Duan, M. (2020). Improvement of the rice "easy-to-shatter" trait via CRISPR/Cas9-mediated mutagenesis of the qSH1 gene. *Frontiers in Plant Science*, 11, 619.

Shi, M., Du, Z., Hua, Q., & Kai, G. (2021). CRISPR/Cas9-mediated targeted mutagenesis of bZIP2 in Salvia miltiorrhiza leads to promoted phenolic acid biosynthesis. *Industrial Crops and Products*, 167, 113560.

Shinozaki, K., & Yamaguchi-Shinozaki, K. (2007). Gene networks involved in drought stress response and tolerance. *Journal of Experimental Botany*, 58(2), 221–227.

Shrivastava, P., & Kumar, R. (2015). Soil salinity: A serious environmental issue and plant growth promoting bacteria as one of the tools for its alleviation. *Saudi Journal of Biological Sciences*, 22(2), 123–131.

Singh, S., Rahangdale, S., Pandita, S., Saxena, G., Upadhyay, S. K., Mishra, G., & Verma, P. C. (2022). CRISPR/Cas9 for insect pests management: A comprehensive review of advances and applications. *Agriculture*, 12(11), 1896.

Smith, W. A., Matsuba, Y., & Dewey, R. E. (2022). Knockout of a key gene of the nicotine biosynthetic pathway severely affects tobacco growth under field, but not greenhouse conditions. *BMC Research Notes*, 15(1), 1–6.

Sohail, A., Shah, L., Liu, L., Islam, A., Yang, Z., Yang, Q., Anis, G. B., Xu, P., Khan, R. M., & Li, J. (2022). Mapping and validation of qHD7b: Major heading-date QTL functions mainly under long-day conditions. *Plants*, 11(17), 2288.

Sommer, A. (2008). Vitamin A deficiency and clinical disease: An historical overview. *The Journal of Nutrition*, 138(10), 1835–1839.

Sun, Y., Jiao, G., Liu, Z., Zhang, X., Li, J., Guo, X., Du, W., Du, J., Francis, F., & Zhao, Y. (2017). Generation of high-amylose rice through CRISPR/Cas9-mediated targeted mutagenesis of starch branching enzymes. *Frontiers in Plant Science*, 8, 298.

Sun, Z., Zang, Y., Zhou, L., Song, Y., Chen, D., Zhang, Q., Liu, C., Yi, Y., Zhu, B., & Fu, D. (2021). A tomato receptor-like cytoplasmic kinase, SlZRK1, acts as a negative regulator in wound-induced jasmonic acid accumulation and insect resistance. *Journal of Experimental Botany*, 72(20), 7285–7300.

Tang, X., Cao, X., Xu, X., Jiang, Y., Luo, Y., Ma, Z., Fan, J., & Zhou, Y. (2017). Effects of climate change on epidemics of powdery mildew in winter wheat in China. *Plant Disease*, 101(10), 1753–1760.

Tang, Y., Abdelrahman, M., Li, J., Wang, F., Ji, Z., Qi, H., Wang, C., & Zhao, K. (2021). CRISPR/Cas9 induces exon skipping that facilitates development of fragrant rice. *Plant Biotechnology Journal*, 19(4), 642.

Tashkandi, M., Ali, Z., Aljedaani, F., Shami, A., & Mahfouz, M. M. (2018). Engineering resistance against tomato yellow leaf curl virus via the CRISPR/Cas9 system in tomato. *Plant Signal Behav* 13 (10): e1525996.

Thomazella, D. P. de T., Brail, Q., Dahlbeck, D., &Staskawicz, B. (2016). CRISPR-Cas9 mediated mutagenesis of a. *DMR6*.

Tripathi, J. N., Ntui, V. O., Ron, M., Muiruri, S. K., Britt, A., & Tripathi, L. (2019). CRISPR/Cas9 editing of endogenous banana streak virus in the B genome of Musa spp. overcomes a major challenge in banana breeding. *Communications Biology*, 2(1), 1–11.

Tripathi, J. N., Ntui, V. O., Shah, T., & Tripathi, L. (2021). CRISPR/Cas9-mediated editing of DMR6 orthologue in banana (Musa spp.) confers enhanced resistance to bacterial disease. *Plant Biotechnology Journal*, 19(7), 1291.

UN World Population Prospects. (2019).

United Nations, Goal 2: Zero Hunger. (2019).

Urnov, F. D., Rebar, E. J., & Holmes, M. C. (2010). Zhang. HS & Gregory, PD. *Nature Rev. Genet*, 11, 636–646.

Varanda, C. M. R., Félix, M. do R., Campos, M. D., Patanita, M., &Materatski, P. (2021). Plant viruses: From targets to tools for CRISPR. *Viruses*, 13(1), 141.

Veley, K. M., Okwuonu, I., Jensen, G., Yoder, M., Taylor, N. J., Meyers, B. C., & Bart, R. S. (2021). Gene tagging via CRISPR-mediated homology-directed repair in cassava. G3, 11(4), jkab028.

Wan, D.-Y., Guo, Y., Cheng, Y., Hu, Y., Xiao, S., Wang, Y., & Wen, Y.-Q. (2020). CRISPR/Cas9-mediated mutagenesis of VvMLO3 results in enhanced resistance to powdery mildew in grapevine (Vitis vinifera). *Horticulture Research*, 7.

Wang, C., Wang, G., Gao, Y., Lu, G., Habben, J. E., Mao, G., Chen, G., Wang, J., Yang, F., & Zhao, X. (2020). A cytokinin-activation enzyme-like gene improves grain yield under various field conditions in rice. *Plant Molecular Biology*, 102(4), 373–388.

Wang, F., Wang, C., Liu, P., Lei, C., Hao, W., Gao, Y., Liu, Y.-G., & Zhao, K. (2016). Enhanced rice blast resistance by CRISPR/Cas9-targeted mutagenesis of the ERF transcription factor gene OsERF922. *PloS One*, 11(4), e0154027.

Wang, L., Sun, S., Wu, T., Liu, L., Sun, X., Cai, Y., Li, J., Jia, H., Yuan, S., & Chen, L. (2020). Natural variation and CRISPR/Cas9-mediated mutation in GmPRR37 affect photoperiodic flowering and contribute to regional adaptation of soybean. *Plant Biotechnology Journal*, 18(9), 1869–1881.

Wang, T., Xun, H., Wang, W., Ding, X., Tian, H., Hussain, S., Dong, Q., Li, Y., Cheng, Y., & Wang, C. (2021). Mutation of GmAITR genes by CRISPR/Cas9 genome editing results in enhanced salinity stress tolerance in soybean. *Frontiers in Plant Science*, 12, 2752.

Wang, W., Ma, S., Hu, P., Ji, Y., & Sun, F. (2021). Genome editing of rice eIF4G loci confers partial resistance to rice black-streaked dwarf virus. *Viruses*, 13(10), 2100.

Wang, W., Wei, X., Jiao, G., Chen, W., Wu, Y., Sheng, Z., Hu, S., Xie, L., Wang, J., & Tang, S. (2020). GBSS-BINDING PROTEIN, encoding a CBM48 domain-containing protein, affects rice quality and yield. *Journal of Integrative Plant Biology*, 62(7), 948–966.

Wang, Y., Wang, J., Guo, S., Tian, S., Zhang, J., Ren, Y., Li, M., Gong, G., Zhang, H., & Xu, Y. (2021). CRISPR/Cas9-mediated mutagenesis of ClBG1 decreased seed size and promoted seed germination in watermelon. *Horticulture Research*, 8.

Wei, Z., Abdelrahman, M., Gao, Y., Ji, Z., Mishra, R., Sun, H., Sui, Y., Wu, C., Wang, C., & Zhao, K. (2021). Engineering broad-spectrum resistance to bacterial blight by CRISPR-Cas9-mediated precise homology directed repair in rice. *Molecular Plant*, 14(8), 1215–1218.

Woo, J. W., Kim, J., Kwon, S. il, Corvalán, C., Cho, S. W., Kim, H., Kim, S.-G., Kim, S.-T., Choe, S., & Kim, J.-S. (2015). DNA-free genome editing in plants with preassembled CRISPR-Cas9 ribonucleoproteins. *Nature Biotechnology*, 33(11), 1162–1164.

Xie, T., Chen, X., Guo, T., Rong, H., Chen, Z., Sun, Q., Batley, J., Jiang, J., & Wang, Y. (2020). Targeted knockout of BnTT2 homologues for yellow-seeded *Brassica napus* with reduced flavonoids and improved fatty acid composition. *Journal of Agricultural and Food Chemistry*, 68(20), 5676–5690.

Xing, S., Chen, K., Zhu, H., Zhang, R., Zhang, H., Li, B., & Gao, C. (2020). Fine-tuning sugar content in strawberry. *Genome Biology*, 21(1), 1–14.

Xu, Z., Xu, X., Gong, Q., Li, Z., Li, Y., Wang, S., Yang, Y., Ma, W., Liu, L., & Zhu, B. (2019). Engineering broad-spectrum bacterial blight resistance by simultaneously disrupting variable TALE-binding elements of multiple susceptibility genes in rice. *Molecular Plant*, 12(11), 1434–1446.

Yan, F., Kuang, Y., Ren, B., Wang, J., Zhang, D., Lin, H., Yang, B., Zhou, X., & Zhou, H. (2018). Highly efficient A· T to G· C base editing by Cas9n-guided tRNA adenosine deaminase in rice. *Molecular Plant*, 11(4), 631–634.

Ye, H., Du, H., Tang, N., Li, X., & Xiong, L. (2009). Identification and expression profiling analysis of TIFY family genes involved in stress and phytohormone responses in rice. *Plant Molecular Biology*, 71(3), 291–305.

Yin, W., Xiao, Y., Niu, M., Meng, W., Li, L., Zhang, X., Liu, D., Zhang, G., Qian, Y., & Sun, Z. (2020). ARGONAUTE2 enhances grain length and salt tolerance by activating BIG GRAIN3 to modulate cytokinin distribution in rice. *The Plant Cell*, 32(7), 2292–2306.

Yu, G., Zou, J., Wang, J., Zhu, R., Qi, Z., Jiang, H., Hu, Z., Yang, M., Zhao, Y., & Wu, X. (2022). A soybean NAC homolog contributes to resistance to Phytophthora sojae mediated by dirigent proteins. *The Crop Journal*, 10(2), 332–341.

Yu, Y., Pan, Z., Wang, X., Bian, X., Wang, W., Liang, Q., Kou, M., Ji, H., Li, Y., & Ma, D. (2022). Targeting of SPCSV-RNase3 via CRISPR-Cas13 confers resistance against sweet potato virus disease. *Molecular Plant Pathology*, 23(1), 104–117.

Yuan, M., Zhu, J., Gong, L., He, L., Lee, C., Han, S., Chen, C., & He, G. (2019). Mutagenesis of FAD2 genes in peanut with CRISPR/Cas9 based gene editing. *BMC Biotechnology*, 19(1), 1–7.

Yue, E., Cao, H., & Liu, B. (2020). OsmiR535, a potential genetic editing target for drought and salinity stress tolerance in *Oryza sativa*. *Plants*, 9(10), 1337.

Yuyu, C., Aike, Z., Pao, X., Xiaoxia, W., Yongrun, C., Beifang, W., Yue, Z., Liaqat, S., Shihua, C., & Liyong, C. (2020). Effects of GS3 and GL3. 1 for grain size editing by CRISPR/Cas9 in rice. *Rice Science*, 27(5), 405–413.

Zeng, X., Luo, Y., Vu, N. T. Q., Shen, S., Xia, K., & Zhang, M. (2020). CRISPR/ Cas9-mediated mutation of OsSWEET14 in rice cv. Zhonghua11 confers resistance to Xanthomonas oryzaepv. oryzae without yield penalty. *BMC Plant Biology*, 20(1), 1–11.

Zeng, Z., Han, N., Liu, C., Buerte, B., Zhou, C., Chen, J., Wang, M., Zhang, Y., Tang, Y., & Zhu, M. (2020). Functional dissection of HGGT and HPT in barley vitamin E biosynthesis via CRISPR/Cas9-enabled genome editing. *Annals of Botany*, 126(5), 929–942.

Zhan, X., Zhang, F., Zhong, Z., Chen, R., Wang, Y., Chang, L., Bock, R., Nie, B., & Zhang, J. (2019). Generation of virus-resistant potato plants by RNA genome targeting. *Plant Biotechnology Journal*, 17(9), 1814–1822.

Zhang, A., Liu, Y., Wang, F., Li, T., Chen, Z., Kong, D., Bi, J., Zhang, F., Luo, X., & Wang, J. (2019). Enhanced rice salinity tolerance via CRISPR/Cas9-targeted mutagenesis of the OsRR22 gene. *Molecular Breeding*, 39(3), 1–10.

Zhang, K., Zhuo, C., Wang, Z., Liu, F., Wen, J., Yi, B., Shen, J., Ma, C., Fu, T., & Tu, J. (2022). BnaA03. MKK5-BnaA06. MPK3/BnaC03. MPK3 module positively contributes to sclerotinia sclerotiorum resistance in *Brassica napus*. *Plants*, 11(5), 609.

Zhang, Y., Bai, Y., Wu, G., Zou, S., Chen, Y., Gao, C., & Tang, D. (2017). Simultaneous modification of three homoeologs of Ta EDR 1 by genome editing enhances powdery mildew resistance in wheat. *The Plant Journal*, 91(4), 714–724.

Zhang, Y., Guo, W., Chen, L., Shen, X., Yang, H., Fang, Y., Ouyang, W., Mai, S., Chen, H., & Chen, S. (2022). CRISPR/Cas9-mediated targeted mutagenesis of GmUGT enhanced soybean resistance against leaf-chewing insects through flavonoids biosynthesis. *Frontiers in Plant Science*, 13, 802716.

Zhang, Z., Ge, X., Luo, X., Wang, P., Fan, Q., Hu, G., Xiao, J., Li, F., & Wu, J. (2018). Simultaneous editing of two copies of Gh14-3-3d confers enhanced transgene-clean plant defense against *Verticillium dahliae* in allotetraploid upland cotton. *Frontiers in Plant Science*, 9, 842.

Zheng, M., Liu, X., Lin, J., Liu, X., Wang, Z., Xin, M., Yao, Y., Peng, H., Zhou, D., & Ni, Z. (2019). Histone acetyltransferase GCN 5 contributes to cell wall integrity and salt stress tolerance by altering the expression of cellulose synthesis genes. *The Plant Journal*, 97(3), 587–602.

Zhou, J., Deng, K., Cheng, Y., Zhong, Z., Tian, L., Tang, X., Tang, A., Zheng, X., Zhang, T., & Qi, Y. (2017). CRISPR-Cas9 based genome editing reveals new insights into microRNA function and regulation in rice. *Frontiers in Plant Science*, 8, 1598.

Zhou, Y., Zhang, Y., Wang, X., Han, X., An, Y., Lin, S., Shen, C., Wen, J., Liu, C., & Yin, W. (2020). Root-specific NF-Y family transcription factor, PdNF-YB21, positively regulates root growth and drought resistance by abscisic acid-mediated indoylacetic acid transport in Populus. *New Phytologist*, 227(2), 407–426.

Zong, Y., Wang, Y., Li, C., Zhang, R., Chen, K., Ran, Y., Qiu, J.-L., Wang, D., & Gao, C. (2017). Precise base editing in rice, wheat and maize with a Cas9-cytidine deaminase fusion. *Nature Biotechnology*, 35(5), 438–440.

List of Abbreviations

ABA	Abscisic acid
ABRE	ABA-responsive elements
CRISPRAK	Clustered regularly interspaced short palindromic repeats
DREB	Dehydration-responsive element-binding protein
Indel	Deletes or adds
GMO	Genetically modified organism
GnEd	Genome editing
HDR	Homologous direct recombination
MeN	Meganuclease
NHEJ	Nonhomologus end joining
PEG	Polyethylene glycol
SDN	Site directed nuclease
sHSps	Small heat shock proteins
SDGs	Sustainable Development Goals
TALEN	Transcription activator-like effector nucleases
TF	Transcription factors
ZFN	Zinc-Finger Nucleases

6 Abiotic and biotic stress tolerance in plants via genome-editing tools

Karam Mostafa, Mohamed Farah Abdulla, and Musa Kavas

6.1 Introduction

External circumstances that devastatingly impact plant growth, development, or output are called stress in plants. Plants constantly face a plethora of environmental challenges since they live a sessile lifestyle (Suzuki et al. 2014). The productivity and survival of plants and crops have significant repercussions due to environmental stresses. Plants are susceptible to various environmental stresses, which may be divided into abiotic (physical environment) and biotic stresses (Redondo-Gómez, 2013). The abiotic stress, which includes radiation, salt, floods, droughts, extremely high temperatures, chemical toxicity, oxidative stress, heavy metals, nutrients, high light intensity, ozone (O_3), and anaerobic stresses, causes unimaginable loss of significant crop plants worldwide (Munns and Tester, 2008; Wang et al. 2003; Chaves and Oliveira, 2004; Agarwal and Grover, 2006; Hirel et al. 2007; Bailey-Serres and Voesenek, 2008). In contrast, biotic stressors comprise attacks by various pathogens, including herbivores, nematodes, viruses, fungi, bacteria, arachnids, and weeds (Suzuki et al. 2014; Mertens et al. 2021; Singla and Krattinger, 2016).

Furthermore, due to fluctuating environmental conditions and global warming these days, its effect is unanimously one of the biggest challenges faced by the world (Nunez et al. 2019). Unfortunately, most climate change scenarios eventually negatively impact the agricultural system and substantially decreases crop productivity (Mahato, 2014). The issue of feeding the ever-growing population is significant; an annual United Nations study in 2019 indicated that an estimated two billion people worldwide did not have regular access to safe, nutritious, and adequate food during the year. Therefore, increasing agricultural sustainability and production is crucial for the entire planet. Technical advancements in crop production and scientific discoveries are critically required to guarantee future global food security. Consequently, innovation and non-traditional solutions have become crucial to maintaining the agricultural system and achieving food security and safety (Gao, 2021).

The regulatory or transcriptional machinery is activated when plants detect biotic stress conditions, and the plant finally produces the necessary response (Iqbal et al. 2021). As a result, plants have unique and complex

DOI: 10.4324/9781003382102-6

stress-tolerance systems. However, plant tolerance to environmental stress varies depending on plant species and the intensity of the stress. Therefore, transitory or constant plants must adapt to new environments that are changing faster than their pace of adaptation. Plant cells respond to stressful circumstances by re-programming their genetic machinery for survival and reproduction. Therefore, plants change their physiologies, metabolic systems, gene expression, and developmental activities in response to stress (Chinnusamy and Zhu, 2009). Agricultural advancement is based on genetic variation. For many scientists, gene editing represents an ideal solution to the future food crises expected to exacerbate.

Genome-editing techniques have introduced precise and predictable genome modifications into DNA to obtain desired traits (Mao et al. 2019). These modifications can result in the knockout or knockdown of one or multiple genes without permanently inserting any foreign DNA. On the other hand, genes from different organisms can be introduced into specific regions in the genome to introduce a new feature. These genes have a role in every step of the stress response, including signaling, transcriptional regulation, membrane protein protection, free radical and toxic chemical scavenging, and more (Wang et al. 2003). Through the years, many promising tools for achieving specific gene edits have been developed, including zinc-finger nucleases (ZFNs), transcription activator-like effector nucleases (TALENs), and CRISPR-associated protein 9 (Cas9) (Barrangou and Doudna, 2016). Although CRISPR/Cas systems have significantly increased the accuracy and effectiveness of making modifications, there is undoubtedly still a role for other gene-editing technologies. Besides modern breeding methods, CRISPR-Cas technologies will play an essential role in crop improvement programs. Gene-editing techniques have created enormous possibilities for developing crops and animals that can better deal with the effects of climate change. In this chapter, we will summarize an update on recent studies focusing on the role of gene-editing approaches and the impact of these tools on abiotic and biotic stress responses. There is a particular focus on the role of CRISPR-Cas9 and the crosstalk between CRISPR-Cas9 and stress response pathways. We will also go through the difficulties that genome-editing faces and how they must be resolved before this technology can fully realize its potential for new crops and food production.

6.2 Clustered regularly interspaced short palindromic repeats/Cas9

6.2.1 History and origin of genome editing

For thousands of years, scientists and breeders have attempted to engineer life to produce and improve valuable traits in diverse creatures (Zhang and Zhou, 2014). Years of research have revealed that altering the coding instructions may alter the creature carrying them. Watson and Crick's 1953 discovery of the DNA molecule's structure, subsequent findings of DNA restriction enzymes, and the

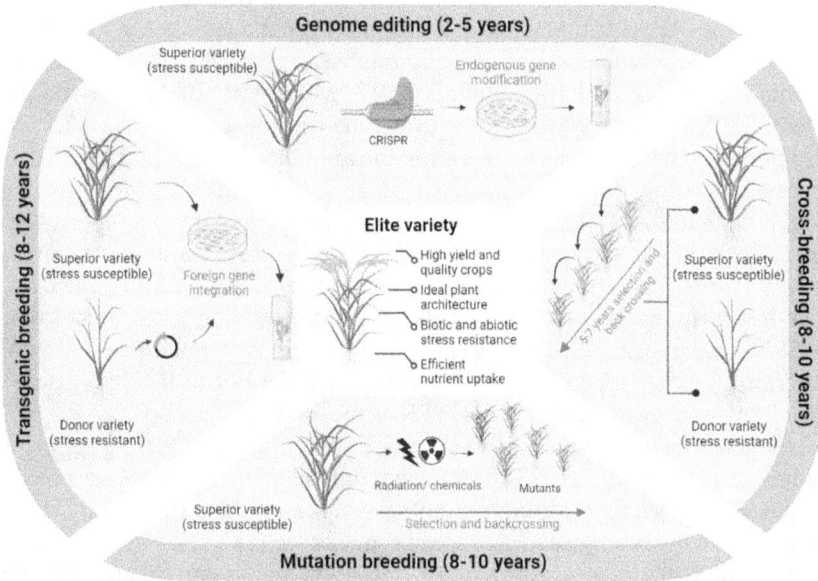

Figure 6.1 Some major modern breeding approaches used in breeding to improve plant traits for enhanced resistance to abiotic and biotic stress.

development of transformation techniques improved how they can engineer organisms (Watson and Crick, 1953). Though over time and effort, scientists produced incredible advancements through plant breeding programs (Hickey et al. 2019). A central purpose of plant breeding relies on a broad genetic variation. Four main new breeding approaches have been used throughout the long history of plant breeding: cross-breeding, mutation breeding, transgenic breeding, and genome-editing breeding (see Figure 6.1) (Gao, 2021).

Cross-breeding (Traditional plant breeding) relies entirely on sexual recombination, which faces many challenges, such as low genetic variability in elite germplasms. It has been performed for a long time and is intrinsically non-specific (Gao, 2021; Hartung and Schiemann, 2014). Physical and chemical mutagenesis is used to induce random mutations genome-wide. Nevertheless, mutation breeding methods can cause numerous random mutations with unknown effects and consequences (Holme et al. 2019). Transgenic breeding, which controls a gene for studying its function as well as increasing yields and quality, reducing pesticide usage, and enhancing nutrition, was a significant advancement in plant breeding (Raman, 2017).

6.2.2 Types of site-specific nucleases technology and method

Through the years, genome-editing technologies to create novel plant types have been used to supplement classical plant breeding techniques. Recent advances in genome-editing technologies enable scientists to generate new crop varieties with desirable features that satisfy the various commercialization demands (Araki and Ishii, 2015). This has been made possible by cellular DNA repair mechanisms and programmable site-specific DNA cleavage reagents (Govindan and Ramalingam, 2016; Miki et al. 2021). Site-specific nucleases (SSNs) technologies may overcome any natural crossing barrier compared to regular breeding, increasing the amount of genetic variation that is accessible and producing plants (or other species) that are not possible through conventional breeding. While transgene technology has opened up many doors, it has also generated much apprehension about its potential effects on human health and the environment (Jung et al. 2018; Nogué et al. 2016). Recently several promising tools for genome-editing technologies that use SSNs have been developed, including zinc-finger nucleases (ZFNs), transcription activator-like effector nucleases (TALENs), RNAi/antisense technology, and more recently, CRISPR-associated protein 9 (Cas9).

These innovative methods have tremendous potential to accelerate the plant breeding process (Sun et al. 2016; Sivanandhan et al. 2016). The precise manipulation of genome-editing technologies relies on the induction of DNA double-strand breaks by sequence-specific nucleases (SSNs) to initiate DNA repair reactions that are based on either non-homologous end joining (NHEJ) or homology-directed repair (HDR) (see Figure 6.2) (Sun et al. 2016).

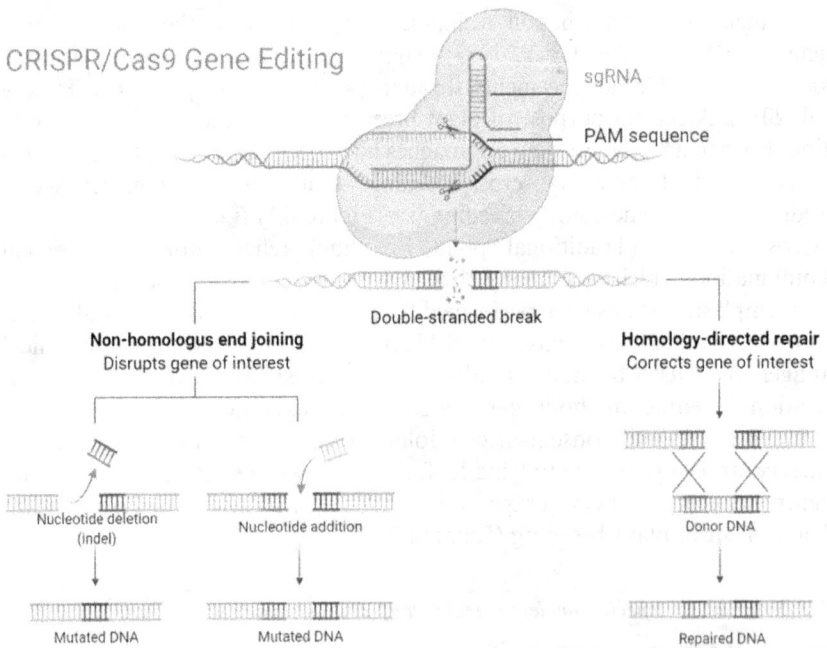

Figure 6.2 Schematic representation of types of double-strand breaks (DSBs) by CRISPR/Cas9 system.

6.2.3 CRISPR/CAS: The luminary gene-editing technology

CRISPR (or, more precisely, CRISPR-Cas9) is a revolutionary, cutting-edge biotechnological tool already used in many crop traits. Gene editing has become elementary and painless with the CRISPR system (Biswas et al. 2021). Nowadays, scientists can rewrite the genetic code in almost any organism effortlessly and inexpensively in days rather than weeks or months.

Today, CRISPR is well-known as a multifunctional genome-editing tool that is highly accurate for targeting and achieving site-specific genome modification than previous gene-editing techniques (ZENs and TALENs) (Javed et al. 2018). Scientists have taken many years to figure out what it was and how to harness its potential. CRISPR was first recognized as an adaptive immune mechanism used by bacteria and archaea to protect themselves against invading viruses by recording and targeting their DNA sequences (Lander, 2016; Mojica et al. 2000, 1993). So, the hazy interface between prokaryotes and viruses leads to a wide diversity of CRISPR-Cas systems. Jennifer and Emmanuel published a groundbreaking paper in 2012, revealing that CRISPR-Cas9 has been repurposed as a gene-editing tool (Jinek et al. 2012). It works as a precise couple of molecular scissors capable of cutting a specific DNA sequence while guided by a customizable guide. Since the emergence of the initial CRISPR/Cas-based system, CRISPR-Cas has quickly evolved into a powerful, versatile genome-editing tool with a wide range of methods and approaches. These include CRISPR interference (CRISPRi) and CRISPR activator (CRISPRa) gene regulators, as well as development as a base editor, prime editor, and epigenetic editor (see Figure 6.3) (Li et al. 2021). All these applications have been groundbreaking

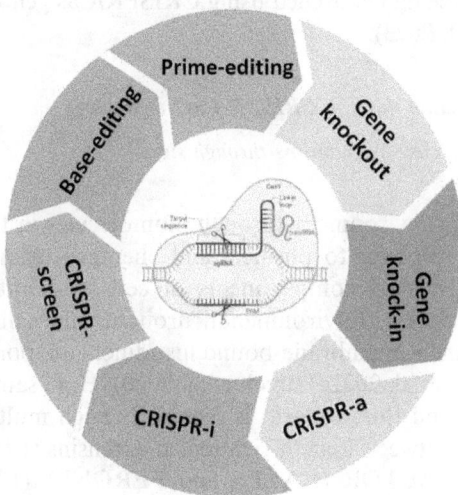

Figure 6.3 Applications of CRISPR technology for crop improvements.

in a variety of biological areas, including biomedical and agricultural progress. Moreover, these approaches have many real-world applications for greatly enhancing tolerance to various environmental stress.

6.3 Functional Investigations of biotic and abiotic stress-related genes

6.3.1 *Plant stressors, major types, and importance*

During the past 150 years, humans have made a deep-rooted effect on the biota and ecosystems of Earth. These effects have resulted in extreme loss of habitat, over-exploitation, pollution of different kinds, the introduction of invasive parasitic species, and, most notably, climate change (Arias et al. 2021). This has consequently resulted in global warming, flooding, widespread droughts, and different biotic stresses such as pathogens, insects, and parasitic weeds, resulting in heavy losses to agricultural produce and threatening global food security (Zandalinas et al. 2021; Lesk et al. 2016). Therefore, it is not surprising to see a number of programs and research focusing on developing superior tolerant plant cultivars. However, it's been reported that plants naturally respond uniquely to single or simultaneous combinations of stresses (Mittler, 2006; Atkinson and Urwin, 2012). Hence, breeding tolerance for single stress may come at the expense of tolerance to another stress. With climate change increasing at an alarming rate, more weather extremes are to be faced, hence increasing the likelihood of experiencing multiple stresses by the crops and stress from biotic diseases (Zandalinas et al. 2022). Therefore, there needs to be a clear understanding of the types of biotic and biotic stresses faced by plants and their similarities and differences regarding response pathways and co-expression to target optimal crop improvement traits. Here we will only focus on abiotic stress caused by salinity, water deficit, and temperature extremes, as they are the major stress factors being researched using CRISPR/Cas genome-editing system (Nascimento et al. 2023).

6.3.2 *Abiotic plant stress and mitigation through CRISPR-Cas9 technology*

6.3.2.1 *Natural plant sensing and defense mechanisms through signal transduction*

Plants can sense changes to their environment through biomolecules in their cells, which triggers a cellular response to physical and chemical changes. Abiotic stresses are sensed in various compartments of cells and initiate molecular changes at different levels. Environmental drought and salinity stress are signaled by Ca^{2+} channels, membrane-bound histidine, aquaporins, and receptor-like kinases (Buoso et al. 2022; Pareek et al. 2020). High salinity disrupts cell wall organization, and this scenario is sensed through multiple routes, including an interaction between Leucin-rich repeat extensins (LRX), RAPID ALKALINIZATION FACTOR (RALF), and FERONIA (FER) (collectively called LRX-RALF-FER regulatory module) (Zhao et al. 2018; Feng et al. 2018). This effect subsequently results in high Cytosolic Ca^{2+}

spikes that signal activation of the SOS3-SOS2-SOS1 pathway, which exports Na+ from roots to soil and maintains homeostasis. Another route through which plants signal hyperosmotic environments is by induction of cell membrane depolarization through the binding of Na+ cations (Jiang et al. 2019). The changes in environmental Na+ concentration generate a salt-dependent intracellular increase in Ca^{2+} through Ca^{2+} transporters (Jiang et al. 2019). In real-time, through modern genetic resources and techniques, researchers are looking for answers to the osmosensing systems of plants through the lens of calcium signatures (Khattar et al. 2022; Wolf, 2022).

Heat and cold stress alter the natural fluidity of phospholipid membranes in plant cellular membranes (Sangwan et al. 2002). Plants sense this alteration, possibly through Ca^{2+} signaling and protein phosphorylation. For instance, when plants are exposed to cold stress, Ca^{2+} ion spikes are induced through Ca^{2+} transporters (Cui et al. 2020). Ca^{2+} signaling induced by cold-sensing activates the MEKK1-MKK2-MPK4 module through protein-protein interactions (Li et al. 2017). Multiple mechanisms in all cellular compartments possibly achieve alleviated heat stress sensing. Heat stress results in protein denaturation and causes the accumulation of misfolded proteins that are then sensed by specific heat shock proteins (HSPs). At average temperatures, HSPs bind to heat stress transcription factors (HSFs), preventing their expression. When heat stress is present, misfolded proteins accumulate, thus, binding to HSPs, then releasing the expression of HSFs into cells and activating heat stress responses. High temperature changes the expression of genes during the transcription/translation process resulting in the generation of heat shock proteins (HSPs) (Khan and Shahwar, 2020). Plants cannot move and are hence equipped with sophisticated mechanisms to sense environmental changes without sacrificing energy. Many studies are conducted on abiotic stress-sensing mechanisms in plants at both cell surfaces and intracellular compartments. These can help us narrow down targets for the preferred response of a given stress condition or a preferred compartment-specific response.

In nature, plants are developed with stress responses in various ways. When plants are exposed to abiotic stresses, they most commonly react by increasing the production of reactive oxygen scavenging activities (ROS), stomatal closure, reduced photosynthesis process, an increase of root length, and decreased leaf growth (Maiti and Satya, 2014). Plants respond to biotic stresses by reduced photosynthesis process, the closing of stomata, the production of toxic secondary compounds, and the induction of localized cell death (Wojtaszek, 1997). Plant phytohormones such as abscisic acid (ABA), jasmonic acid (JA), salicylic acid (SA), and ethylene play a significant role in harmonizing these processes. ABA and JA are responsible for regulating tolerance to abiotic stresses, whereas SA, JA, and ethylene primarily participate in signaling during the pathogenic attack.

6.3.3 Abiotic stress and current GE studies

6.3.3.1 Salt

The term salinity refers to the presence of elevated levels of salts resulting in retarded growth rate and decreased productivity in crops. Saline soils are primarily composed of salts like sodium chloride, calcium sulfate, magnesium sulfates, magnesium bicarbonates, and calcium bicarbonates (Alexandre Bosco de et al. 2013). Salinity or osmotic stress is one of the most difficult challenges for agriculture production in this century. It is estimated that more than 20% of arable lands worldwide are affected by soil salinity. It is also estimated that by the year 2050, due to alarming changes in global climate, almost 50% of arable lands will be under salinization (Machado and Serralheiro, 2017). Plants in nature are developed with endogenous mechanisms to cope with salinity stress. To name a few, it includes accumulation and exclusion of ions from cells, changes in photosynthetic pathways, induction of enzymatic and non-enzymatic antioxidant and plant hormone, decreases in stomatal conductance, and reduction of water potential in leaf and internal CO_2 concentration (Chauhan et al. 2022).

Genome editing using the CRISPR/Cas system has been employed to produce mutant lines tolerant to saline soils or lines with alleviated sensitivity to salt to understand the underlining gene expression mechanisms. For instance, in *Cucurbita moschata,* the knockout of *CmoPIP,* an aquaporin protein, produced lines with hypersensitivity to salt stress displayed by weakened growth and development at both physiological and molecular levels (Sohail et al. 2022). Liu et al. (2022) reported hypersensitive rice mutants when *OsGLYI* genes were silenced; they concluded that expression of this gene is essential for salt tolerance in rice seeds. Loss of function of *OsNAC45,* a transcription factor related to ABA signaling, resulted in more resistance to NaCl through repression of ABA and co-expression of other salt resistance genes like *OsDREB1F, OsEREBP2*, and *OsSIK1* (Zhang et al. 2020). In tomatoes, *SlABIG1* knockout lines were reported to be significantly more tolerant to a saline environment than WT lines. This result was achieved through the decreased accumulation of ROS, MDA, and sodium ions in cells (Ding et al. 2022). *SlACO2* knockout mutants also showed significant tolerance to salt but at the cost of decreased pollen viability (Secgin et al. 2022). In another study on tomatoes, the *SlHyPRP1* gene was shown to regulate salt stress response negatively. Through CRISPR/Cas9 genome editing, higher salinity tolerance mutant lines were produced at both germination and vegetative stages (Tran et al. 2021).

6.3.3.2 Drought

Variations in precipitation patterns are another impact of climate change that affects plant growth and development. The ongoing threat of drought has significantly impacted global food production in recent years. To address this challenge, numerous studies have been conducted to better understand the mechanisms of plant drought tolerance and the potential use of the CRISPR/Cas system to enhance these processes. It has been observed that drought stress, similar to salinity stress, leads to hyperosmotic stress in plant cells.

Scientists have found that mutating *Peroxidase 63-like* in tobacco (*NtPOD63*) CRISPR performed significantly better than wild-type (WT) under drought stress conditions (Xu et al. 2023). This result was linked with the gene functioning as an up regulator of the ABA pathway and other stress-related genes and the closure of stomata. Another group of scientists led by Ying, Scheible (Ying et al. 2022) reported that knockout mutants of the *BdRFS* gene using the CRISPR/Cas9 mutation system exhibited sensitivity to drought instead of their overexpression. In *Arabidopsis*, the knockout of the *trehalase* (*TRE1*) gene resulted in a higher level of trehalose, which saw more robust drought tolerance than WT lines (Nuñez-Muñoz et al. 2021). Employing CRISPR/Cas system to target the ethylene receptor signaling genes is common to produce tolerant plants to abiotic stress. For instance, the *ARGOS8* gene in maize, a negative regulator of ethylene responses at early stages, was modified to enhance its expression using CRISPR/Cas breeding (Shi et al. 2017). In short, this was achieved through the swap of the endogenous promoter controlling *ARGOS8* to a more constitutive endogenous promoter (*GOS2* promoter) by the HDR repair system. *RGA* genes are negative regulators of GA signaling, and their expression is enhanced in the presence of drought and ABA. In *Brassica napus*, the loss-of-function of this gene by CRISPR/Cas9 showed enhanced water deficit stress tolerance and less sensitivity to ABA treatments. In rice, targeting two sites in both *SRL1* and *SRL2* genes produced mutant lines with rolled leaf structure in addition to the deformed plant structure. However, mutant lines were recorded to perform better in drought stress with higher ABA expression and antioxidant activities (Liao et al. 2019). *ENHANCED RESPONSE TO ABA1* (*ERA1*) genes were targeted in rice plants using CRISPR/Cas9 to produce frame-shift mutations. This resulted in increased tolerance to drought and ABA signaling (Ogata et al. 2020). Liu, Zhang (Liu et al. 2020) found that mutation of *LATERAL ORGAN BOUNDARIES DOMAIN* (*LBD*) in tomato plants is linked to a Jasmonic acid (JA) signaling pathway through targeted CRISPR/Cas9-mediated mutation resulting in drought tolerance plants. This phenomenon was further confirmed by the sensitivity of the over-expressing line to drought stress. Using multiplexed CRISPR/Cas9 knockout system, five homologs of *TaSal1* in hexaploidy wheat were targeted simultaneously. The resulting mutant lines had rolled leaf phenotype and closed elongated stomata, increasing seed germination in drought environments (Abdallah et al. 2022). In cotton plants, it was reported that knockout of *GhHB12* genes conferred higher tolerance to drought by increased sensitivity to ABA (He et al. 2020).

6.3.3.3 Temperature extremes

Temperature extremes cause heat or cold stress and are becoming severe agricultural problems worldwide. The primary reason for this phenomenon can be pointed to global warming and climate change. When plant cells face temperature stress, the reaction can be seen indirectly as poor seed germination rate, sub-optimal growth, rescued photosynthesis rate, loss of floret fertility, and increased leaf senescence. Cold stress in plants is given limited attention compared to heat stress. Low temperatures cause seed imbibition, slowed seed germination rate, damaged cell membrane, less growth, and low yield of crops (Zhang et al. 2019). Scientists are making efforts to improve the understanding of mechanisms of physiological and molecular traits that lead to tolerance to temperature extremes, for instance, identification of heat tolerance pathways, chilling tolerant genotypes, and their use in genomic breeding through CRISPR systems (Posch et al. 2019; Li et al. 2022). Low-temperature stress, based on their physiological characteristics, can be divided into two levels; chilling stress (0–18°C) faced by tropical and sub-tropical plants and freezing stress (<0°C) faced by temperate plants (Thomashow, 1999). Using CRISPR/Cas9 systems, Zeng, Wen (Zeng et al. 2019) and colleagues reported simultaneously targeted mutation of three genes related to cold sensitivity, including *OsPIN5b*, *GS3,* and *OsMYB30*. By simultaneously targeting these genes, they found that rice yielded significantly better than WT in double and triple mutants. In tomato plants, to better understand genetic regulators of cold sensitivity, *Slcbf1* mutant lines were generated using CRISPR/Cas9; it was reported that mutant tomato lines showed increased cold sensitivity by increased contents of malondialdehyde (MDA) and electrolyte leakage indicating sensitivity to cold-induced membrane damage (Li et al. 2018). In rice, knockout of *OsCNGC9, OsSAPK8,* and *OsDREB1A* conferred more sensitivity to cold stress. Based on this study, the relation and interaction of these genes in cold stress-induced triggering Ca^{2+} spike mechanism and the activation of cold stress-related genes were further demonstrated (Wang et al. 2021).

Effects of heat stress can also be observed directly as damaged membrane, denaturation of protein, a spike in ROS, and induction of metabolic disturbance that can further lead to a decrease in plant growth and development, in addition to reduced productivity and crop yield. To maintain plant integrity and protect plants from heat-stress-mediated effects, studies have found that ROS-scavenging enzymes like SOD, APX, CAT, Prx, and GR play an essential role in maintaining proper redox status (Mittler, 2006). Tomato plants, like other horticultural crops, are heat sensitive. *BZR1* gene, a plant signaling gene to heat stress, was mutated in tomato plants, and the study demonstrated a retarded growth and higher sensitivity to heat stress compared to WT (Yin et al. 2018). Another study on tomatoes studied *MAPK* genes' role in regulating heat-induced plant response using CRISPR/Cas9 mutation system (Yu et al. 2019). The loss-of-function of *SlMAPK3,* a serine-threonine protein kinase, conferred tolerance phenology to heat stress at 42°C. This result suggests the role of this gene as a negative regulator of heat stress response in tomato plants by regulating levels of ROS homeostasis. *MULTIPROTEIN BRIDGING FACTOR* 1 (*MBF1*) is a transcriptional coactivator mediating the transcription activation and regulating plant developmental processes. Its function was shown to handle abiotic stress, including heat stress. As reported by Tian et al. (2022), *TAMBF1,* which is highly conserved to wheat species, contributed positively to heat stress tolerance. Knockout (KO) mutant lines were shown to perform poorly under heat stress at 42°C, partially due to their role in regulating heat-responsive genes. Transcription factors (TFs) have been shown to play an essential role in controlling abiotic stress responses during plant growth; this includes heat stress. *OsNAC006a* TF in rice was mutated using the CRISPR/Cas9 system; this resulted in rice plants with decreased antioxidative enzyme activity (POD and SOD) and higher levels of MDA, indicating membrane damage. This ultimately resulted in sensitive mutant lines to heat stress in rice plants and showed the importance of *NAC* TF in regulating plant growth and their function in stress tolerance regulation (Wang et al. 2020).

6.3.4 Biotic plant stress and acquisition of resistance through CRISPR-Cas genome editing

Biotic stress is an adverse condition resulting from the interaction of plants with deleterious microorganisms such as viruses, bacteria, fungi, nematodes, insects and arachnids, and weeds, which are a significant constraint on agricultural productivity (20% to 40% of global agricultural productivity) (Sapre et al. 2021; Savary et al. 2012). Infections caused by a combination of fungus, bacteria, and viruses are more prevalent and induce severe disease symptoms than infections caused by individual pathogens (Pandey et al. 2017). However, only a small percentage of these attacks and invasions succeed and cause disease. Plants have evolved sophisticated mechanisms in order to overcome these threats of biotic stress (Hammond-Kosack and Jones, 1997). The first way is known as physical barriers, including waxy or thick cuticles and specific trichomes that restrict insects or diseases from settling, entering plant surfaces, and effectively invading plants. The second approach to disease control depends on resistance (R) genes (resistant varieties) (Kaloshian, 2004). Resistance genes induce various signal transduction pathways, which lead to differential transcriptional changes making the plant tolerant to stress. Intriguingly, many characteristics are shared by the proteins expressed by R genes from various plants against different infections (Audil et al. 2019).

The complex interaction between a host plant and a pathogen plays a vital role in disease, and the resistance/susceptibility response can affect different components. In plants, several identified genes may change the interaction and increase the possibility of inhibiting specific steps in the mechanism of infection. Fortunately, several contemporary gene-editing technologies have advanced quickly and demonstrated promise in providing plant disease resistance. The following part focuses on the effectiveness of CRISPR/Cas9 technology for enhancing plants in response to biotic stresses.

6.3.4.1 Bacteria

Only a few hundred bacterial species are involved in crop damage, frequently exhibiting multiple disease symptoms (Schloss and Handelsman, 2004). The CRISPR knockout of *OsSWEET13* is the most compelling example of using the CRISPR tool. Bacterial resistance in rice crops was conducted to fight bacterial blight disease induced by –proteobacterium caused by the pathogen *Xanthomonas oryzae* pv. *Oryzae (*Zhou et al. 2015; Blanvillain-Baufumé et al. 2017). Two new studies have shown the use of CRISPR/Cas9 to develop citrus plants resistant to citrus bacterial canker (CBC) (Jia et al. 2016). *Pseudomonas syringae* pv. *lycopersicum* causes bacterial speck disease in tomato plants (*Solanum lycopersicum*). Tomato DC 3000 produces coronatine, which, in combination with the co-receptor *JAZ2*, causes stomata to open and permits bacteria to colonize the leaf. *SlJAZ2* was edited using CRISPR/Cas9 to create a dominant allele expressing a version of *JAZ2* lacking the C-terminal Jas domain, which limits stomatal closure and offers resistance to bacterial speck disease (Ortigosa et al. 2019).

6.3.4.2 Fungi

Fungi cause many diseases; around 30% of the emerging plant diseases include mildew, smut, rust, and rot. These diseases cause significant crop losses worldwide each year and affect the quality of the harvested items (Borrelli et al. 2018). RNA-guided Cas9 endonuclease has targeted and disrupted *MLO* loci in bread wheat, tomato, and grapevine (Wang et al. 2014; Malnoy et al. 2016; Nekrasov et al. 2017). *MLO* was reported as susceptibility (S) genes, with homozygous loss-of-function mutants, significantly boosting resistance to powdery mildew in barley, *Arabidopsis*, and tomato (Piffanelli et al. 2004; Consonni et al. 2006). Bread wheat plants modified by CRISPR/Cas9 in one of three *MLO* homeoalleles (TaMLO-A1) showed increased resistance to *Blumeria graminis* f. sp. tritici infection (Wang et al. 2014). *SlMlo1*, previously recognized as the most significant of 16 *SlMlo* genes, was targeted at two places in tomato, yielding a 48-bp deletion. The modified plants were self-pollinated to produce CRISPR/Cas cassette-free individuals. "Tomelo," a novel non-transgenic cultivar, proved completely resistant to *Oidium neolycopersici*. Furthermore, off-target research revealed no influence on genomic areas other than the *SlMlo1* gene (Nekrasov et al. 2017). CRISPR/Cas9 has also been used to target *Edr1*, which encodes a Raf-like mitogen-activated protein, resulting in a significant decrease in powdery mildew in wheat (Wang et al. 2014). These findings show that the CRISPR/Cas9 technology may be used effectively and efficiently to boost crop resistance to fungal diseases.

6.3.4.3 *Viruses*

Many economically important staple and specialty crops are under threat from plant viruses. Plant viruses are typically handled in two ways: (1) to address the virus genome directly or (2) directly target plant S genes critical for viral disease development (Zaidi et al. 2018). The initial method of engineering viral resistance directly targeted the virus genome inside plant cells (Zaidi et al. 2016). Recently, improving CRISPR-edited plants for virus resistance by targeting ssDNA geminivirus genomes is also a popular approach (Borrelli et al. 2018; Ali et al. 2015). *Geminiviridae* is a significant plant virus family that causes widespread agricultural losses in numerous vital groups, including *Cucurbitaceae, Euphorbiaceae, Solanaceae, Malvaceae,* and *Fabaceae* (Zaidi et al. 2016). Gilbertson illustrated that the most important genus of geminiviruses is Begomovirus, which infects several dicotyledonous plants and is mainly associated with infected plants' phloem (Gilbertson et al. 2015). To reduce crop losses due to geminivirus infection, Baltes, Hummel (Baltes et al. 2015) targeted the bean yellow dwarf virus (*BeYDV*) genome for destruction with the CRISPR–Cas system. This research demonstrated a novel strategy for engineering resistance to geminiviruses. Transgenic plants expressing CRISPR–Cas reagents and challenged with *BeYDV* had reduced virus accumulation in inoculated leaves at varying levels. Furthermore, transient tests in *Nicotiana benthamiana* employing the geminivirus, beet severe curly top virus (*BSCTV*), show that the sgRNA-Cas9 constructs suppress virus accumulation and introduce mutations at the target sites. Transgenic *Arabidopsis* and *Nicotiana benthamiana* plants that express sgRNA-Cas9 are also extremely resistant to viral infection (Ji et al. 2015). Because the traditional SpCas9 from *Streptococcus pyogenes* exclusively detects dsDNA, protection against RNA viruses appeared to be more challenging. However, searching for and characterization of comparable nucleases has resulted in finding enzymes that can bind to and cut RNA, such as *FnCas9* from *Francisella novicida* and *LwaCas13a* from *Leptotrichia wadei* (Zhang et al. 2018).

6.4 Limitations

6.4.1 *The accuracy of genome editing*

Despite recent breakthroughs in plant genome-editing technology, making all necessary alterations in a genome is still not practicable. In order to improve the crop's traits, more precise genome editing is urgently needed, such as gene indels, gene replacements, and targeting base substitutions. In theory, Homology Directed repair (HDR) mediated genome editing might rewrite any genome and precisely create the desired modifications. Several studies have been used to increase HDR reliability in plant cells (Baltes et al. 2014; Fauser et al. 2012; Ali et al. 2020). To circumvent these constraints, the recently developed base editing and prime editing systems could make all needed genome editing (Chen et al. 2019; Kurt et al. 2021; Zhao et al. 2020).

6.4.2 *Off-target effects*

Genome editing using CRISPR/Cas system still suffers from the off-target effects creating deleterious phenotypic effects in plant species and hence hindering the widespread of the system. To resolve this issue, sgRNA selection with less predicted off-target mutation is selected based on the robust reference genome sequence. As defined by Moreb, Hutmacher (Moreb et al. 2020), an off-target site in CRISPR/Cas system is all sites other than the targeted site, possessing an experimental method by which Cas RNP can interact by transient or permanent binding. This effect can change gene function, which would cause the likelihood of the production of undesirable changes in the expression of specific genes in plants. Theoretically, in CRISPR/Cas genome-editing system, an off-target is caused when multiple undesired mutations are introduced into different positions on the genome by degenerated gRNAs which can edit plant DNA at the multitude of proto-spacers in the DNA of plants, similar to a degenerated primers that show non-specific annealing on unintended sites (Gerashchenkov et al. 2020) (see Figure 6.4).

Numerous researchers are making efforts to minimize the off-target cleavage of gRNA and to maximize its efficiency (Alkan et al. 2018; Hiranniramol et al. 2020; Secgin et al. 2021; Aksoy et al. 2022). These are necessary to protect human health and the environment without compromising the advantages this system provides to humanity (Chapman et al. 2017). However, off-target effects can also be considered to improve agriculturally essential traits by introducing mild editing of individual nitrogenous bases. However, this approach will be regarded as more laborious and costly than using known mutagens like chemicals or radiation on plant seeds (Gerashchenkov et al. 2020). Hence, it is always necessary to pay utmost attention while designing gRNA, which will circumvent all off-target cleavage.

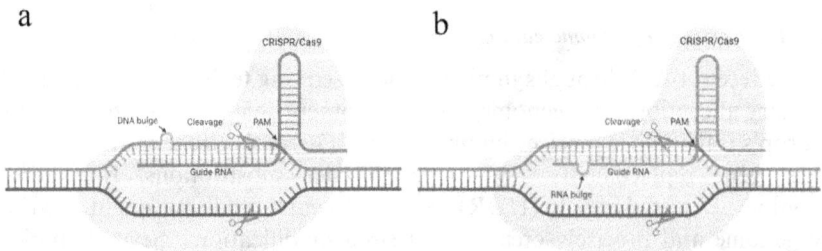

Figure 6.4 Schematic diagram illustrating different mechanisms of off-target effect in CRISPR/Cas9 system with the formation of R loop (bulge) between the gRNA and the DNA loci (a) or the sgRNA (b) potential target site as a result introducing a mismatch recognition.

6.4.3 *Improving the delivery method and regeneration conditions*

The delivery of editing reagents to plant cells and subsequent plant regeneration of editing cells are critical processes in genome editing. CRISPR/Cas reagents can be delivered to plants using several methods: *Agrobacterium*-mediated transformation, biolistic (particle bombardment), polyethylene glycol (PEG), microinjection, electroporation, and nanoparticles. The method chosen depends on the specific plant species, the size of the genome-editing construct, and the desired transformation efficiency. Despite the diversity of delivery methods, tissue culture-free, genotype independent, and directly applied to particular tissues are still the ideal plant genome-editing reagent in plant delivery systems (Ellison et al. 2020).

6.5 Future prospects/Conclusion

Climate change is a global threat to crop growth and productivity, and almost every aspect of our life in the present time is affected by it. Changes in global temperature, population growth, and water scarcity are increasing at an alarming rate and are likely to worsen in the future. Besides political and cultural pressure to limit pesticide use, crop protection via genetic modification offers a promising option with no discernible impact on human health or the environment. Although CRISPR/Cas systems have significantly increased the accuracy and effectiveness of making modifications, there is undoubtedly still a role for other gene-editing technologies. Besides modern breeding methods, CRISPR-Cas technologies will play an essential role in crop improvement programs. Gene-editing techniques have created enormous possibilities for developing crops and animals that can better deal with the effects of climate change.

Even though many countries have been developing genome-edited crops for several years, only a few have clarified their opinion toward genome editing. In countries like Canada and the USA and developing countries such as Pakistan and India, genome-edited crops can quickly be passed through their regulatory procedures. However, New Zealand and Europe are strict with their old GMO regulatory procedures. In Turkey, the GMO regulation procedures are the same as in Europe. With the predictions that genome-edited products will no longer be recognized as GMOs soon in the United Kingdom after it leaves the European Union (EU). Nowadays, the EU court is planning to change current rules so that gene-edited plants using the CRISPR system are treated differently than GMOs.

References

Abdallah, N.A., et al., *Multiplex CRISPR/Cas9-mediated genome editing to address drought tolerance in wheat.* GM Crops & Food, 2022: p. 1–17.

Agarwal, S. and A. Grover, *Molecular biology, biotechnology and genomics of flooding-associated low O2 stress response in plants.* Critical Reviews in Plant Sciences, 2006. **25**(1): p. 1–21.

Aksoy, E., et al., *General guidelines for CRISPR/Cas-based genome editing in plants.* Molecular Biology Reports, 2022. **49**(12): p. 12151–12164.

Alexandre Bosco de, O., A. Nara Lidia Mendes, and G.-F. Enéas, *Comparison Between the Water and Salt Stress Effects on Plant Growth and Development*, in *Responses of Organisms to Water Stress*, A. Sener, Editor. 2013, IntechOpen: Rijeka. p. Ch. 4.

Ali, Z., et al., *CRISPR/Cas9-mediated viral interference in plants.* Genome Biology, 2015. **16**(1): p. 238.

Ali, Z., et al., *Fusion of the Cas9 endonuclease and the VirD2 relaxase facilitates homology-directed repair for precise genome engineering in rice.* Communications Biology, 2020. **3**(1): p. 44.

Alkan, F., et al., *CRISPR-Cas9 off-targeting assessment with nucleic acid duplex energy parameters.* Genome Biology, 2018. **19**(1): p. 177.

Araki, M. and T. Ishii, *Towards social acceptance of plant breeding by genome editing.* Trends in Plant Science, 2015. **20**(3): p. 145–149.

Arias, P., et al., *Climate Change 2021: The Physical Science Basis. Contribution of Working Group14 I to the Sixth Assessment Report of the Intergovernmental Panel on Climate Change; Technical Summary.* 2021.

Atkinson, N.J. and P.E. Urwin, *The interaction of plant biotic and abiotic stresses: from genes to the field.* Journal of Experimental Botany, 2012. **63**(10): p. 3523–3543.

Audil, G., L. Ajaz Ahmad, and W. Noor Ul Islam, *Biotic and Abiotic Stresses in Plants*, in *Abiotic and Biotic Stress in Plants*, O. Alexandre Bosco de, Editor. 2019, IntechOpen: Rijeka. p. Ch. 1.

Bailey-Serres, J. and L.A. Voesenek, *Flooding stress: acclimations and genetic diversity.* Annual Review of Plant Biology, 2008. **59**: p. 313–339.

Baltes, N.J., et al., *DNA replicons for plant genome engineering.* The Plant Cell, 2014. **26**(1): p. 151–163.

Baltes, N.J., et al., *Conferring resistance to geminiviruses with the CRISPR–Cas prokaryotic immune system.* Nature Plants, 2015. **1**(10): p. 15145.

Barrangou, R. and J.A. Doudna, *Applications of CRISPR technologies in research and beyond.* Nature Biotechnology, 2016. **34**(9): p. 933–941.

Biswas, S., D. Zhang, and J. Shi, *CRISPR/Cas systems: opportunities and challenges for crop breeding.* Plant Cell Reports, 2021. **40**(6): p. 979–998.

Blanvillain-Baufumé, S., et al., *Targeted promoter editing for rice resistance to Xanthomonas oryzae pv. oryzae reveals differential activities for SWEET14-inducing TAL effectors.* Plant Biotechnology Journal, 2017. **15**(3): p. 306–317.

Borrelli, V.M.G., et al., *The enhancement of plant disease resistance using CRISPR/Cas9 technology.* Frontiers in Plant Science, 2018. **9**.

Buoso, S., et al., *Infection by phloem-limited phytoplasma affects mineral nutrient homeostasis in tomato leaf tissues.* Journal of Plant Physiology, 2022. **271**: p. 153659.

Chapman, J.E., D. Gillum, and S.J.A.B. Kiani, *Approaches to reduce CRISPR off-target effects for safer genome editing.* 2017. **22**(1): p. 7–13.

Chauhan, P.K., et al., *Understanding the salinity stress on plant and developing sustainable management strategies mediated salt-tolerant plant growth-promoting rhizobacteria and CRISPR/Cas9.* Biotechnology and Genetic Engineering Reviews, 2022: p. 1–37.

Chaves, M.M. and M.M. Oliveira, *Mechanisms underlying plant resilience to water deficits: prospects for water-saving agriculture.* Journal of Experimental Botany, 2004. **55**(407): p. 2365–2384.

Chen, K., et al., *CRISPR/Cas genome editing and precision plant breeding in agriculture*. Annual Review of Plant Biology, 2019. **70**(1): p. 667–697.

Chinnusamy, V. and J.-K. Zhu, *Epigenetic regulation of stress responses in plants*. Current Opinion in Plant Biology, 2009. **12**(2): p. 133–139.

Consonni, C., et al., *Conserved requirement for a plant host cell protein in powdery mildew pathogenesis*. Nature Genetics, 2006. **38**(6): p. 716–720.

Cui, Y., et al., *Cyclic nucleotide-gated ion channels 14 and 16 promote tolerance to heat and chilling in rice*. Plant Physiology, 2020. **183**(4): p. 1794–1808.

Ding, F., et al., *Knockout of a novel salt responsive gene SlABIG1 enhance salinity tolerance in tomato*. Environmental and Experimental Botany, 2022. **200**: p. 104903.

Ellison, E.E., et al., *Multiplexed heritable gene editing using RNA viruses and mobile single guide RNAs*. Nature Plants, 2020. **6**(6): p. 620–624.

Fauser, F., et al., *In plantaIn planta gene targeting*. Proceedings of the National Academy of Sciences, 2012. **109**(19): p. 7535–7540.

Feng, W., et al., *The FERONIA receptor kinase maintains cell-wall integrity during salt stress through Ca2+ signaling*. Current Biology, 2018. **28**(5): p. 666–675.e5.

Gao, C., *Genome engineering for crop improvement and future agriculture*. Cell, 2021. **184**(6): p. 1621–1635.

Gerashchenkov, G., et al., *Design of guideRNA for CRISPR/Cas plant genome editing*. 2020. **54**(1): p. 24–42.

Gilbertson, R.L., et al., *role of the insect supervectors Bemisia tabaci and Frankliniella occidentalis in the emergence and global spread of plant viruses*. Annual Review of Virology, 2015. **2**(1): p. 67–93.

Govindan, G. and S. Ramalingam, *Programmable site-specific nucleases for targeted genome engineering in higher eukaryotes*. Journal of Cellular Physiology, 2016. **231**(11): p. 2380–2392.

Hammond-Kosack, K.E. and J.D. Jones, *Plant disease resistance genes*. Annual Review of Plant Biology, 1997. **48**(1): p. 575–607.

Hartung, F. and J. Schiemann, *Precise plant breeding using new genome editing techniques: opportunities, safety and regulation in the EU*. The Plant Journal, 2014. **78**(5): p. 742–752.

He, X., et al., *GhHB12 negatively regulates abiotic stress tolerance in Arabidopsis and cotton*. Environmental and Experimental Botany, 2020. **176**: p. 104087.

Hickey, L.T., et al., *Breeding crops to feed 10 billion*. Nature Biotechnology, 2019. **37**(7): p. 744–754.

Hiranniramol, K., et al., *Generalizable sgRNA design for improved CRISPR/Cas9 editing efficiency*. Bioinformatics, 2020. **36**(9): p. 2684–2689.

Hirel, B., et al., *The challenge of improving nitrogen use efficiency in crop plants: towards a more central role for genetic variability and quantitative genetics within integrated approaches*. Journal of Experimental Botany, 2007. **58**(9): p. 2369–2387.

Holme, I.B., P.L. Gregersen, and H. Brinch-Pedersen, *Induced genetic variation in crop plants by random or targeted mutagenesis: convergence and differences*. Frontiers in Plant Science, 2019. **10**.

Iqbal, Z., et al., *Plant defense responses to biotic stress and its interplay with fluctuating dark/light conditions*. Frontiers in Plant Science, 2021. **12**.

Javed, M.R., et al., *CRISPR-Cas system: history and prospects as a genome editing tool in microorganisms*. Current Microbiology, 2018. **75**(12): p. 1675–1683.

Ji, X., et al., *Establishing a CRISPR–Cas-like immune system conferring DNA virus resistance in plants.* Nature Plants, 2015. **1**(10): p. 15144.

Jia, H., et al., *modification of the PthA4 effector binding elements in Type I CsLOB1 promoter using Cas9/sgRNA to produce transgenic Duncan grapefruit alleviating XccΔpthA4:dCsLOB1.3 infection.* Plant Biotechnology Journal, 2016. **14**(5): p. 1291–1301.

Jiang, Z., et al., *Plant cell-surface GIPC sphingolipids sense salt to trigger Ca2+ influx.* Nature, 2019. **572**(7769): p. 341–346.

Jinek, M., et al., *A programmable dual-RNA–Guided DNA endonuclease in adaptive bacterial immunity.* Science, 2012. **337**(6096): p. 816–821.

Jung, C., et al., *Recent developments in genome editing and applications in plant breeding.* Plant Breeding, 2018. **137**(1): p. 1–9.

Kaloshian, I., *Gene-for-gene disease resistance: bridging insect pest and pathogen defense.* Journal of Chemical Ecology, 2004. **30**(12): p. 2419–2438.

Khan, Z., and Shahwar, D., *Role of heat shock proteins (HSPs) and heat stress tolerance in crop plants.* Sustainable agriculture in the era of climate change, 2020. p. 211–234.

Khattar, V., L. Wang, and J.-B. Peng, *Calcium selective channel TRPV6: structure, function, and implications in health and disease.* Gene, 2022. **817**: p. 146192.

Kurt, I.C., et al., *CRISPR C-to-G base editors for inducing targeted DNA transversions in human cells.* Nature Biotechnology, 2021. **39**(1): p. 41–46.

Lander, Eric S., *The heroes of CRISPR.* Cell, 2016. **164**(1): p. 18–28.

Lesk, C., P. Rowhani, and N. Ramankutty, *Influence of extreme weather disasters on global crop production.* Nature, 2016. **529**(7584): p. 84–87.

Li, H., et al., *MPK3- and MPK6-mediated ICE1 phosphorylation negatively regulates ICE1 stability and freezing tolerance in arabidopsis.* Developmental Cell, 2017. **43**(5): p. 630–642.e4.

Li, R., et al., *Reduction of tomato-plant chilling tolerance by CRISPR–Cas9-mediated SlCBF1 mutagenesis.* Journal of Agricultural and Food Chemistry, 2018. **66**(34): p. 9042–9051.

Li, C., et al., *CRISPR/Cas: a Nobel Prize award-winning precise genome editing technology for gene therapy and crop improvement.* Journal of Zhejiang University Science B, 2021. **22**(4): p. 253–284.

Li, X., et al., *CRISPR/Cas9 technique for temperature, drought, and salinity stress responses.* Current Issues in Molecular Biology, 2022. **44**(6): p. 2664–2682.

Liao, S., et al., *CRISPR/Cas9-induced mutagenesis of semi-rolled leaf1,2 confers curled leaf phenotype and drought tolerance by influencing protein expression patterns and ROS scavenging in rice (Oryza sativa L.).* Agronomy, 2019. **9**(11): p. 728.

Liu, L., et al., *CRISPR/Cas9 targeted mutagenesis of SlLBD40, a lateral organ boundaries domain transcription factor, enhances drought tolerance in tomato.* Plant Science, 2020. **301**: p. 110683.

Liu, S., et al., *OsGLYI3, a glyoxalase gene expressed in rice seed, contributes to seed longevity and salt stress tolerance.* Plant Physiology and Biochemistry, 2022. **183**: p. 85–95.

Machado, R.M.A. and R.P. Serralheiro, *Soil salinity: effect on vegetable crop growth. management practices to prevent and mitigate soil salinization.* Horticulturae, 2017. **3**(2): p. 30.

Mahato, A., *Climate change and its impact on agriculture.* International Journal of Scientific and Research Publications, 2014. **4**(4): p. 1–6.

Maiti, R.K. and P. Satya, *Research advances in major cereal crops for adaptation to abiotic stresses.* GM Crops & Food, 2014. **5**(4): p. 259–279.

Malnoy, M., et al., *DNA-free genetically edited grapevine and apple protoplast using CRISPR/Cas9 ribonucleoproteins.* Frontiers in Plant Science, 2016. **7**.

Mao, Y., et al., *Gene editing in plants: progress and challenges.* National Science Review, 2019. **6**(3): p. 421–437.

Mertens, D., et al., *Predictability of biotic stress structures plant defence evolution.* Trends in Ecology & Evolution, 2021. **36**(5): p. 444–456.

Miki, D., et al., *Gene targeting facilitated by engineered sequence-specific nucleases: potential applications for crop improvement.* Plant and Cell Physiology, 2021. **62**(5): p. 752–765.

Mittler, R., *Abiotic stress, the field environment and stress combination.* Trends in Plant Science, 2006. **11**(1): p. 15–19.

Mojica, F.J., et al., *Biological significance of a family of regularly spaced repeats in the genomes of Archaea, Bacteria and mitochondria.* Molecular Microbiology, 2000. **36**(1): p. 244–246.

Mojica, F.J., G. Juez, and F. Rodríguez-Valera, *Transcription at different salinities of Haloferax mediterranei sequences adjacent to partially modified PstI sites.* Molecular Microbiology, 1993. **9**(3): p. 613–621.

Moreb, E.A., M. Hutmacher, and M.D. Lynch, *CRISPR-Cas "non-target" sites inhibit on-target cutting rates.* The CRISPR Journal, 2020. **3**(6): p. 550–561.

Munns, R. and M. Tester, *Mechanisms of salinity tolerance.* Annual Review of Plant Biology, 2008. **59**: p. 651–681.

Nascimento, F.d.S., et al., *Gene editing for plant resistance to abiotic factors: a systematic review.* Plants, 2023. **12**(2): p. 305.

Nekrasov, V., et al., *Rapid generation of a transgene-free powdery mildew resistant tomato by genome deletion.* Scientific Reports, 2017. **7**(1): p. 482.

Nogué, F., et al., *Genome engineering and plant breeding: impact on trait discovery and development.* Plant Cell Reports, 2016. **35**(7): p. 1475–1486.

Nunez, S., et al., *Assessing the impacts of climate change on biodiversity: is below 2 °C enough?* Climatic Change, 2019. **154**(3): p. 351–365.

Nuñez-Muñoz, L., et al., *Plant drought tolerance provided through genome editing of the trehalase gene.* Plant Signaling & Behavior, 2021. **16**(4): p. 1877005.

Ogata, T., et al., *CRISPR/Cas9-targeted mutagenesis of OsERA1 confers enhanced responses to abscisic acid and drought stress and increased primary root growth under nonstressed conditions in rice.* PLOS ONE, 2020. **15**(12): p. e0243376.

Ortigosa, A., et al., *design of a bacterial speck resistant tomato by CRISPR/Cas9-mediated editing of SlJAZ2.* Plant Biotechnology Journal, 2019. **17**(3): p. 665–673.

Pandey, P., et al., *Impact of combined abiotic and biotic stresses on plant growth and avenues for crop improvement by exploiting physio-morphological traits.* Frontiers in Plant Science, 2017. **8**.

Pareek, A., O.P. Dhankher, and C.H. Foyer, *Mitigating the impact of climate change on plant productivity and ecosystem sustainability.* Journal of Experimental Botany, 2020. **71**(2): p. 451–456.

Piffanelli, P., et al., *A barley cultivation-associated polymorphism conveys resistance to powdery mildew.* Nature, 2004. **430**(7002): p. 887–891.

Posch, B.C., et al., *Exploring high temperature responses of photosynthesis and respiration to improve heat tolerance in wheat*. Journal of Experimental Botany, 2019. **70**(19): p. 5051–5069.

Raman, R., *The impact of Genetically Modified (GM) crops in modern agriculture: a review*. GM Crops & Food, 2017. **8**(4): p. 195–208.

Redondo-Gómez, S., *Abiotic and Biotic Stress Tolerance in Plants*, in Molecular Stress Physiology of Plants, G.R. Rout and A.B. Das, Editors. 2013, Springer India: India. p. 1–20.

Sangwan, V., et al., *Opposite changes in membrane fluidity mimic cold and heat stress activation of distinct plant MAP kinase pathways*. The Plant Journal, 2002. **31**(5): p. 629–638.

Sapre, S., et al., *Chapter 20 - Molecular techniques used in plant disease diagnosis*, in *Food Security and Plant Disease Management*, A. Kumar and S. Droby, Editors. 2021, Woodhead Publishing. p. 405–421.

Savary, S., et al., *Crop losses due to diseases and their implications for global food production losses and food security*. Food Security, 2012. **4**(4): p. 519–537.

Schloss, P.D. and J. Handelsman, *Status of the microbial census*. Microbiology and Molecular Biology Reviews, 2004. **68**(4): p. 686–691.

Secgin, Z., et al., *Genome-wide identification of the aconitase gene family in tomato (Solanum lycopersicum) and CRISPR-based functional characterization of SlACO2 on male-sterility*. International Journal of Molecular Sciences, 2022. **23**(22): p. 13963.

Secgin, Z., M. Kavas, and K. Yildirim, *Optimization of Agrobacterium-mediated transformation and regeneration for CRISPR/Cas9 genome editing of commercial tomato cultivars*. Turkish Journal of Agriculture and Forestry, 2021. **45**(6): p. 704–716.

Shi, J., et al., *ARGOS8 variants generated by CRISPR-Cas9 improve maize grain yield under field drought stress conditions*. Plant Biotechnology Journal, 2017. **15**(2): p. 207–216.

Singla, J. and S.G. Krattinger, Biotic Stress Resistance Genes in Wheat, in *Encyclopedia of Food Grains (Second Edition)*, C. Wrigley, et al., Editors. 2016, Academic Press: Oxford. p. 388–392.

Sivanandhan, G., et al., *Targeted genome editing using site-specific nucleases, ZFNs, TALENs, and the CRISPR/Cas9 system Takashi Yamamoto (ed.)*. Annals of Botany, 2016. **118**(2): p. vii–viii.

Sohail, H., et al., *Genome-wide identification of plasma-membrane intrinsic proteins in pumpkin and functional characterization of CmoPIP1-4 under salinity stress*. Environmental and Experimental Botany, 2022. **202**: p. 104995.

Sun, Y., J. Li, and L. Xia, *Precise genome modification via sequence-specific nucleases-mediated gene targeting for crop improvement*. Frontiers in Plant Science, 2016. **7**.

Suzuki, N., et al., *Abiotic and biotic stress combinations*. New Phytologist, 2014. **203**(1): p. 32–43.

Thomashow, M.F., *Plant cold acclimation: freezing tolerance genes and regulatory mechanisms*. Annual Review of Plant Physiology and Plant Molecular Biology, 1999. **50**(1): p. 571–599.

Tian, X., et al., *Stress granule-associated TaMBF1c confers thermotolerance through regulating specific mRNA translation in wheat (Triticum aestivum)*. New Phytologist, 2022. **233**(4): p. 1719–1731.

Tran, M.T., et al., *CRISPR/Cas9-based precise excision of SlHyPRP1 domain(s) to obtain salt stress-tolerant tomato.* Plant Cell Reports, 2021. **40**(6): p. 999–1011.

Wang, Y., et al., *Simultaneous editing of three homoeoalleles in hexaploid bread wheat confers heritable resistance to powdery mildew.* Nature Biotechnology, 2014. **32**(9): p. 947–951.

Wang, B., et al., *Knockout of the OsNAC006 transcription factor causes drought and heat sensitivity in rice.* International Journal of Molecular Sciences, 2020. **21**(7): p. 2288.

Wang, J., et al., *Transcriptional activation and phosphorylation of OsCNGC9 confer enhanced chilling tolerance in rice.* Molecular Plant, 2021. **14**(2): p. 315–329.

Wang, W., B. Vinocur, and A. Altman, *Plant responses to drought, salinity and extreme temperatures: towards genetic engineering for stress tolerance.* Planta, 2003. **218**(1): p. 1–14.

Watson, J.D. and F.H.C. Crick, *Molecular structure of nucleic acids: a structure for deoxyribose nucleic acid.* Nature, 1953. **171**(4356): p. 737–738.

Wojtaszek, P., *Oxidative burst: an early plant response to pathogen infection.* Biochemical Journal, 1997. **322**(3): p. 681–692.

Wolf, S., *Cell wall signaling in plant development and defense.* Annual Review of Plant Biology, 2022. **73**(1): p. 323–353.

Xu, L., et al., *Physiological and phosphoproteomic analyses revealed that the NtPOD63 L knockout mutant enhances drought tolerance in tobacco.* Industrial Crops and Products, 2023. **193**: p. 116218.

Yin, Y., et al., *BZR1 transcription factor regulates heat stress tolerance through FERONIA receptor-like kinase-mediated reactive oxygen species signaling in tomato.* Plant and Cell Physiology, 2018. **59**(11): p. 2239–2254.

Ying, S., W.-R. Scheible, and P.K. Lundquist, *A stress-inducible protein regulates drought tolerance and flowering time in Brachypodium and Arabidopsis.* Plant Physiology, 2022. **191**(1): p. 643–659.

Yu, W., et al., *Knockout of SlMAPK3 enhances tolerance to heat stress involving ROS homeostasis in tomato plants.* BMC Plant Biology, 2019. **19**(1): p. 354.

Zaidi, S.S.-e.-A., et al., *Engineering plant immunity: using CRISPR/Cas9 to generate virus resistance.* Frontiers in Plant Science, 2016. **7**.

Zaidi, S.S.-e.-A., M.S. Mukhtar, and S. Mansoor, *Genome editing: targeting susceptibility genes for plant disease resistance.* Trends in Biotechnology, 2018. **36**(9): p. 898–906.

Zandalinas, S.I., et al., *Plant responses to climate change: metabolic changes under combined abiotic stresses.* Journal of Experimental Botany, 2022. **73**(11): p. 3339–3354.

Zandalinas, S.I., F.B. Fritschi, and R. Mittler, *Global warming, climate change, and environmental pollution: recipe for a multifactorial stress combination disaster.* Trends in Plant Science, 2021. **26**(6): p. 588–599.

Zeng, Y., et al., *Rational improvement of rice yield and cold tolerance by editing the three genes OsPIN5b, GS3, and OsMYB30 with the CRISPR-Cas9 system.* Front Plant Sci, 2019. **10**: p. 1663.

Zhang, J., et al., *Crop improvement through temperature resilience.* Annual Review of Plant Biology, 2019. **70**(1): p. 753–780.

Zhang, X., et al., *OsNAC45 is involved in ABA response and salt tolerance in Rice* Rice, 2020. **13**(1): p. 79.

Zhang, L. and Q. Zhou, *CRISPR/Cas technology: a revolutionary approach for genome engineering*. Science China Life Sciences, 2014. **57**(6): p. 639–640.

Zhang, Y.-Z., M. Shi, and E.C. Holmes, *Using metagenomics to characterize an expanding virosphere*. Cell, 2018. **172**(6): p. 1168–1172.

Zhao, C., et al., *Leucine-rich repeat extensin proteins regulate plant salt tolerance in ArabidopsisArabidopsis*. Proceedings of the National Academy of Sciences, 2018. **115**(51): p. 13123–13128.

Zhao, D., et al., *New base editors change C to A in bacteria and C to G in mammalian cells*. Nature Biotechnology, 2020.

Zhou, J., et al., *Gene targeting by the TAL effector PthXo2 reveals cryptic resistance gene for bacterial blight of rice*. The Plant Journal, 2015. **82**(4): p. 632–643.

List of Abbreviations

Cas	CRISPR-associated (enzyme)
CRISPR	Clustered regularly interspaced short palindromic repeats
CRISPRa	CRISPR-mediated gene activation
CRISPRi	CRISPR interference
dCas9	Dead Cas9 (catalytically dead Cas9)
GMO	Genetically modified organisms
DSB	Double-strand break (double strand DNA break)
HDR	Homology-directed repair
HSFs	Heat stress transcription factors
HSPs	Heat shock proteins
NHEJ	Non-homology end joining
PAM	Protospacer-adjacent motif
RNAi	RNA interference
ROS	Reactive oxygen species
SSNs	site-specific nucleases
TALENs	Transcription activator-like effector nucleases
TFs	Transcription factors
ZFNs	Zinc finger nucleases
KO	Knock-out

7 Recent advances in genome editing towards sustainable agriculture

Kaushik Kumar Panigrahi, Ayesha Mohanty,
Smruti Ranjan Padhan, Priya brata Bhoi,
Syed Mohammad Bashir Ali, and
Purandar Mandal

7.1 Introduction

Understanding gene function and revealing biological mechanisms requires the ability to modify and edit genetic information. Since 1971, when the production of specific DNA fragments using restriction enzymes was demonstrated for the first time, scientists have been controlling and using prokaryotic molecules for genome engineering. The science and applications of DNA-binding proteins that modify specific loci are extremely advanced. However, developing modular DNA-binding proteins that bind to a specific target is a very complex process that mandates protein engineering expertise. This problem has been solved by the clustered regularly interspaced short palindromic repeats (CRISPR) and CRISPR-associated protein 9 (Cas9) technology, as the target specificity of CRISPR-Cas9 is completely reliant on nucleic acid base pairing rather than DNA-protein interaction (Pickar-Oliver and Gersbach, 2019). In the past few years, CRISPR-Cas9, a highly versatile genome-editing technology, has transformed genome engineering, allowing researchers to make precise sequence-specific alterations to the genetic material of various cell types and organisms (Gaj et al. 2016). Homing endonucleases (HENs) or meganucleases, zinc-finger nucleases (ZFNs), transcription activator-like effector nucleases (TALENs), and CRISPR-Cas9 are the most frequently employed genome engineering/editing technologies. The advancement of genome editing technologies such as HENs, ZFNs, and TALENs allowed for more precise targeting of any gene of interest. While these first-generation genome editing techniques require difficult steps for protein engineering, making them expensive, time-consuming, and labour-intensive, CRISPR-Cas9 technology encompasses simple designing and cloning procedures that enable the use of equivalent Cas9 with diverse guide RNAs (gRNAs) that target multiple locations in the genome (Jaganathan et al. 2018).

Plant genome editing techniques using site-directed nucleases (SDNs) have evolved into potent tools. By inducing specific DNA double-strand breaks (DSBs), tools like mega nucleases, zinc-finger nucleases (ZFNs), transcription activator-like effector nucleases (TALENs), and clustered regularly interspaced short palindromic repeats/CRISPR-associated protein (CRISPR/Cas)

DOI: 10.4324/9781003382102-7

are used to modify genetic material precisely. The DSB caused by these methods can be repaired by either non-homologous end-joining (NHEJ) or homology-directed repair, which is influenced by the cell cycle phase and the availability of repair template with homologous terminal regions (HDR) (Tripathi et al. 2020; Ntui et al. 2021). The crops covered here, such as cassava and maize, have successfully used HDR processes (Hummel et al. 2018; Veley et al. 2021). Moreover, some cereal crops, like maize, have undergone base and prime editing (Zong et al. 2017).

Meganucleases are inherently present DNA-cleaving enzymes that can recognise lengthy DNA targets (20 base pairs) encoded by mobile genetic elements or introns.Cys2-His2 zinc-finger protein and the FokI restriction endonuclease cleavage domain are the two components that make up ZFNs. Bacterial TALEs are called TALENs to have a carboxy-terminal FokI cleavage domain linked to an amino-terminal TALE DNA-binding domain. CRISPR/Cas9 originates from the adaptive immune system of *Streptococcus pyogenes*. Due to its simplicity, effectiveness, specificity, adaptability, and capacity to multiplex characteristics, CRISPR/Cas9 has become the most extensively used genome editing technique.

7.2 The CRISPR structure emerges

The origin of CRISPR can be traced back a generation to Atsuo Nakata's lab at the Research Institute of Microbial Diseases, Osaka University in Japan. A postdoctoral researcher named Yoshizumi Ishino sequenced the *iap* gene of *Escherichia coli*, which produced an enzyme that contributes to alkaline phosphatase metabolism, a critical procedure for disassembling complex compounds into simpler ones (Isozyme of alkaline phosphatase is represented by the abbreviation *iap*). Since the 1960s, the K-12 *E. coli* strain has been the workhorse of molecular genetics, and Ishino was using it. To comprehend the function of the gene, he sequenced the DNA regions on either side of the gene. In these flanking sections, regulatory sequences were expected to be discovered, as is typical for bacterial genes like the lac operon. These DNA segments are regulatory sequences, and other molecules can bind to them to activate or deactivate a gene. Ishino and his co-workers found an interesting pattern in the flanking region downstream of the gene rather than a regulatory sequence: five virtually identical repetitive segments, each made up of twenty-nine DNA bases, and spaced by thirty-two bases of varied DNA (dubbed "spacers"). These repeating sequences also had palindromic sequences:

CGGTTTA<u>TCCCCGCT</u>**<u>CGCGGGGA</u>ACTC,

A: T and C: G base pair complementarity makes the seven-base portion of the sequence palindromic. The "*" denotes that there was some variation (possibly the presence of either G or A) among the five repeats.

By the middle of the 1980s, the investigation of bacteria and all the genomes at the sequence level had led to the recognition of a variety of repeating DNA sequences. However, a novelty in this structure warranted a mention in the study that was eventually published. After spotting it, Ishino and his co-authors refrained from speculating on its purpose. They simply stated that "to date, no sequence similar to these have been found anywhere in prokaryotes, and the biological relevance of these sequences is unclear."

It has been now known that a genome may contain somewhere between 2 and 100 CRISPR arrays. Each CRISPR repeat consists of 25 to 35 base pairs, which vary between species and among various genome arrays within a species. The number of repeats can differ and can increase, in response to a single prokaryotic cell introducing more spacers. A "leader" sequence and two *cas* genes are present at the beginning of each CRISPR array. Cas proteins have a wide range of applications, so classifying all CRISPR systems into two main categories, each with several subcategories, was a time-consuming process. In 2005, the CRISPR spacer sequences were independently recognised by three different groups as being identical to or strikingly similar to DNA sequences found in bacteriophages, archaeal viruses, and other prokaryotic pathogens.

7.3 Towards gene editing

The foundation for the development of CRISPR-based gene editing was laid in 2011 by three papers. There were two experimental papers; the third one was a theoretical analysis based on bioinformatics done by Koonin and Makarova with help from CRISPR researchers worldwide. The laboratory of Emmanuelle Charpentier in Ume, Sweden, was the source of the initial experimental paper. Charpentier and her co-workers discovered a novel step in the transformation of pre-crRNA into crRNA while working with the pathogenic bacterium *Streptococcus pyogenes*. A trans-activating crRNA (tracrRNA), which was coded by a different gene and had a 24-nucleotide base sequence that was in sync with the repeat segments of the crRNAs, was required to participate in this step. In addition to the typical cellular machinery found in bacteria, it became now clear that interference required three types of molecules: Cas proteins, tracrRNA, and crRNA.

The second experimental paper, which came from Virginijus Siksnys's lab at Vuknius University in Lithuania, was very important in the development of CRISPR-based gene editing. This paper presented two significant findings. The collaborators, including Barrangou and Horvath, two Danisco biologists, were successful in transferring DNA with a CRISPR locus from one species to another. They established that the new species' CRISPR mechanism sustained to perform adequately. The transfer occurred from *S. thermophiles* to *E. coli*, the laboratory workhorse, which explains Danisco's interest. This finding was substantial because it suggested CRISPR-based system could potentially be utilised the likelihood that a CRISPR-based system could be utilised in different settings with no compromise in its functionality. The study also demonstrated that the CRISPR system can function properly with just one Cas protein, now known as Cas9. This observation strongly suggested that a CRISPR-based gene editing method might be fairly straightforward to develop. Additionally, this paper marked the beginning of an investigation into the functions of various components—or "domains"—of the Cas9 protein in DNA cleavage. These two papers laid the groundwork for the development of a CRISPR-based gene editing technology experimentally. The third paper, was a theoretical and significant one, as it acted as a catalyst. The majority of these papers consisted of a review of preceding work on the CRISPR system; the list of authors contained nearly all prominent CRISPR researchers, with Makarova and Koonin being the most prominent. The paper's main accomplishment was classifying the various CRISPR systems that had emerged over the past decade. Furthermore, the paper also summarised the findings uncovered regarding the functions of different RNA and Cas9 types, especially in DNA cleavage. This allowed the reconfiguration of the CRISPR system as a universal gene (and genome) editing technology that could target nearly any DNA sequence within any species.

At the University of California, Berkeley, Charpentier, and renowned RNA biologist Jennifer Doudna began working together. The paper describing how the CRISPR system functions as a fully programmable gene editing technology was published in 2012. In vitro, or a test tube, the team constructed an astonishingly straightforward CRISPR system, using only the Cas9 protein and a single "guide" RNA (sgRNA), a chimera of a crRNA and tracrRNA. Programming the spacer sequence of the crRNA segment could be used to induce cleavage at any DNA sequence. The system's competence was demonstrated by creating a chimeric sgRNA aimed at green fluorescent protein. They demonstrated that the two segments of the targeted DNA were cleaved by distinct domains of the Cas9 molecule. The team members' successes were fully acknowledged. Acknowledging the latest accomplishments of ZFNs and TALENs in genome manipulation, the report ended by recommending "an alternative technology based on RNA-programmed Cas9 that could offer substantial potential for gene targeting and genome-editing applications."

A typical CRISPR-based genome editing was developed as a result of the research conducted, using a molecular construct comprising of a guide RNA targeted at a specific DNA sequence and the genes that needed to be inserted. After Cas9 cleaves the targeted DNA, the construct is put into the chromosome. This technique permits rapid genome editing in both germline and somatic cells.

The CRISPR system's remarkable programming simplicity set it apart from ZFNs and TALENs. For insertion into the sgRNA, all that is required is the creation of a complementary RNA sequence to that DNA sequence. This is not only conceptually simpler than engineering protein segments for

ZFNs and TALENs, but it is also experimentally much simpler and considerably less expensive. It is not surprising that the CRISPR system completely supplanted its predecessors right away. After the target DNA is cleaved at the intended location, insertable DNA can be induced at that location because it was included in the sgRNA. To replace a gene that isn't working, for instance, a functional version can be added.

The DNA cleaver Cas9 is what the normal CRISPR system uses, however, other nucleases have been tested and could someday take Cas9's place. The Cas9 protein requires a particular PAM sequence located upstream of the spacer to cleave DNA. The prevalent Cas9 molecule is produced by the *Streptococcus pyogenes*. Its PAM sequence is NGG, with "N" denoting any nucleotide other than the normative quadruplet (A, T, C, or G). In case the protospacer being targeted doesn't have this triplet just upstream of it, Cas9 for *S. pyogenes* wouldn't function. Numerous solutions to the issue have been developed over time. The simplest method is to take advantage of Cas9's inherent variation between species. PAMs are required by the various Cas9 versions. Cas9 protein from *S. aureus*, for instance, employs the PAM sequence NGRRT or NGRRN, with "N" denoting any nucleotide base and "R" denoting either an A or a G; The PAM sequence NAAN is used by Cas9 from *Treponema denticola*, a bacterium that causes dental disease in humans. Today, laboratories use about thirty-odd Cas9 variants, and the number keeps going up. In addition, apart from Cas9, other nucleases employ other PAMs, and it is possible to modify nucleases to fit the appropriate region close to any targeted DNA space—although it presents some difficulty. This leads to the conclusion that anyone can program the CRISPR system.

To be effective, the CRISPR system, like all other gene editing systems, must be delivered to the cells being targeted. Delivery is relatively simple when germline genes are targeted. A microscale needle can be used to inject the CRISPR construct—sgRNA and Cas9, into a single-cell embryo or zygote. In humans, this technique has been used to repair point mutations with some success. The sgRNA and the Cas9 protein were injected into the cell during these experiments. However, it is also possible for Cas9 RNA to enter the cytoplasm to undergo ribosome-mediated protein translation; alternatively, transcription and translation can be initiated by inserting Cas9 DNA into the nucleus. The CRISPR system initiates its function right upon the introduction of the Cas9 protein itself into any cell, resulting in rapid gene editing.

7.4 Crop plant editing with CRISPR/CAS9

Several plant species have benefited from the CRISPR/Cas9 system techniques. Besides model plants like Arabidopsis, these include agricultural products including rice, maize, tobacco, tomato, sorghum, wheat, apple, soybean, poplar, potato, and banana. Applications of CRISPR gene editing system are discussed briefly in some of the crops.

7.4.1 Rice

With the help of CRISPR technology, new rice varieties with better nutritional and economic qualities have been successfully developed. The genes in Table 7.1 that are enhanced by these traits are listed. One of the quality indications is the amylose content; rice with a high amylose level digests more slowly and therefore has higher nutritious value. The genes SBEI and SBEIIb govern the synthesis of starch. Targeted mutations in these genes are produced by the CRISPR-Cas9 system, and the mutant rice generates 25% more amylose and 9.8% more resistant starch, respectively (Sun et al. 2017). Resistant starch is crucial to human health and reduces the likelihood of non-communicable disorders (Romero and Gatica-Arias, 2019). Some hybrid types, including indica hybrids, have a low market value because of their high amylose content. Reducing the amylose content, in this case, can result in sticky rice, which is a very desired quality.

One of the rice derivatives is called rice bran oil (RBO). Oleic acid, one of its elements with strong health-promoting characteristics, aids in the prevention of several lifestyle disorders like excessive cholesterol and blood pressure. The Rice genome contains four fatty acid desaturase 2 (FAD2) genes that are expressed at high levels in rice seeds and convert oleic acid to linoleic acid. Oleic acid concentration increases the yield of more valuable RBO. Oleic acid is converted into linoleic acid by the enzyme FAD2. So, by Agrobacterium-mediated transformation, FAD2-1 knockout via the CRISPR/Cas system

Table 7.1 List of genes responsible for the commercially and nutritionally important characters modified by CRISPR/CAS system. (Erum Shoeb et al. 2021)

No	Name	Target gene	Characters
1	*Oryza sativa*	SBEIIbSBEI	Boost amylase and Amylose content
2	*Oryza sativa*	DEP1 NHEJ Gn1a, GS3,	Enhanced grain number, larger grain size, and dense erect panicles
3	*Oryza sativa*	FAD2	Rise oleic acid
4	*Oryza sativa*	NRAMP5	Lessen the accretion of cadmium
5	*Triticum aestivum*	TAMLO	resistance to fungal powdery mildew
6	*Triticum aestivum*	α-Gliadin	Low-gluten
7	*Triticum aestivum*	TaDREB2, TaERF3	Abiotic stress tolerance
8	*Gossypium hirsutum*	GhMYB25, GhMYB25	Transcription factors and Development of trichome
9	*Zea mays*	ARGOS8	Drought tolerance and resistance
10	*Zea mays*	LG1	Reduced leaf angle making it more vertical oriented
11	*Glycine max*	GMFT2	Deferred flowering sequences
12	*Solanum lycopersicum*	SlORRM4	Late ripening
13	*Solanum tuberosum*	GBSS	Deficient in amylose entirely
14	*Malus domestica*	PPOs	Reduced the browning
15	*Citrus sinensis*	CsLOB1	Resistance to citrus canker

might produce a rice variety with increased oleic acid. Linoleic acid levels are below detection limits, and the oleic acid quantity in the homozygous FAD2-1 mutants is double that of the wild type (Abe et al. 2018).

All living things are at risk from toxic heavy metals, presence of high levels of cadmium in rice can cause serious health issues to consumers (Bertin and Averbeck, 2006; Clemens et al. 2013). Tang et al. (2017) used CRISPR/Cas9 to modify the metal transporter gene NRAMP5 to develop a novel indica rice variety with minimal cadmium accumulation. The roots' absorption of cadmium is decreased by this transporter And the Mutant plants thus produced possess less cadmium than wild-type plants, while maintaining the same agronomic characteristics, like grain yield, strawweight, and grain quality.

One of the most desirable and complicated traits for crop improvement is high yield, which is controlled by several factors including agronomic practices and various genes, including quantitative trait loci (QTLs) (Shen et al. 2018). Some of these have been identified as a potential target for CRISPR/Cas, to create new rice lines with increased yield. Li et al. (2016) altered the Gn1a, DEP1, GS3, and IPA1 genes in the Zhonghua 11 rice line, which control grain number, panicle architecture, size, and plant architecture. The sgRNAs were engineered to target high-yield mutations in the first exons of Gn1a and GS3, the third exons of DEP1 and IPA1, and the fourth exon of GS3. The Gn1a mutation results in an approximately 90% increase in plant height, panicle size, and flower count per panicle. The grain size and husk awn length were bigger in the GS3 mutant lines. However, they had more flowers per panicle (an increase of around 50% compared to the control), while having shorter panicles and shorter plant heights (about 20% less than the wild type). Finally, three unique phenotypes are produced by IPA1 mutations, depending on the type of mutation (Bae et al. 2014; Chari et al. 2017; Doench et al. 2016; Zhao et al. 2017). Xu et al. (2016) modified three QTLs linked to grain weight, GW2, GW5, and TGW6, using a CRISPR/Cas9-mediated method, increasing grain weight. Additionally, compared to the wild type, CRISPR elimination of 625 bp from the DEP1 gene causes dense, upright panicles with more grains and shorter plants (Huang et al. 2009). Genes HD2, HD4, and HD5 negatively regulate rice heading date, which has significance on crop productivity and distribution (Matsubara et al. 2014; Li et al. 2015). Early-maturing sgRNAs based on CRISPR were created by Li et al. (2017).

7.4.2 *Wheat*

A CRISPR-mediated targeted genetic editing technique has already been employed to enhance desired crop traits in *Triticum aestivum*, or bread wheat. Genome editing techniques are seriously hindered by the enormous allohexaploid (2n = 42) genome of homologous genes in wheat. The three homoeoalleles for a dominant gene at the TAMLO locus that causes the fungus powdery mildew susceptibility were eliminated, using CRISPR technology

(Shan et al. 2013). To create wheat lines with low-gluten and transgene-free properties for patients with autoimmune Celiac disease, Sanchez-Leon et al. (2018), targeted the 33-mer in the -gliadin genes and reported a reduction in gliadins in wheat. An enzyme called lipoxygenase (LOX) influences the color and quality of items made from wheat. According to Shan et al. (2014), CRISPR/Cas9 system was used to target TaLOX2. The CRISPR/Cas9 method was also used to modify TaERF3 (ethylene-responsive factor 3) and TaDREB2 (dehydration-responsive element binding protein 2) to create wheat variants with enhanced drought and frost resistance (Kim et al. 2018). Kernel size is another element that affects wheat yield. Using RNAi, TaDA1, a negative regulator of kernel size is upregulated, resulting in increased weight, length, and width in wheat kernels (Liu et al. 2019).

7.4.3 Cotton

Among the most important natural sources of vegetable oil and premium fiber is cotton (*Gossypium hirsutum L.*). Cotton has an allotetraploid (2n = 52) DNA. The GhCLA1 (*Cloroplastos Alterados* 1 gene) in cotton, which is associated with chloroplast formation has reportedly been modified via the CRISPR/Cas9 system (Wang et al. 2018). It produced the albino phenotype; GhMYB25, transcription factors for the development of trichomes and fiber (Li et al. 2017); GhVP, which stands for "vacuolar H+-pyrophosphatase," 2017) for resistance to salt stress; Phytoene desaturase, or GhPDS; The aminoacyl-tRNA binding is catalyzed by GhEF1, an elongation factor-1 protein (Gao et al. 2017); Arginase gene (GhARG).

7.4.4 Soybean

Cai et al. (2018) delayed the soybean flowering cycle using the CRISPR/Cas9 system (Glycine max). GMFT2 genes were knocked out to create transgene-clean mutants that were stable, homozygous, and free of any signs of the transgenic element.

7.4.5 Potato

The potato (Solanum tuberosum), a significant staple crop, is widely consumed around the world. According to Andersson et al. (2017), the GBSS (granule-bound starch synthase) gene was specifically targeted in the development of the waxy genotype in hexaploid potatoes. GBSS catalyses the production of amylose in starch granules. Amylose-free mutants are created by knocking out the GBSS gene, and these mutant variants can be suggested for people with amylose-related conditions such as cystic fibrosis. Several fruits and vegetables, such as potatoes, develop oxidative browning as a result of the polyphenol oxidases (PPOs) gene. The PPO gene was suppressed using the RNAi method to create potatoes that brown less quickly.

7.4.6 Maize

One of the most significant cereal crops is maize (*Zea mays*), and the seeds of this plant contain e antinutritional compound phytic acid. Liang et al. (2014) employed CRISPR technology to block the production of phytic acid by focusing on the genes ZmIPK1A, ZmIPK, and ZmMRP4. To make seeds easier for consumers to digest, it is preferred that they have a low phytic acid level. A favorable characteristic in maize is a reduced leaf angle because it boosts yield by increasing plant density per unit of area. Li et al. (2017) used CRISPR-based guide RNA to target the Liguleless1 (LG1) gene and created heritable mutations that resulted in erect leaves with 50% fewer leaf angles than the non-mutant genotype. According to Shi et al. (2017), the native maize GOS2 promoter was employed in place of the ARGOS8 gene's native promoter to boost maize grain yield in drought-like conditions.

7.4.7 Sorghum

In cells of developing sorghum embryos, the Cas9 gene, green fluorescent protein (GFP) gene, sgRNA, and out-of-frame RFP gene were used to detect the CRISPR/Cas system activity for the first time (DsRED2). The DsRED2 gene was intended to be deleted by the sgRNA/Cas9 complex, and be subsequently functionally repaired via NHEJ. Jiang et al. (2013) found that 5 of the 18 groups of transformed cells that exhibited positive GFP expression, contained regions expressing DsRED2. Che et al. (2018) used the CRISPR system to mute the endogenous H3 gene (b-CENH3), which controls chromosomal segregation, in an early embryo to induce haploid induction. Li et al. (2018b) used CRISPR/Cas technology to improve the digestibility and protein quality of sorghum, thus increasing the nutritional value of the grain, by targeting the k1C gene families, which are responsible for the majority of kafirins. The resultant mutant had a decreased kafirin level and exhibited a 1.3–1.5-fold increase in protein digestibility.

7.4.8 Tomato

Ripening regulation is one of the most important tomato properties, which is why the ripening inhibitor gene (RIN) was created using the gene-editing technique CRISPR/Cas9 because tomatoes (Solanum lycopersicum) are a significant crop around the world. RIN-protein mutants exhibited a poor ripening phenotype as well as suppression of ethylene and carotenoid production. (Ito et al. 2015; Li et al. 2018b). Knocking out of SlORRM4 gene results in delayed ripening of tomato fruit (Yang et al. 2017). Similarly, Yu et al. (2017) claimed that CRISPR/Cas9 could be used to extend shelf life by replacing the dominant ALC (Alcobaca) gene with the recessive alc gene (Alcobaca) in tomato.

7.4.9 Citrus

Citrus is an important economical crop. Targeted genome editing technology has the potential to accelerate the development of certain traits, such as disease resistance.in Wanjincheng orange (*Citrus sinensisOsbeck*) and the Duncan grapefruit (Citrus × paradisi Macfad) (Jia et al. 2017) against citrus canker. Both have the lateral Organ Boundaries (CsLOB1) gene modified to introduce citrus canker resistance at the 50 regulatory sequences (Peng et al. 2017). The mutated plants showed no side effects or off-target mutations.

7.4.10 Apple

Conventional breeding methods on fruit trees are quite challenging because of their lengthy breeding cycles and heterozygosity. Rapid targeting and modification of desirable traits are made possible by CRISPR. Slices of apple (*Malus domestica*) turns brown as a result of polyphenol oxidation to quinones, which is catalyzed by PPOs (Mellidou et al. 2014). This alters the flavor and quality of the apple. Employing RNAi resulted in a considerable decrease in PPO protein production, which, in turn, prevented the browning of apples (Waltz, 2015). Apples that don't oxidise could decrease fruit waste while also improving their usability.

7.5 Analysis of CRISPR mutation design

The clustered regularly interspaced short palindromic repeats (CRISPR)/ CRISPR-associated protein 9 (Cas9) system represents a cutting-edge advance in precise mutagenesis technology. Nevertheless, screening a large number of original samples for CRISPR/Cas9-induced mutations is a time-consuming and expensive process. The majority of mutations induced by genome editing with CRISPR in many diploid species are non-chimeric mutations, comprising biallelic, heterozygous, and homozygous mutations. Direct Sanger sequencing of PCR amplicons harboring non-homozygous mutations reveals overlapping peaks starting at the mutation sites by superimposing sequencing chromatograms.

Double-strand breaks (DSBs) at the targeted DNA locations are produced by the CRISPR/Cas9 system by combining a single-guide RNA (sgRNA) and the CRISPR-associated endonuclease Cas9. Non-homologous end-joining (NHEJ) repair is then used to create genetic alteration (Jinek et al. 2012; Cong et al. 2013; Mali et al. 2013). The CRISPR/Cas9 system typically causes insertion or deletion (indel) mutations in 3 base pairs upstream of the protospacer adjacent motif (PAM) (Cong et al. 2013). Due to the rapidly expanding use of genome editing in biological research, the number of studies on CRISPR/Cas9-generated mutants has considerably risen in the past few years, especially for screening a large number of mutants.

PCR is a popular method that can quickly and accurately screen a large number of samples with high specificity. There are three steps in a single PCR cycle: extension, annealing, and denaturation Effectiveprimer–template pairing is determined by the annealing temperature, so choosing the right temperature is essential for successful PCR. At the optimal temperature, mismatched annealing is suppressed, which lowers the output of non-specific products. We created the "annealing at critical temperature PCR" (ACT-PCR) approach which allows for rapid, ready, cheap, and accurate detection of CRISPR/Cas9-induced mutations based on this idea (Wang et al. 2017). This approach involves three steps:

1 Designing of primers,
2 Identification of the crucial annealing temperature using the preliminary gradient PCR, and
3 Screening of mutants.

Initially, the target gene-specific primer pairs are created. The forward primer, which is known as the DSB site-specific primer (primer DS), has a 4-bp overhang at its 30th end relative to the DSB site to guarantee specificity and sensitivity of PCR amplification and attachment of wild-type (WT) genes. To guarantee DNA template binding at the critical annealing temperature, the reverse primer (primer R) has a higher Tm value than the DS primer because it is outside the DSB site. Preliminary gradient PCR determines the critical annealing temperature. Lastly, conventional PCR is carried out at the critical annealing temperature that was previously established. If a mutation occurs, the DS primer fails to attach to the mutated sequence, and no amplicons are produced. As a result, the lack of amplicons differentiates the mutants from the wild-type samples.

7.5.1 Pathways for the repair of plant DNA and how they can be used in genome engineering

Exceptional advances in the creation of technology for sequence-specific DNA modification of fundamental DNA sequences have made it possible to precisely create plants possessing novel properties. Such programmable sequence-specific modifiers comprise base editors and site-directed nucleases. (At present, these genome editing tools can alter the sequence by focusing on certain chromosomal regions. However, the results of sequence mutations depend significantly on the nature of DNA damage that is being induced, the condition of the host's DNA repair apparatus, and the structure and presence of the DNA repair template. The insertion or deletion of DNA sequences with different lengths is typically the outcome of the sequence alteration that comes about as a result of repairing DNA double-strand breaks (DSBs). The effectiveness of base editing (BE) is contingent upon the type of BE utilised or on the activity of the host's DNA repair systems although the accuracy of planned alterations is much higher.

Plants are exposed to a wide range of biological and environmental conditions that have the potential to harm genomic DNA. For instance, the genomic DNA of leaf cells are constantly subjected to extremely harmful UV radiation when they are exposed to sunlight (Britt, 2004). Disintegrated nucleotides or DNA breaks are also produced by essentially biological processes like DNA replication, recombination, and transcription (Spampinato, 2017; Manova and Gruszka, 2015). Free radicals that are capable of causing DNA base damage are also produced by the body's metabolic processes and genotoxic stresses like heat and infection by pathogens. To keep the genome stable, these various types of DNA damage must be repaired correctly and promptly. Cell death occurs when there is an excessive amount of DNA damage in cells. Plant cells use several major mechanisms to repair different kinds of DNA damage (Spampinato, 2017; Manova and Gruszka, 2015; Hu et al. 2016). Photo reactivation, base excision repair (BER), and nucleotide excision repair (NER) pathways are used to rectify the damaged bases and nucleotides. Mismatched and unpaired nucleotides are recognised and corrected via mismatch repair (MMR) pathways. If homologous donors are provided, the BER and the homology-directed repair (HDR) pathways can recognise and repair nicks or single-strand breaks (SSBs). To repair the most damaging DSBs, mechanisms such as non-homologous ends joining (NHEJ) and homology recombination (HR) are used (Spampinato 2017, Manova and Gruszka, 2015; Puchta and Fauser, 2014).

Base repair caused by damage due to exposure to UV light is a major function of plant DNA repair machinery. Typical DNA damage induced by UV radiation is Cyclobutane pyrimidine dimers (CPDs) and pyrimidine (6-4) pyrimidones (6-4 photoproducts). According to Ueda and Nakamura (2011), these pyrimidine dimers in plants are eliminated through the photo-reactivation process, just as it is in other organisms. The damaged bases are removed by two related yet distinct photolyases, CPD photolyase and 6-4 photolyase (Britt, 2004; Spampinato 2017; Ueda and Nakamura, 2011). All photolyases have a photolyase-homologous region (PHR), according to Faraji and Dreuw (2017), which attaches the chromophore flavin adenine dinucleotide (FAD). This chromophore split the pyrimidine dimer lesion to give two repaired pyrimidines using the energy by absorbing visible or blue light.

Nucleotide excision repair (NER) in bulky helix-distorting CPDs and pyrimidine (6-4) pyrimidones brought on by UV exposure is a crucial process (Schearer, 2013). According to Alekseev and Coin (2015), NER is also responsible for identifying and eliminating a wide variability of structurally dissimilar DNA lesions. The bulky helix-distorting lesions can be detected using two distinct methods to commence NER: the transcription-coupled NER (TC-NER) and the global genome NER (GG-NER) (Spampinato, 2017; Schearer, 2013; Alekseev and Coin, 2015). The heterotrimeric RAD4/XPC-RAD23-CEN2 complex and the heterodimeric damaged DNA-binding (DDB) protein complex are used to detect DNA damage in the GG-NER pathway (Spampinato, 2017; Schearer, 2013; Alekseev and Coin 2015). With the assistance of the proteins CSA, CSB, and XAB2 in TC-NER, the identification procedure originate from an arrested RNA polymerase (Spampinato 2017; Schearer, 2013; Alekseev and Coin, 2015). A stable pre-incision complex which comprises transcriptional factor II H (TFIIH), XPA (*Xeroderma pigmentosum* group A), RPA (replication protein A), XPG, and ERCC1 (excision repair cross-complementing 1)-XPF) is formed once GG-NER and TC-NER integrate to form a common pathway and recruit additional proteins. The pre-incision complex's endonucleases, ERCC1/XPF and XPG, collaborate to remove the damaged site from a single-strand oligonucleotide fragment that is 24–32 nucleotides in length. According to Spampinato (2017) and Schearer (2013), DNA synthesis by DNA polymerases, nick ligation by DNA ligase 1 or 3, or subject to the chromatin accessibility in the damaged region completes the repair.

According to Krokan and Bjoras (2013), the basic (apurinic or apyrimidinic, AP) sites, as well as base damages from oxidation, deamination, and alkylation, are recognised and corrected by the base excision repair (BER) process. DNA glycosylase clears the damaged base to form an AP site. In the cell, various DNA glycosylases target specific kinds of damaged bases. An AP endonuclease or the AP lyase activity of the DNA glycosylase then cleaves the sugar-phosphate backbone at the AP site (Spampinato, 2017; Manova and Gruszka, 2015; Krokan and Bjoras, 2013). The nick in the DNA backbone is then joined by DNA polymerase and ligase.

Since plants lack homologs for DNA polymerase and ligase 3, it is probable for ligase 1 to participate in both the "short" and "long" patch forms of BER (Manova and Gruszka, 2015). BER also has a crucial function to control gene expression in plants by DNA demethylation, where 5-methylcytosine (5-meC) is specifically eliminated by a specialised glycosylase/lyase mechanism (Morales-Ruiz et al. 2006).

Mismatch repair (MMR), according to Crouse (2016) and Kunkel and Erie (2015), is in charge of fixing mismatches in healthy or damaged bases as well as insertion/deletion loops brought on by strand misalignment. Examples of these include the removal of the amine group from 5-methylcytosine, recombination in dissimilar sequences, single base-base mismatches, and unmatched nucleotides brought on by replication mistakes (Kunkel and Erie, 2015). According to Crouse (2016), MMR is crucial in the suppression of insertion/deletion loops (IDL) caused by a phenomenon known as slipped mispairing. MMR not only acts as a barrier for speciation and rearrangement but plays a role in foiling recombination between homoeologous sequences in plant cells and bacteria (Spampinato, 2017; Crouse, 2016). MutS proteins, which are made up of akin but not identical heterodimeric MutS homolog (MSH) subunits, can identify a DNA mismatch. These MSH subunits recognise various types of lesions in plants by forming functionally distinct complexes like MutS (MSH2-MSH6), MutS (MSH2-MSH3), and MutS (MSH2-MSH7) (Spampinato 2017; Manova and Gruszka, 2015). After MutS proteins recognise a lesion, a DNA repair complex is formed by recruiting heterodimeric MutL and endonuclease PMS1, resulting in a nick formation in the DNA strand containing the aberration. Exonuclease I (ExoI) then reselect the damaged DNA strand for subsequent repair by PCNA, the replication fork complex (RFC), RFA, and DNA polymerase (Spampinato, 2017; Manova and Gruszka, 2015).

Single-strand breaks (SSBs) are the most common types of DNA damage in cells. It may be produced as a result of spontaneous DNA decay, exposure to intracellular compounds like reactive oxygen species (ROS), or due to DNA topoisomerase 1 abortive activity; it can also result from incorrect base alteration by APOBEC and TET family proteins or the correction of ribonucleotide damage or mismatch (Caldecott, 2008; Caldecott, 2014). Several mechanisms including base excision repair (BER), nucleotide excision repair (NER), mismatch repair (MMR), and ribonucleotide excision repair (RER) are responsible for the effective repair of SSBs (Caldecott, 2008; Caldecott, 2014; Abbotts and Wilson, 2017). The readily available Cas9 nickase has greatly facilitated the study of genomic SSB repair mechanisms (Fauser et al. 2014; Kan et al. 2017; Davis and Maizels, 2016; Komor et al. 2016). Deep sequencing analysis in Arabidopsis indicated that repairing of targeted nicked targeted sequences resulted in only insertions and deletions (indels) (Fauser et al. 2014). For HDR, SSB also makes use of homologous sequences that are present in both trans and cis (Fauser et al. 2014; Kan et al. 2017; Davis and Maizels, 2016). Nickase-generated site-specific SSB is viable to direct site-specific editing using a homologous DNA template (Fauser et al. 2014; Kan et al. 2017; Davis and Maizels, 2016) and also influence the outcomes of base editing in terms of repair (Komor et al. 2016). The cells' DSBs, or double-strand breaks, may be the most harmful and mutagenic. There are several ways to repair double-strand breaks such as: nonhomologous end joining (cNHEJ), alternative end joining (altEJ), single-strand annealing (SSA), and homologous recombination (HR) mechanisms (Spampinato, 2017; Manova and Gruszka, 2015; Gorbunova and Levy, 1999; Deriano and Roth 2013; Chang et al. 2017, Puchta and Fauser 2014; Puchta, 2005; Ceccaldi et al. 2016; Steinert et al. 2016; Verma and Greeberg, 2017). An initial episode of End processing selection serves as a determinant in the DSB repair pathways and outcomes (Ceccaldi et al. 2016). The cNHEJ pathway that relies on Ku and the altEJ pathway that is independent of Ku, both rapidly repair DSBs in plants, particularly in non-germline cells which are often treated as substrates for genome manipulation studies (Spampinato 2017; Manova and Gruszka, 2015; Hu et al. 2016; Puchta and Fauser 2014). Backup NHEJ (b-NHEJ) and microhomology-mediated end joining (MMEJ) are other names for AltEJ (Chang et al. 2017). The Ku70-Ku80 heterodimer recognises as well as tightly binds DSBs in cNHEJ. After that, DNA polymerases and additional cNHEJ factors like DNA-PKcs, XRCC4-ligase IV-XLF (Cernunnos), and Artemis nuclease are called upon to perform end processing and ligation at the site of DNA break (Deriano and Roth, 2013; Chang et al. 2017; Ceccaldi et al. 2016). According to Deriano and Roth (2013), the cNHEJ pathway brings about small indels (1–4 nucleotides) as a part of end processing, leading to a minor DNA (Deriano and Roth, 2013; Chang et al. 2017; Ceccaldi et al. 2016), whereas the poly ADP-ribose polymerase (PARP) proteins in the altEJ pathway bind the DSB. The MRE11-RAD50-NBS1 (MRN) complex is called upon to initiate end resection when PARP binds to the broken ends. This makes it easier to generate microhomology among the two DNA strands that have unpaired DNA termini. After that, MRN processes the micro-homology to facilitate end joining aided by DNA ligase 3/XRCC1 complex (Deriano and Roth, 2013; Chang et al. 2017; Ceccaldi et al. 2016). The low-fidelity DNA polymerase uses templates containing microhomology in both cis and trans to extend the broken DNA ends in the altEJ pathway. This leads to larger-scale deletions as well as insertions of 'donor' sequences, which sometimes results in the reversing the sequence orientation and translocating of the chromosome (Deriano and Roth, 2013; Chang et al. 2017; Ceccaldi et al. 2016).

7.5.2 *Utilising CRISPR/Cas9 for intron-targeted gene replacement*

The CRISPR-Cas9 technology gained immense popularity for modifying genomes and is being employed by expanding variety of organisms, especially for the aim of inducing site-specific precise gene alterations, achieved by non-homologous end joining (NHEJ) of DNA double-strand breaks (DSBs). Additionally, the ability to apply homology-directed repair (HDR) for accurate gene modification, such as changing one amino acid with another in a specific gene expression by replacing a short segment of DNA, would be highly desirable. This is still a big problem for plants, though. The type II prokaryotic CRISPR/Cas9 system is found in bacteria and archaea as an adaptive immune defense mechanism (Barrangou et al. 2007; Jinek et al. 2012), and it has been tailored for eukaryotic genome editing (Mali et al. 2013; Cong et al. 2013). It is now the most predominantly employed tool for accurately customising eukaryotic genomes due to its high efficiency, affordability, simplicity, as well as versatility (Voytas and Gao, 2014; Kim and Kim 2014; Weeks et al. 2016; Wang et al. 2016). Cas9 generates targeted DNA double-strand breaks (DSBs) by utilising a single guide RNA (sgRNA) that recognises target DNA sequences via Watson-Crick base pairing.

The DNA DSBs can be repaired by nonhomologous end joining (NHEJ) or homology-directed repair (HDR) pathways (Symington and Gautier 2011). In several cell types and organisms, precise NHEJ repair is employed to induce non-functionality in the targeted gene (Westra et al. 2014). The genomic sequence is modified in a predetermined manner by High-fidelity HDR using a homologous DNA template (Puchta, 2005; Bibikova et al. 2003).

According to Reddy (2007), the splicing of precursor mRNAs (pre-mRNAs) removes untranslated introns from the majority of eukaryotic open reading frames (ORFs). The branch point sequence and the GU and AG dinucleotides that make up the 50 and 30 splice sites, respectively, as well as the splicing signals that define non-coding regions remain unaltered (Lydia et al. 2017).

Alterations to a non-coding sequence in regions other than the signal sequences may have little effect on transcription or differential splicing of the gene under consideration.

For studies of functional genomics in plants, point mutations, and gene substitution are very useful because they may aid in the development of traits that are beneficial to agriculture. However, utilising HDR to produce precise genome modifications remains a significant challenge for the majority of plants (Cermak et al. 2015). Usually, Cas9 precisely deletes or inserts 1 base pair located just upstream of the DNA double-strand break, at the fourth base from the protospacer-adjacent motif (PAM) site (Zhang et al. 2014). Li et al. (2016) employed Cas9 in rice to aim adjoining introns for the creation of gene replacement events because small intron changes can be tolerated (Li et al. 2016). The gene replacement's effectiveness is mostly determined by two factors: the quantity of the donor fragment that is available and the effectiveness of DSB induction. A pHUN411 vector that has two sgRNAs that specifically target neighbouring introns of an exon alongside a donor vector containing a donor fragment which includes the sgRNA target sequences at both ends is both injected into the callus cells of rice. Expression of Cas9 and the sgRNAs in plant cells bring about cleavage in genomic sequence as well as the donor plasmid at both the sgRNA sites, which in turn produces four double-stranded breaks (DSBs). These DSBs are comprised of DSB1 and DSB2, which release the donor sequence from the donor plasmid, and DSB3 and DSB4, which occur in the endogenous locus. When the repair process that joins the free ends is done correctly, the donor sequence is substituted with the appropriate region in the host gene.

Contrary to the HDR route, this replacement strategy uses the DNA repair machinery that predominates in NHEJ and does not require the donor fragment to have any extra homology arms. It is error-prone and frequently leads to insertion, omission, or inversion of genomic fragments at target locations, incorporation and orientation reversal of donor fragments, as well as the targeted substitution. Although indels are regularly created in specific intronic regions, modest modifications are acceptable as long as the splicing signal and transcription are not adversely damaged. A rice callus is used to regenerate a plantlet that has a site-specific gene substitution, and this callus is subsequently inherited without any alterations.

7.5.3 *Multiplex genome editing in plants using a CRISPR-Cas9 system with a single transcript*

The CRISPR-Cas9 system is extensively employed to carry out genome editing. Altering the 20 bp guide sequence makes it easier to change any sequence in a genome that is close to a protospacer adjacent motif (PAM). To perform multiplex genome editing, several single-guide RNAs (sgRNAs) are simultaneously expressed. Because sgRNAs are expressed by Pol III promoters, several complete sgRNA expression cassettes are required for multiplex genome editing. Vector creation, however, tends to be challenging.

Nature's evolution is fundamentally based on DNA mutations. There are four basic types of mutations based on the responsible factors, viz., mutation due to physical factors, mutation due to chemical factors, spontaneous mutations, insertional mutation, and site-specific mutation aided by genome editing. Physical, chemical, and insertional mutagenesis, specifically T-DNA insertion, are all stochastic and unpredictable processes (Krysan et al. 1999). Random mutagenesis occurs at a low efficiency as well (Saxer et al. 2012). Genome editing, on the other hand, can introduce mutations at specific locations.

Zinc-finger nucleases (ZFN), TALE nucleases (TALEN), and CRISPR nucleases are the genome editing tools (Gaj et al. 2013). In a variety of species, CRISPR/Cas9 system demonstrated high efficacy across multiple species when used as a genome editing tool. Cas9 protein and single-guide RNA (sgRNA) are components of this system. A sgRNA directs the Cas9 protein to accurately insert DNA double-strand breaks. SpCas9 is a popularly employed Cas9 protein, which relies on NGG PAM to identify the target genome for editing (Jinek et al. 2012). Small insertions and deletions (indels) are the majority of the mutations that CRISPR-Cas9 produces (Hsu et al. 2013). To delete a chromosomal region, it may be necessary to simultaneously express two sgRNAs and Cas9 in some situations, such as non-coding sequence knockout. When compared to conventional breeding, this type of multiplex genome editing is extremely beneficial in crop improvement that can edit multiple trait genes simultaneously, resulting in significant time and effort savings. SgRNAs are typically transcribed by a CRISPR-Cas9 system using Pol III promoters.

7.6 Sustainable agriculture with CRISPR

CRISPR offers plant breeders many opportunities to advance breeding programs towards ambitious goals since it can be utilised in a wide variety of crop species and is a versatile genetic tool that is rapidly advancing. CRISPR-based gene manipulation tools have the potential to produce inheritable genetic mutations like In/Dels, directed insertions, point mutations, and nucleotide substitutions, which are the predominant types. Furthermore, this technology can also produce directed chromosomal rearrangements that allow for genetic or epigenetic regulation of genes. Additionally, it has the benefit of lessening pleiotropic and off-target impacts, and transgene-free modified plants are still a possibility. CRISPR emerged as a potent tool in crop breeding for enhancing yield, quality, safety, resilience to ecological stresses, plant-based biopharming, etc. (Table 7.2).

After the successful engineering of the bacterial CRISPR system by Jinek and colleagues in 2012, it did not take long for scientists to showcase the use of CRISPR technology in plants (Jinek et al. 2012; Li et al. 2013; Feng et al. 2013;

Table 7.2 List of crops where CRISPR has been used. Species of plants, targeted genes, and trait(s) for Sustainable Agriculture (Camerlengo et al. 2022)

Application	Crop	Target Genes	Trait(s)
Genetic Variability	Tomato	RECQ4	Raised the genetic recombination frequency
	Wheat	ZIP4-B2	Amplified crossover frequency
	Pea	Zep1	Improved recombination among the homologous chromosomes
Abiotic stress resistance	Rice	ARGOS8	Improved the tolerance and resistance to drought
	Maize	MLO	Boosted resistance to powdery mildew
	Wheat	Qsd1	Extended dormancy
	Soybean	F3H1, F3H2, FNSII-1	Improved isoflavone content and resistance to the soybean mosaic virus
Yield	Rice	PYL1, PYL4, PYL6	Better-quality growth and productivity
	Maize	CLE	Greater kernel count
	Wheat	GW2	Enhanced the grain size and test weight
	Soybean	GASR7	Improved grain size
Quality	Rice	IPK1	Lowered the phytic acid content
	Maize	SBEIIa	Raised the amylose content
	Wheat	a-gliadin genes	Lessen gluten content
	Sweet potato	ASN2	Reduced the free asparagine
Synthetic Biology	Tomato	SBEI	Boosted the amylose content
	Rice	CrtI, PSY	Addition of large DNA fragments
	Potato	>GBSS	Reduce the amylose content
	Arabidopsis	Chromosome 1, Chromosome 2	Reciprocal translocation

Holton et al. 2015), as evident from growing research publications in the last decade. In 2013, Shan and colleagues utilised CRISPR/Cas-assisted genome modification to disable the OsPDS and OsBADH2 genes in rice. Soon after, in 2013 and 2014, respectively, Upadhyay et al. and Shan et al. also used this technique to modify the inositol oxygenase (inox) and phytoene desaturase (pds) genes in a cultured cell suspension in wheat. In 2014, modifications were introduced to the ZmIPK gene in maize protoplast (Liang et al. 2014). ARGONAUTE7 (SlAGO7) gene-focused modifications were induced by employing CRISPR/Cas9 system to create homozygous edited tomatoes in T0 generations Brooks et al. (2014). Since then, the CRISPR system has been used to edit many crop species. The year 2015 evidences the use of CRISPR to edit potato genes by two different groups (Wang et al. 2015; Butler et al. 2015), as well as multiple scholarly publications describing CRISPR/Cas9-mediated genome changes in soybean (Michno et al. 2015; Jacobs et al. 2015; Sun et al. 2015; Cai et al. 2015). As a consequence of this, CRISPR developed into a potent tool in crop breeding to enhance several traits of crops (Table 7.1), such as yield, quality, safety, resistance to ecological stressors, biopharming, etc. (Dey, 2021). Even though CRISPR is being extensively utilised for numerous purposes over the past decade (such as enhancing yield performance and quality features through biofortification) (Menz et al. 2020), there are currently very few genetically-modified products in the market circulation. The CRISPR/Cas9 technology is being utilised to modify numerous agriculturally important plants not only to add desirable traits but also heritable features like higher production and improved stress adaptability.

7.6.1 *Expanding the variability of genetics*

A deeper comprehension of the capabilities of the CRISPR/Cas toolkit and its possible epigenetic implementations may lead to the formation of novel genetic variability suitable for the creation of novel kinds with novel allele blends. Due to the rarity of meiotic crossovers, crop production from new alleles and their combination is constrained. The introduction of crossover events among chromosomal regions that are homologous or non-homologous in this context (Filler hayut et al. 2021), can increase genetic variability by manipulating meiotic recombination (Taagen et al. 2020; Blary and Jenczewski, 2019). Editing some genes that prevent meiotic recombination has increased the frequency of crossovers in plants. Rice, pea, and tomato, in particular, experienced an increase in crossovers as a result of mutations in RecQ Like Helicase 4 (RECQ4) (Mieulet et al. 2018). According to Liu et al. 2021 mutations in the zeaxanthin epoxidase, ZEP1 increased the frequency of genetic recombination in rice. Homologous recombination (HR) is made easier by the ZIP4-B2 protein in wheat, but crossovers between homeologous chromosomes are prevented. Martin et al. (2021) obtained Zip4-ph1d, a mutant of ZIP4-B2 that ordinarily takes part in HR and besides allowing recombination across homeologous chromosomes. Indeed, these methods make it simpler to add a promising novel assortment of desired characteristics to crops (Schaart et al. 2016; Ahmar et al. 2020; Qaim, 2020; Hartung and Schiemann, 2014).

7.6.2 High yield

Several investigations have demonstrated that some specific genes' expressions have an adverse effect on crop yield and that CRISPR/Cas9-mediated deletion of these genes may have a positive influence. Many investigations have demonstrated that some genes' expression has a negative impact on yield and that their deletion using CRISPR/Cas9 technology may have a favourable effect. CRISPR/Cas9 technology has been used to silence several rice genes, namely GN1a, DEP1, GS3, GW2, GW5, and TGW6, leading to a significant improvement in grain size and weight, grain number, and dense and erect panicles. (Li et al. 2016; Xu et al. 2016). In a similar study, hexaploid wheat with the genes GW2, GW7, and GASR7 knocked out increased seed size and weight (Wang et al. 2018; Zhang et al. 2016; Wang et al. 2019). CRISPR/Cas-derived cultivars with increased yield are not yet available for purchase. Nonetheless, a few studies have entered the trial phase. For instance, in rice, the combined alterations of the PYL1, PYL4, and PYL6 genes, increased grain yield as well as improved growth. Additionally, the CLV-WUS feedback signaling pathway controls the inflorescence meristem, which in turn affects the number of maize kernels per ear. Increasing meristem activity and grain yield can be achieved through cis-regulatory region gene editing (Liu et al. 2021). The alteration in CLV-WUS pathway's cis-regulatory component also resulted in a bigger tomato fruit (Rodríguez-Leal et al. 2017).

Alternately, crop yield can be influenced by genome editing in other ways. By eliminating the waxy gene, high amylopectin variants of superior cultivars were produced using CRISPR/Cas9 (Zhang et al. 2018). The edited maize cultivars outperformed conventionally bred high amylopectin varieties in terms of yield by 5.5 bushels per acre. Additionally, they could also be created more quickly, proving the potential of genome editing for particular specialised functions (Zhang et al. 2018). The yield can also be increased by decreasing the ABA sensitivity in rice crops. Rice plants modified using CRISPR/Cas9 outperformed the control (Beying et al. 2020). These rice plants have simultaneous mutations in class I PYL genes, which encode ABA receptors.In optimal-hydrated conditions, triple elimination of PYLs 1, 4, and 6 increased yield by 30% in well-watered conditions (Beying et al. 2020). It is attractive to perceive how the yield is affected by these PYL genes that encode for ABA under less-than-ideal conditions. According to a recent study (Schachtsiek and Stehle, 2019), wheat PYL1-1B (TaPYL1-1B), which showed better ABA sensitivity, photosynthetic capacity, and water use efficiency under drought conditions, is accountable for augmented production and drought resistance.

Using CRISPR/Cas9, you can also alter the flower repressor gene to increase tomato yield. Tomato plants that lack the SELF-PRUNING 5G

(SP5G), a flowering repressor gene exhibit rapid flowering, yield earlier, and compact, determined growth (Morris and Spillane, 2008). On the other hand, variations in the SELF-PRUNING (SP) gene altered the plant architecture by causing plants to become bushier and to bear additional branches (Podevin et al. 2012). When juxtaposed with the control, the resulting mutants with two mutations had quicker blooming timings and early fruit ripening. In another experiment, tomato SlAGL6 was eliminated using CRISPR to increase yield when exposed to high temperatures. Heap (2013) reported that the tomato agl6 mutants had facultative parthenocarpy with no pleiotropic effects, and produced seedless fruits of the same size and shape as WT. Plant height, pod count, and seed weight had a considerably lesser impact on the yield of the CRISPR-edited soybean mutants (GmAITRs) under salinity stress than in the wild type (Hartung and Schiemann, 2014). The volume of research aiming to boost crop yield and resistance is projected to increase due to the quick development of genome editing techniques.

7.6.3 *Quality improvement*

Considering this environment, vitamin A deficiency (VAD) is among the most widespread contributors to nutritional disorders. It mostly affects children and pregnant women, resulting in several health problems like vision loss, high infection risk, fetal malformations, and neonatal death (Sommer, 2008). Various methods for carotenoid bio-fortification have been used to successfully apply CRISPR/Cas9 genome editing to tomatoes and rice. Rice with a high dry weight of beta-carotene was generated by CrtI and PSY overexpression as marker-free mutants (Waltz, 2016), while in tomato, silencing five genes (SGR1, LCY-E, Blc, LCY-B1, and LCY-B2) related to carotenoid metabolic pathway promoted the synthesis of lycopene, which is a bioactive substance employed in the treatment of chronic illnesses and reduce the likelihood of cancer as well as cardiovascular disorders (Li et al. 2018).

Phytic acid (PA), a primary phosphorus depository in plant seeds is difficult to be digested by monogastric animals, including humans because their digestive tract lacks phytase. PA is regarded as an anti-nutritional compound since it restricts the bioavailability of minerals and phosphorus (Sparvoli et al. 2015). Three BnITPK gene paralogs were silenced in *Brassica napus* to produce low phytic acid (lpa) mutants. The mutants simultaneously showed a 35% reduction in phytic acid and a rise in Pi (Sashidhar et al. 2020). TaIPK1. A substantial decrease in the levels of phytic acid in common wheat, a disruption using CRISPR/Cas9 resulted in an increase in Fe concentration of 1.5 to 2.1 times and an increase in Zn absorption of 1.6 to 1.9 times (Ibrahim et al. 2022).

Cereals, the most important source of carbohydrates for human nutrition can be improved nutritionally by altering their output and carbohydrate contents. Due to their positive benefits on human health and reduced likelihood of chronic diseases that are diet-related and not caused by infections, cereals high in resistant starch and amylose are particularly attractive (Regina et al. 2006).

In many crop products, proteins are another important nutrient. The technological as well as nutritive value of processed foods can be affected by their quantity and composition. However, some proteins are undesirable due to their toxicity and allergenicity. Consuming foods made from wheat and other cereals can lead to several health problems. Wheat-dependent exercise-induced anaphylaxis (WDEIA), as well as Celiac disease (CD), are caused by gluten proteins, which also play a role in determining the technological characteristics of doughs (Tatham and Shewry, 2008). In 2018, Sánchez-León et al. reduced the -gliadins in wheat grain using the CRISPR/Cas9 technology in a bid to develop low-gluten, transgene-free wheat (Sánchez-León et al. 2018).

According to Mansueto et al. (2019) and Geisslitz et al. (2021), metabolic as well as structural proteins including amylase/trypsin inhibitors (ATI), are the underlying cause of both wheat allergies and non-celiac wheat sensitivity (NCWS) (Mansueto et al. 2019; Geisslitz et al. 2021). The CRISPR/Cas9 multiplexing strategy was employed to manipulate ATI subunits WTAI-CM3 as well as WTAI-CM16 of durum wheat which subsequently facilitated the development of transgene-free wheat lines having fewer allergens (Camerlengo et al. 2020). Finally, the D-hordein gene deletion causes a significant rise in glutenins and a decline in prolamines in barley, enabling the adjustment of the gluten content (Yang et al. 2020).

7.6.4 Abiotic stress resistance

Numerous abiotic stresses brought on by climate change are a significant threat to global agricultural food production (Neupane et al. 2022). Abiotic stresses like salinity, scorching heat waves, and water stress affect about 90% of all arable lands (Lau et al. 2021). Plants have developed a variety of mechanisms to deal with and respond to these stresses to survive (Mohd Amnan et al. 2021). However, because the plant stress-responsive and adaptation mechanisms are complicated and a variety of genes regulate them, it is difficult to develop novel cultivars using conventional methods (Vats et al. 2019). Therefore, the CRISPR/Cas9 system might be employed to focus genetic modifications on a single or numerous target positions to create crop lines that are resistant to abiotic adversities (Wada et al. 2020).

Crop viability under environmental adversities has increased through employing the CRISPR/Cas9 strategy. As an illustration, Zhang et al. (2019) developed salinity-resistant rice using the CRISPR/Cas9 technique. The authors discovered that under salinity conditions, the generated rice exhibited superior plant growth to the wild-type OsNAC041 was found to be a crucial transcription factor in rice's salt stress response in a recent study. The CRISPR/Cas9-targeted osnac041 mutant was reported to be taller than the control (Bo et al. 2019). Salinity stress adaptation was implicated in other studies by members of the RAV (related to ABI3/VP1) transcription factor family belonging to the AP2/ERF domain (Xie et al. 2019; Faraji et al. 2020).

The OsRAV2 gene, for instance, was activated when salt stress was applied to the rice. To investigate the function of the GT-1 element in the OsRAV2 gene, Duan et al. (2016) created a sgRNA that specifically targets the GT-1 region of the promoter. They discovered that under salinity conditions, the mutant lines were unable to exhibit the OsRAV2 gene when exposed to salinity, demonstrating the gene's importance in responding under saline adversities. Liu et al. (2020), have reported a similar finding when salt hypersensitivity was seen in the CRISPR/Cas9-mediated OsGTg-2 silenced lines. The CRISPR/Cas9 technology for editing the genome has also been used on wheat (Nazir et al. 2022), soybean (Wang et al. 2021), maize (Zhang et al. 2018), and tomato (Tran et al. 2021).

Plant growth and yield are constrained by the disruption of biochemical and physiological functions caused by drought stress (Amnan et al. 2022). It has been established that many genes and plant hormone signaling and regulation mechanisms are crucial for drought stress reactions. Among them, abscisic acid (ABA) is a pivotal player in regulating water usage and coordinating the plant's behavior to drought stress. As a result, numerous studies have been conducted to specifically target the genes involved in ABA signaling and enhance the drought tolerance of crops. Take for instance, the ABA 80-hydroxylase gene OsABA8ox2, which was identified by Zhang et al. (2020), and was found to be important for rice's resistance to drought. It was discovered that the varieties with silenced OsABA8ox2 gene mediated by CRISPR/Cas9 have a higher concentration of ABA in the roots induced by exposure to drought which eventually increases its ability to withstand drought. On the other hand, overexpressing OsABA8ox2 in rice, reduced root elongation and made it more sensitive to drought stress (Zhang et al. 2020). The β-subunit of the protein farnesyltransferase, encoded by the enhanced response to ABA1 (ERA1), was mutated in Japonica rice cv. utilising the CRISPR/Cas9 system, Nipponbare (Ogata et al. 2020). Through stomatal regulation, the rice osera1 mutant strains demonstrated improved drought tolerance and ABA sensitivity, indicating that ERA1 may be a promising candidate gene for crop drought resistance. One investigation by Yin et al. (2017) demonstrated that CRISPR/Cas9-edited rice plants had OsEPFL9 (Epidermal Patterning Factor like-9) mutants with an SD that was more than eight times lower. Because of the reduced SD, the modified rice lines can resist drought stress under ideal conditions, a 50% drop in SD in barley and wheat resulted in an increase in water usage efficiency (WUE) and a considerable decrease in carbon absorption and conductance (Caine et al. 2019; Hughes et al. 2017). Similarly, in sufficiently-hydrated conditions, CRISPR assisted a knockout of VvEPFL9-1 in grapevine, lowered the carbon absorption, and reduced SD by 60% (Clemens et al. 2022). The CRISPR/Cas9-produced slmapk3 mutants, which similarly exhibited greater wilting signs and cell membrane impairment during drought stress, showed that SlMAPK3 is implicated in response to drought in tomatoes (Wang et al. 2017).

To lessen the toxicity of minerals, some studies utilised the CRISPR/Cas9 technology. Nieves-Cordones et al. (2017) created rice plants with low levels of Cesium by using the CRISPR/Cas9 system to disable the K+ transporter OsHAK1. Arsenic tolerance in rice was enhanced by knocking out OsARM1 and OsNramp5 (Wang et al. 2017) and tolerable cadmium accumulation (Tang et al. 2017) in each case. Shao et al. (2020) provided yet another illustration of increasing plant stress resistance in which a semi-dwarf banana with disrupted gibberellin genes was developed using the CRISPR/Cas9 system. This has resulted in mature bananas being more resistant to storms and strong winds. Both knock-in mutations of desired genes and knockout mutations of genes that are susceptible can be successfully achieved using genome editing techniques. Examples include Shimatani et al. (2017) inserted a maize promoter before the ARGOS8 gene, which is associated with drought tolerance using CRISPR/Cas9. As a result, under water stress, the edited maize crops produced more grain.

These findings showed that the CRISPR/Cas system effectively alters the plant genome, enabling investigators to examine the roles of genes implicated in abiotic stressor response. There have been limited publications focusing modification of genes associated with abiotic stress tolerance, because of the intricacy of abiotic stress tolerance, which typically requires the manipulation of numerous genes to change the trait of interest.

7.6.5 Biotic stress resistance

Using CRISPR/Cas9, genome editing has produced remarkable results in the fight against disease in rice. Many plants encode sucrose transporters for the SWEET gene family, which are used by the majority of pathogens (Jiang et al. 2013). The promoter regions of several OsSWEET genes were altered by employing CRISPR/Cas9 to develop tolerance to bacterial leaf blight (Xu et al. 2019; Oliva et al. 2019). The OsERF922 gene, which controls plant response to ethylene, was knocked down using CRISPR/Cas9, resulting in improved plant tolerance to leaf blast disease (Wang et al. 2016). Additionally, the eukaryotic elongation factor eIF4G in rice was edited employing CRISPR/Cas9 resulting in the development of plants resistant to the rice tungro virus (Macovei et al. 2018). The CRISPR-modified plants that were infected produced higher yields compared to the wild-type plants and lacked any detectable viral proteins.

The development of the CRISPR/Cas9 system has made it easier to develop simultaneous resistance to multiple diseases. Large-scale engineering of a wide range of disease resistance could provide a unified solution to address multiple crop diseases that impact agricultural production (Xu et al. 2019). Alteration of bsr-k1, a gene in rice that attaches to defense-related genes and boosts their turnover. Zhou et al. (2018) serve as an illustration of this tactic. In engineered rice plants, "turning off" these crucial defense genes prevented bacterial leaf blight and leaf blast. The transgenic lines exhibited a 50% increase in yield when confronted with rice leaf blast under field conditions without affecting other yield attributes (Zhou et al. 2018). Other crops have also benefited from the same method to enhance disease resistance. For example, tomatoes have achieved broad-spectrum resistance by modifying a single locus (Thomazella et al. 2021). CRISPR/Cas9-mediated SlDMR6-1 mutations in the edited lines confer tolerance to *Pseudomonas syringae, Phytophthora capsici*, and *Xanthomonas spp.*, maintaining an elevated presence of salicylic acid in the plant while significantly reducing disease symptoms and pathogen abundance (Thomazella et al. 2021). CRISPR/Cas9-based gene modification of MORC1, a gene involved in defense mechanisms, enhanced barley's tolerance to *Fusarium graminearum* and powdery mildew of barley (Kumar et al. 2018). The study also demonstrated that there were fewer lesions and less fungal DNA in the altered barley plants.

Some species' resistance to fungi infections was boosted by focusing on the MLO and other loci's homologs. Three MLO homologs, TaMLO-A, TaMLO-B, and TaMLO-D, can be simultaneously modified using CRISPR/Cas9 to increase wheat resilience to powdery mildew. The Tomelo transgene-free tomato, which was developed by using CRISPR/Cas9 assisted SlMlo1 gene modification, is resistant to powdery mildew disease, and serves as another example (Nekrasov et al. 2017). Using CRISPR/Cas9, Zhang et al. (2017) modified three wheat TaEDR1 homologs at once to increase their defense against the powdery mildew disease. While the altered grapevine line demonstrated a decrease in powdery mildew spore production by a factor of two, the targeting MLO homologs resulted in increased powdery mildew resistance in grapevine (Wan et al. 2020). Other studies found that knocking out the 14-3-3 c and 14-3-3 d proteins, both of which are negative regulators of disease response, increased immunity against Verticillium *dahlia* (Wang et al. 2018). The engineered cotton exhibited reduced infection signs and a lower level of pathogens than in the WT (Wang et al. 2018).

7.6.6 Herbicide resistance

Herbicide-tolerant invasive plants represent a danger to global hunger prevention and incur large losses since they can lower crop productivity. Growing herbicide-resistant crops turns out to be a successful weed-control tactic in the absence of innovative herbicides due to decreased crop phytotoxicity and an enlarged herbicidal range. CRISPR/Cas, a genome-editing system, has the potential to create certain plants that are herbicide-resistant and enable effective targeted alteration.

7.6.7 Synthetic biology

The term "plant synthetic biology" broadly encompasses the creation of modified plant species through the alteration, deletion, or introduction of biological systems and components to achieve specific objectives. Therefore, all of the successes utilising CRISPR systems that have been previously outlined could be considered as synthetic biology methods. CRISPR editing undoubtedly has the potential to help in the advancement of synthetic biology. Although there has been a lot of work done on crop species, model plants have shown how synthetic gene circuits can be created not only by the insertion of engineered DNA sequences but also through the manipulation of metabolic processes that are involved in gene action. Studies have demonstrated that chromosomal changes like deletions, insertions, and rearrangements can be induced using CRISPR systems (Schmidt et al. 2020; Schindele et al. 2020; Zhou et al. 2014). In addition, Dong et al. (2020) demonstrated that CRISPR-based genetic modification editing can be used to achieve site-specific integration of large gene cassettes to precisely insert a 5.2 kb carotenoid biosynthesis cassette into two rice chromosomal locations that were previously identified. These particular regions are referred to as genomic safe harbors (GSHs) because they can accommodate transgenes without causing harm to the recipient's biological systems.

Furthermore, the CRISPR system also facilitates the production of chromosomal rearrangements and crossovers by inducing DSBs in the chromosome, which favor homologous recombination (HR) among target gene alleles (Taagen et al. 2020; Blary and Jenczewski, 2019). Employing CRISPR/Cas9 technology to specifically induce DSBs in the PHYTOENE SYNTHASE (PSY1) gene and Carotenoid isomerase (CRTISO) locus resulted in inter-homologous somatic recombination in tomato plants (Hayut et al. 2017; Shlush et al. 2021).

Generally, the portions of chromosomes do not undergo crossing over because chromosomal rearrangements such as inversions and translocations inhibit meiotic recombination. CRISPR-induced chromosomal rearrangements, such as the induction or reversion of chromosomal inversions, has the potential to weaken or reinforce genetic connections and bring about reciprocal translocations. CRISPR allows for the induction of deletions, inversions, and reciprocal translocations by introducing several DSBs on homologous or non-homologous chromosomal loci. Schwartz and colleagues (Schwartz et al. 2020) recently demonstrated that CRISPR/Cas9 effectively induced a specific chromosomal inversion of 75.5-Mb in maize, resulting in two DSBs on either side of the inversion. Beying et al. (2020) induced a 1 Mb reciprocal translocation between Chromosomes 1 and 2 in *Arabidopsis thaliana*. They targeted intergenic regions situated at a distance of 0.5 Mb from the ends of both chromosomes' q-arms.

Additionally, the development of gene circuits is increasingly using synthetic promoters and transcription factor engineering. Through the use of these techniques, specific chemicals that could be used for commercial or medical applications are synthesised. The best strategy to influence gene expression in this situation at the transcriptional or epigenetic level is by using CRISPR orthologs and variants (Huang et al. 2021). Reports on the transcriptional control and epigenetic modification of plant genes have primarily centered on model plant species (Karlson et al. 2021), demonstrating their extensive crop application potential. Additionally, the translation of a particular protein is frequently regulated by upstream open reading frame (uORF) modification-mediated translation regulation (Si et al. 2020). Nucleotide insertions and deletions in the upstream open reading frames (uORFs) of genes involved in the development and biosynthesis of antioxidants in lettuce, tomato, and *Arabidopsis thaliana* have been reported by Zhang et al. (2018).

Despite the loss of cultivable land and the imperative requirement for sustainable food production being linked to climate change, biofuel demand is severely impacted by constant fuel consumption. To develop improved varieties suitable for biodiesel production, oil seed crops need to have a higher oil content and the best fatty acid composition. The fatty acid desaturase 2 (FAD2) gene in *Brassica napus*, responsible for the production of an enzyme that catalyzes the desaturation of oleic acid, was altered using CRISPR/Cas9. This modification led to an increase in the concentration of oleic acid in the modified plant seeds than that in wild-type seeds (Okuzaki et al. 2018). Through deletion of the NtAn1 transcription factor, *Nicotiana tabacum* seed lipid accumulation increased (Tian et al. 2021). These tobacco plants can be used to make biodiesel.

In addition, the production of more digestible forage and biofuels, as well as the production of paper and textiles, could all benefit from the lower lignin content of biomass crops. Park et al. (2017) reported the decrease in lignin content caused by the 4-coumarate knockout: switchgrass (*Panicum virgatum*), a perennial grass that is a lignocellulosic source for bioenergy production, possesses the coenzyme A ligase (4CL) gene. Switchgrass that had undergone mutation showed a drop in lignin and produced more sugar. CRISPR-mediated modification of the caffeic acid O-methyltransferase 1 (COMT-1) gene dramatically increased the bioethanol content in the mutant line while reducing the amount of lignin in the barley (Lee et al. 2021).

7.7 Conclusion

Genome editing tools are ideal for developing more sustainable agricultural systems. They provide novel strategies for raising food production while preserving non-renewable resources by creating altered crops with higher yields or more efficient water use (such as soil, energy, and water). Additionally, a lot of successful initiatives have been documented to boost crops' resistance to heat and drought, which lower crop output and jeopardise food security. Regarding biotic stressors, GE crops may present a chance to cut back on the usage of phytochemicals while also defending the environment and saving farmers money. Additionally, CRISPR-mediated editing offers a fantastic chance to develop plant synthetic biology. The adaptability of CRISPR systems, which enables the introduction of significant chromosome modifications as well as the modulation or production of synthetic gene circuits, can be advantageous for applications ranging from plant pharming to improving crop performance and the quality of food products. With unprecedented accuracy and efficiency, scientists can now tweak genes thanks to the CRISPR/Cas9 genome editing system. Crops can be improved in countless ways, including productivity, nutrient content, stress resistance, and a host of other factors. It is being utilised passionately to increase crop quality and yield.

RNA-guided endonuclease (RGEN) RNPs and/or donor DNA can be directly transferred into protoplast cells, allowing for the generation of genetically stable mutations without leaving any vector DNA in the plant genome. Such extraneous-DNA-free genetically modified crops have a bright future in agriculture and food supply. The US Department of Agriculture (USDA) has stated that GE maize and mushrooms shouldn't be classified as genetically modified crops because CRISPR technology was used to generate them. However, CRISPR/Cas technology critics contend that these genome-edited crops should be treated as generic GMOs, which would thwart the application of this cutting-edge and promising technology.

Additional investigation into genome editing and its impacts is necessary to address public concerns such as phenotypic effects on plants, genome-wide off-target effects produced by guide RNAs, and remedial challenges. We conclude that CRISPR/Cas9-based genome editing will depend largely not only on scientific capacity but also on public confidence in science to go even further in adopting and strengthening the targets related to a sufficient food supply for humanity. CRISPR/Cas9 has been hailed as one of the most potent tools for the GE of numerous significant crops because of its high efficiency, low cost, and ease of use in comparison to other GE technologies like ZFNs and TALENs. CRISPR/Cas9 has started to alter biological research as the preferred technique for focusing on particular genome sequences in simple or complex organisms. With the help of this technique, germplasm can be created that has higher yields, better nutrition is more resistant to biotic and abiotic challenges, and has fewer off-target impacts. Researchers may be helped by the availability of whole-genome sequencing data from various kinds of cereal in their quest to identify novel genes that compassionately and precisely govern vital agricultural processes. Despite substantial advancements in efficiency and target specificity, CRISPR technology still needs to be improved. The novel CRISPR advancements CRISPR/Cas12, Cas13, and Base editing need more study to become dependable and readily accessible for crop improvement programs. The genome editing will also assist farmers in realising zero hunger by cultivating better-quality cultivars.

References

Abbotts, R., Wilson, D.M. III 2017. Coordination of DNA single strand break repair. *Free Radic. Biol. Med.* 107:228–244.

Abe, K., Araki, E., Suzuki, Y., Toki, S., Saika, H. 2018. Production of high oleic/low linoleic rice by genome editing. *Plant Physiol. Biochem.* 131, 58–62.

Ahmar, S., Gill, R.A., Jung, K.H., Faheem, A., Qasim, M.U., Mubeen, M., Zhou, W. 2020. Conventional and molecular techniques from simple breeding to speed breeding in crop plants: Recent advances and future outlook. *Int. J. Mol. Sci.* 21, 2590.

Alekseev, S., Coin, F. 2015. Orchestral maneuversat the damaged sites in nucleotide excision repair. *Cell Mol. Life Sci.* 72:2177–2186.

Amnan, M.A.M., Aizat, W.M., Khaidizar, F.D., Tan, B.C. 2022. Drought stress induces morpho-physiological and proteome changes of Pandanus amaryllifolius. *Plants* 11:221.

Andersson, M., Turesson, H., Nicolia, A., Fa¨lt, A.-S., Samuelsson, M., Hofvander, P. 2017. Efficient targeted multiallelic mutagenesis in tetraploid potato Solanum tuberosum by transient CRISPR-Cas9 expression in protoplasts. *Plant Cell. Rep.* 36 (1):117–128.

Bae, S., Park, J., Kim, J.S. 2014. Cas-OFFinder: A fast and versatile algorithm that searches for potential off-target sites of Cas9 RNA-guided endonucleases. *Bioinformatics* 30 (10):1473–1475.

Barrangou, R., Fremaux, C., Deveau, H. et al. 2007. CRISPR provides acquired resistance against viruses in prokaryotes. *Science* 315:1709–1712.

Bertin, G., Averbeck, D. 2006. Cadmium: Cellular effects, modifications of biomolecules, modulation of DNA repair and genotoxic consequences a review. *Biochimie* 88 (11):1549–1559.

Beying, N., Schmidt, C., Pacher, M., Houben, A., Puchta, H. 2020. CRISPR–Cas9-mediated induction of heritable chromosomal translocations in arabidopsis. *Nat. Plants* 6:638–645.

Bibikova, M., Beumer, K., Trautman, J. et al. 2003. Enhancing gene targeting with designed zinc finger nucleases. *Science* 300:764.

Blary, A., Jenczewski, E. 2019. Manipulation of crossover frequency and distribution for plant breeding. *Theor. Appl. Genet* 132:575–592.

Bo, W., Zhaohui, Z., Huanhuan, Z., Xia,W., Binglin, L., Lijia, Y., Xiangyan, H., Deshui, Y., Xuelian, Z., Chunguo,W., et al. 2019. Targeted mutagenesis of NAC transcription factor gene, OsNAC041, leading to salt sensitivity in rice. *Rice Sci.* 26:98–108.

Britt, A.B. 2004. Repair of DNA damage induced by solar UV. *Photosynth. Res.* 81:105–112.

Brooks, C., Nekrasov, V., Lipppman, Z.B., Van Eck, J. 2014. Efficient gene editing in tomato in the first generation using the clustered regularly interspaced short palindromic repeats/CRISPR-associated9 system. *Plant Physiol.* 166:1292–1297.

Butler, N.M., Atkins, P.A., Voytas, D.F., Douches, D.S. 2015. Generation and inheritance of targeted mutations in potato (Solanum tuberosum L.) using the CRISPR/Cas system. *PLoS ONE* 10:1–12.

Cai, Y., Chen, L., Liu, X., Guo, C., Sun, S., Wu, C., et al. 2018. CRISPR/Cas9-mediated targeted mutagenesis of GmFT2a delays flowering time in soya bean. *Plant Biotechnol. J.* 16:176–185.

Cai, Y., Chen, L., Liu, X., Sun, S., Wu, C., Jiang, B., Han, T., Hou, W. 2015. CRISPR/Cas9-mediated genome editing in soybean hairy roots. *PLoS ONE* 10:1–13.

Caine, R.S., Yin, X., Sloan, J., Harrison, E.L., Mohammed, U., Fulton, T., Biswal, A.K., Dionora, J., Chater, C.C., Coe, R.A., et al. 2019. Rice with reduced stomatal density conserves water and has improved drought tolerance under future climate conditions. *NewPhytol* 221:371–384.

Caldecott, K.W. 2008. Single-strand break repair and genetic disease. *Nat. Rev. Genet.* 9:619–631.

Caldecott, K.W. 2014. DNA single-strandbreak repair. *Exp. Cell Res.* 329:2–8.

Camerlengo, F., Frittelli, A., Pagliarello, R. 2022. CRISPr towards a sustainable agriculture. *Encyclopaedia* 2:538–558.

Camerlengo, F., Frittelli, A., Sparks, C., Doherty, A., Martignago, D., Larré, C., Lupi, R., Sestili, F., Masci, S. 2020. CRISPR-Cas9 multiplex editing of the -amylase/trypsin inhibitor genes to reduce allergen proteins in durum wheat. *Front. Sustain. Food Syst.* 4:104.

Ceccaldi, R., Rondinelli, B., D'Andrea, A.D. 2016. Repair pathway choices and consequences at the double-strand break. *TrendsCell Biol.* 26:52–64.

Cermak, T., Baltes, N., Cegan, R. et al. 2015. High-frequency, precise modification of the tomato genome. *Genome Biol.* 16:232

Chang, H.H.Y., Pannunzio, N.R., Adachi, N. et al. 2017. Non-homologous DNA end joiningand alternative pathways to double-strandbreak repair. *Nat. Rev. Mol. Cell Biol.* 18:495–506.

Chari, R., Yeo, N.C., Chavez, A., Church, G.M. 2017. sgRNA scorer 2.0: a species-independent model to predict CRISPR/Cas9 activity. *ACS Synth. Biol.* 6:902–904.

Che, P., Anand, A., Wu, E., Sander, J.D., Simon, M.K., Zhu, W., et al. 2018. Developinga flexible, high-efficiency Agrobacterium-mediated sorghum transformation systemwith broad application. *Plant. Biotechnol.* 16:1388–1395.

Clemens, M., Faralli, M., Lagreze, J., Bontempo, L., Piazza, S., Varotto, C., Malnoy, M., Oechel, W., Rizzoli, A., Dalla Costa, L. 2022. VvEPFL9-1 knock-out via CRISPR/Cas9 reduces stomatal density in grapevine. *Front. Plant Sci.* 13:878001

Clemens, S., Aarts, M.G., Thomine, S., Verbruggen, N. 2013. Plant science: the key topreventing slow cadmium poisoning. *Trends Plant Sci.* 18:92–99.

Cong, L., Ran, F.A., Cox, D. et al. 2013. Multiplexgenome engineering using CRISPR/Cas systems. *Science* 339:819–823.

Crouse, G.F. 2016. Non-canonical actions ofmismatch repair. *DNA Repair* 38:102–109.

Davis, L., Maizels, N. 2016. Two direct pathways support gene correction by single strand ed donors at DNA nicks. *Cell Rep.* 17:1872–1881.

Deriano, L., Roth, D.B. 2013. Modernizing the nonhomologous end-joining repertoire: Alternative and classical NHEJ share the stage. *Annu. Rev. Genet.* 47:433–455.

Dey, A. 2021. CRISPR/Cas genome editing to optimize pharmacologically active plant natural products. *Pharmacol. Res.* 164:105359.

Doench, J.G., Fusi, N., Sullender, M., Hegde, M., Vaimberg, E.W., Donovan, K.F., Root, D.E. 2016. Optimized sgRNA design to maximize activity and minimize off-target effects of CRISPR-Cas9. *Nat. Biotechnol.* 34:184–191.

Dong, O.X., Yu, S., Jain, R., Zhang, N., Duong, P.Q., Butler, C., Li, Y., Lipzen, A., Martin, J.A., Barry, K.W., et al. 2020. Marker-free carotenoid-enriched rice generated through targeted gene insertion using CRISPR-Cas9. *Nat. Commun.* 11: 1–10.

Duan, Y.B., Li, J., Qin, R.Y., Xu, R.F., Li, H., Yang, Y.C., Ma, H., Li, L., Wei, P.C., Yang, J.B. 2016. Identification of a regulatory element responsible for salt induction of rice OsRAV2 through ex situ and in situ promoter analysis. *Plant Mol. Bio.* 90:49–62.

Faraji, S., Dreuw, A. 2017. Insights into light-driven DNA repair by photolyases: Challenges and opportunities for electronic structure theory. *Photochem. Photobiol.* 93:37–50.

Faraji, S., Filiz, E., Kazemitabar, S.K., Vannozzi, A., Palumbo, F., Barcaccia, G., Heidari, P. 2020. The AP2/ERF gene family in Triticumdurum: Genome-wide identification and expression analysis under drought and salinity stresses. *Genes* 11:1464.

Fauser, F., Schiml, S., Puchta, H. 2014. Both CRISPR/Cas-based nucleases and nickases can be used efficiently for genome engineering in Arabidopsis thaliana. *Plant J*79:348–359.

Feng, Z., Zhang, B., Ding, W., Liu, X., Yang, D.L., Wei, P., Cao, F., Zhu, S., Zhang, F., Mao, Y., et al. 2013. Efficient genome editing in plants using a CRISPR/Cas system. *Cell Res.* 23:1229–1232.

Filler-hayut, S., Kniazev, K., Melamed-bessudo, C., Levy, A.A. 2021. Targeted inter-homologs recombination in arabidopsis euchromatin and heterochromatin. *Int. J. Mol. Sci.* 22:12096.

Gaj, T., Gersbach, C.A., Barbas, C.F. 3rd 2013. ZFN, TALEN, and CRISPR/Cas-based methodsfor genome engineering. *Trends Biotechnol.* 31(7):397–405.

Gaj, T., Sirk, S.J., Shui, S.L., Liu, J. 2016 Dec 1. Genome-editing technologies: Principles and applications. *Cold Spring Harb Perspect Biol* 8(12):a023754. doi: 10.1101/cshperspect.a023754. PMID: 27908936; PMCID: PMC5131771.

Gao, W., Long, L., Tian, X., Xu, F., Liu, J., Singh, P.K., et al. 2017. Genome editing in cotton with the CRISPR/Cas9 system. *Front. Plant Sci.* 8:1364.

Geisslitz, S., Shewry, P., Brouns, F., America, A.H.P., Caio, G.P.I., Daly, M., D'Amico, S., De Giorgio, R., Gilissen, L., Grausgruber, H., et al. 2021. Wheat ATIs: Characteristics and role in human disease. *Front. Nutr.* 8:1–16.

Gorbunova, V., Levy, A.A. 1999. How plants make ends meet: DNA double-strand break repair. *Trends Plant Sci.* 4:263–269.

Hartung, F., Schiemann, J. 2014. Precise plant breeding using new genome editing techniques: Opportunities, safety and regulation in the EU. *Plant J.* 78:742–752.

Hayut, S.F., Bessudo, C.M., Levy, A.A. 2017. Targeted recombination between homologous chromosomes for precise breeding in tomato. *Nat. Commun.* 8:1–9.

Heap, B. 2013. Europe should rethink its stance on GM crops. *Nature* 498:409.

Holton, N., Nekrasov, V., Ronald, P.C., Zipfel, C. 2015. The phylogenetically-related pattern recognition receptors EFR and XA21 recruit similar immune signaling components in monocots and dicots. *PLoS Pathog* 11:1–22.

Hsu, P.D., Scott, D.A., Weinstein, J.A., Ran, F.A., Konermann, S., Agarwala, V., Li, Y., Fine, E.J., Wu, X., Shalem, O., Cradick, T.J., Marraffini, L.A., Bao, G., Zhang, F. 2013. DNA targeting specificity of RNA-guided Cas9 nucleases. *Nat. Biotechnol.* 31(9):827–832.

Hu, Z., Cools, T., De Veylder, L. 2016. Mechanisms used by plants to cope with DNA damage. *Annu. Rev. Plant Biol.* 67:439–462.

Huang, D., Kosentka, P.Z., Liu, W. 2021. Synthetic biology approaches in regulation of targeted gene expression. *Curr. Opin. PlantBiol.* 63:102036

Huang, X., Qian, Q., Liu, Z., Sun, H., He, S., Luo, D., et al. 2009. Natural variation at the DEP1 locus enhances grain yield in rice. *Nat. Genet.* 41:494–497.

Hughes, J., Hepworth, C., Dutton, C., Dunn, J.A., Hunt, L., Stephens, J., Waugh, R., Cameron, D.D., Gray, J.E. 2017. Reducing stomatal density in barley improves drought tolerance without impacting on yield. *Plant Physiol.* 174:776–787.

Hummel, A.W., Chauhan, R.D., Cermak, T., Mutka, A.M., Vijayaraghavan, A., Boyher, A., et al. 2018. Allele Exchange at the EPSPS Locus ConfersGlyphosate Tolerance in Cassava. *Plant Biotechnol. J.* 16:1275–1282. doi:10.1111/pbi.12868

Ibrahim, S., Saleem, B., Rehman, N., Zafar, S.A., Naeem, M.K., Khan, M.R. 2022. CRISPR/Cas9 mediated disruption of inositol pentakisphosphate 2-kinase 1 (TaIPK1) reduces phytic acid and improves iron and zinc accumulation in wheat grains. *J. Adv. Res.* (37):33–41, in adaptive bacterial immunity. Science 337: 816–821

Ito, Y., Nishizawa-Yokoi, A., Endo, M., Mikami, M., Toki, S. 2015. CRISPR/Cas9-mediated mutagenesis of the RIN locus that regulates tomato fruit ripening. *Biochem. Biophys. Res. Commun.* 467:76–82.

Jacobs, T.B., LaFayette, P.R., Schmitz, R.J., Parrott,W.A. 2015. Targeted genome modifications in soybean with CRISPR/Cas9. *BMC Biotechnol.* 15:1–10.

Jaganathan, D., Ramasamy, K., Sellamuthu, G., Jayabalan, S., Venkataraman, G. 2018. CRISPR for crop improvement: An update review. *Front. Plant Sci.* 9:985.

Jia, H., Xu, J., Orbovi´c, V., Zhang, Y., Wang, N. 2017. Editing citrus genome via SaCas9/ sgRNA system. *Front. Plant Sci.* 8:2135.

Jiang, W., Zhou, H., Bi, H., Fromm, M., Yang, B., Weeks, D.P. 2013. Demonstration of CRISPR/Cas9/sgRNA-mediated targeted gene modification in Arabidopsis, tobacco, sorghum and rice. *Nucleic Acids Res.* 41:e188.

Jinek, M., Chylinski, K., Fonfara, I., Hauer, M., Doudna, J.A., Charpentier, E. 2012. A programmable dual-RNA-guided DNA endonuclease in adaptive bacterial immunity. *Science*, 337:816–822.

Kan Y., Ruis B., Takasugi T. et al. 2017. Mechanisms of precise genome editing using oligonucleotide donors. *Genome Res.* 27:1099–1111.

Karlson, C.K.S., Mohd-noor, S.N., Nolte, N., Tan, B.C. 2021. Crispr/Dcas9-based Systems: Mechanisms and applications in plant sciences. *Plants*, 10:2055

Kim, H., Kim, J. 2014. A guide to genome engineering with programmable nucleases. *Nat. Rev. Genet.* 15(5):321–334.

Kim, D., Alptekin, B., Budak, H. 2018. CRISPR/Cas9 genome editing in wheat. *Funct. Integr. Genomic.* 18:31–41.

Komor, A.C., Kim, Y.B., Packer, M.S. 2016. Programmable editing of a target base in genomic DNA without double-stranded DNA cleavage. *Nature* 533:420–424.

Krokan, H.E., Bjoras, M. 2013. Base excision repair. *Cold Spring Harb. Perspect. Biol.* 5:a012583

Krysan, P.J., Young, J.C., Sussman, M.R. 1999. T-DNA as an insertional mutagen in Arabidopsis. *Plant Cell.* 11(12):2283–2290.

Kumar, N., Galli, M., Ordon, J., Stuttmann, J., Kogel, K.-H., Imani, J. 2018. Further analysis of barley MORC1 using a highly efficientRNA-guided Cas9 gene-editing system. *Plant Biotechnol. J.* 16:1892–1903.

Kunkel, T., Erie, D. 2015. Eukaryotic mismatch repair in relation to DNA replication. *Annu. Rev. Genet.* 49:291–313.

Lau, S.E., Hamdan, M.F., Pua, T.L., Saidi, N.B., Tan, B.C. 2021. Plant nitric oxide signaling under drought stress. *Plants*, 10:360.

Lee, J.H., Won, H.J., Hoang Nguyen Tran, P., Lee, S.M., Kim, H.Y., Jung, J.H. 2021. Improving lignocellulosic biofuel production by CRISPR/Cas9-mediated lignin modification in barley. *GCB Bioenergy* 13:742–752.

Li, J., Meng, X., Zong, Y. et al. 2016. Gene replacements and insertions in rice by intron targeting using CRISPR–Cas9. *Nat. Plants* 12(2):16139.

Li, A., Jia, S., Yobi, A., Ge, Z., Sato, S.J., Zhang, C., et al. 2018b. Editing of an alphakafiringene family increases digestibility and protein quality in sorghum. *Plant. Physiol.* 77:1425–1438.

Li, X., Liu, H., Wang, M., Tian, X., Zhou, W., Lu, T., et al. 2015. Combinations of Hd2 and Hd4 genes determine rice adaptability to Heilongjiang Province, northern limit of China. *J. Integr. Plant Biol.* 57:698–707.

Li, X., Zhou, W., Ren, Y., Tian, X., Lv, T., Wang, Z., et al. 2017. High-efficiency breeding of early-maturing rice cultivars via CRISPR/Cas9mediated genome editing *J. Genet. Genom.* 44:175–178.

Li, X., Wang, Y., Chen, S., Tian, H., Fu, D., Zhu, B., Luo, Y., Zhu, H. 2018. Lycopene is enriched in tomato fruit by CRISPR/Cas9- mediated multiplex genome editing. *Front. Plant Sci.* 9:1–12.

Li, W., Teng, F., Li, T., Zhou, Q. 2013. Simultaneous generation and germline transmission of multiple gene mutations in rat using CRISPR-Cas systems. *Nat. Biotechnol.* 31:684–686.

Liang, Z., Zhang, K., Chen, K., Gao, C. 2014. Targeted mutagenesis in *Zea mays* using TALENs and the CRISPR/Cas system. *J. Genet. Genom.* 41:63–68.

Liu, H., Li, H., Hao, C., Wang, K., Wang, Y., Qin, L., et al. 2019. TaDA1, a conserved negative regulator of kernel size, has an additive effect with TaGW2 in common wheat Triticum aestivum L. *Plant. Biotechnol. J.* 18:1330–1342.

Liu, L., Gallagher, J., Arevalo, E.D., Chen, R., Skopelitis, T., Wu, Q., Bartlett, M., Jackson, D. 2021. Enhancing grain-yield-related traits by CRISPR–Cas9 promoter editing of maize CLE genes. *Nat. Plants* 7:287–294.

Liu, X., Wu, D., Shan, T., Xu, S., Qin, R., Li, H., Negm, M., Wu, D., Li, J. 2020. The trihelix transcription factor OsGT g-2 is involved adaption to salt stress in rice. *Plant Mol. Biol.* 103:545–560.

Lydia, H., Diana, S., Tara, A. et al. 2017. Splicing and transcription touch base: Co-transcriptional spliceosome assembly and function. *Nat RevMol Cell Biol* 8:637–650.

Macovei, A., Sevilla, N.R., Cantos, C., Jonson, G.B., Slamet-Loedin, I., Cˇermák, T., Voytas, D.F., Choi, I.R., Chadha-Mohanty, P. 2018. Novel alleles of rice eif4g generated by CRISPR/Cas9-targeted mutagenesis confer resistance to rice tungro spherical virus. *PlantBiotechnol. J.* 16:1918–1927.

Mali, P., Yang, L., Esvelt, K. et al. 2013 RNA-guided human genome engineering via Cas9. *Science* 339(6121):823–826.

Manova, V., Gruszka, D. 2015. DNA damage and repair in plants-from models to crops. *Front. Plant Sci.* 6:885

Mansueto, P., Soresi, M., Iacobucci, R., La Blasca, F., Romano, G., D'Alcamo, A., Carroccio, A. 2019. Non-celiac wheat sensitivity: A search for the pathogenesis of a self-reported condition. *Ital. J. Med.* 13:15–23.

Martín, A.C., Alabdullah, A.K., Moore, G. 2021. A separation-of-function ZIP4 wheat mutant allows crossover between related chromosomes and is meiotically stable. *Sci. Rep.* 11(1): 21811.

Matsubara, K., Hori, K., Ogiso-Tanaka, E., Yano, M. 2014. Cloning of quantitative trait genes from rice reveals conservation and divergence of photoperiod flowering pathways in Arabidopsis and rice. *Front. Plant Sci.* 5:193.

Mellidou, I., Buts, K., Hatoum, D., Ho, Q.T., Johnston, J.W., Watkins, C.B., et al. 2014. Transcriptomic events associated with internal browning of apple during postharvest storage. *BMC Plant Biol.* 14:328

Menz, J., Modrzejewski, D., Hartung, F., Wilhelm, R., Sprink, T. 2020. Genome edited crops touch the market: A view on the global development and regulatory environment. *Front. Plant Sci.* 11:1–17.

Michno, J.M., Wang, X., Liu, J., Curtin, S.J., Kono, T.J., Stupar, R.M. 2015. CRISPR/Cas mutagenesis of soybean and medicago truncatula using a new web-tool and a modified Cas9 enzyme. *GM Crops Food* 6:243–252.

Mieulet, D., Aubert, G., Bres, C., Klein, A., Droc, G., Vieille, E., Rond-Coissieux, C., Sanchez, M., Dalmais, M., Mauxion, J.P., et al. 2018. Unleashing meiotic crossovers in crops. *Nat. Plants* 4:1010–1016.

Mohd Amnan, M.A., Pua, T.L., Lau, S.E., Tan, B.C., Yamaguchi, H., Hitachi, K., Tsuchida, K., Komatsu, S. 2021. Osmotic stress in banana is relieved by exogenous nitric oxide. *PeerJ* 9:e10879

Morales-Ruiz, T., Ortega-Galisteo, A.P., Ponferrada-Marin M.I. et al. 2006. DEMETER and REPRESSOR OF SILENCING1 encode 5-methylcytosine DNA glycosylases. *Proc. Natl. Acad. Sci. USA* 103:6853–6858.

Morris, S.H., Spillane, C. 2008. GM directive deficiencies in the European Union. The current framework for regulating GM Crops in the EU weakens the precautionary principle as a policy tool. *EMBO Rep.* 9:500–504.

Nazir, R., Mandal, S., Mitra, S., Ghorai, M., Das, N., Jha, N.K., Majumder, M., Pandey, D.K., Dey, A. 2022. Clustered regularly interspaced short palindromic repeats (CRISPR)/CRISPR-associated genome-editing toolkit to enhance salt stress tolerance in rice and wheat. *Physiol. Plant.* 174:12.

Nekrasov, V., Wang, C., Win, J., Lanz, C., Weigel, D., Kamoun, S. 2017. Rapid generation of a transgene-free powdery mildew resistant tomato by genome deletion. *Sci. Rep.* 7:482.

Neupane, D., Adhikari, P., Bhattarai, D., Rana, B., Ahmed, Z., Sharma, U., Adhikari, D. 2022 Does climate change affect the yield of the top three cereals and food security in the world? *Earth* 3:45–71.

Nieves-Cordones, M., Mohamed, S., Tanoi, K., Kobayashi, N.I., Takagi, K., Vernet, A., Guiderdoni, E., Perin, C., Sentenac, H., Very, A.A. 2017. Production of low-Cs + rice plants by inactivation of the k+ transporter OsHAK1 with the CRISPR-Cas system. *Plant J.* 92:43–56.

Ntui, V.O., Uyoh, E.A., Ita, E.E., Markson, A.A., Tripathi, J.N., Okon, N.I., et al. 2021. Strategies to combat the problem of yam anthracnose disease: Status and prospects. *Mole. Plant Pathol.* 22(10):1302–1314. doi:10.1111/mpp.13107

Ogata, T., Ishizaki, T., Fujita, M., Fujita, Y. 2020. CRISPR/Cas9-targeted mutagenesis of osera1 confers enhanced responses to abscisic acid and drought stress and increased primary root growth under nonstressed conditions in rice. *PLoS ONE,* 15:e0243376.

Okuzaki, A., Ogawa, T., Koizuka, C., Kaneko, K., Inaba, M., Imamura, J., Koizuka, N. 2018. CRISPR/Cas9-mediated genome editing of the fatty acid desaturase 2 gene in Brassica napus. *Plant Physiol. Biochem.* 131:63–69.

Oliva, R., Ji, C., Atienza-Grande, G., Huguet-Tapia, J.C., Perez-Quintero, A., Li, T., Eom, J.-S., Li, C., Nguyen, H., Liu, B., et al. 2019. Broad-spectrum resistance to bacterial blight in rice using genome editing. *Nat. Biotechnol.* 37:1344–1350.

Park, J.J., Yoo, C.G., Flanagan, A., Pu, Y., Debnath, S., Ge, Y., Ragauskas, A.J., Wang, Z.Y. 2017. Defined tetra-allelic gene disruption of the 4-coumarate: Coenzyme A Ligase 1 (Pv4CL1) gene by CRISPR/Cas9 in switchgrass results in lignin reduction and improved sugar release. *Biofuels BioMed. Cent.* 10:1–11.

Peng, A., Chen, S., Lei, T., Xu, L., He, Y., Wu, L., et al. 2017. Engineering canker resistant plants through CRISPR/Cas9-targeted editing of the susceptibility geneCsLOB1 promoter in citrus. *Plant Biotechnol. J.* 15:1509–1519.

Pickar-Oliver, A., Gersbach, C.A. 2019. The next generation of CRISPR-Cas technologies and applications. *Nat Rev Mol Cell Biol.* 20(8):490–507.

Podevin, N., Devos, Y., Davies, H.V., Nielsen, K.M. 2012. Transgenic or not? No simple answer! new biotechnology-based plant breeding techniques and the regulatory landscape. *EMBO Rep.* 13:1057–1061.

Puchta, H. 2005. The repair of double-strand breaks in plants: Mechanisms and consequences for genome evolution. *J. Exp. Bot.* 56:1–14.

Puchta, H., Fauser, F. 2014 Synthetic nucleases for genome engineering in plants: Prospects for a bright future. *Plant J* 78:727–741.

Qaim, M. 2020. Role of new plant breeding technologies for food security and sustainable agricultural development. *Appl. Econ. Perspect. Policy* 42:129–150.

Reddy A. 2007. Alternative splicing of pre-messenger RNAs in plants in the genomicera. *Annu Rev Plant Biol* 58:267–294.

Regina, A., Bird, A., Topping, D., Bowden, S., Freeman, J., Barsby, T., Kosar-Hashemi, B., Li, Z., Rahman, S., Morell, M. 2006. High-amylose wheat generated by RNA interference improves indices of large-bowel health in rats. *Proc. Natl. Acad. Sci. USA* 103:3546–3551.

Rodríguez-Leal, D., Lemmon, Z.H., Man, J., Bartlett, M.E., Lippman, Z.B. 2017. Engineering quantitative trait variation for crop improvement by genome editing. *Cell* 171:470–480.e8.

Romero, F.M., Gatica-Arias, A. 2019. CRISPR/Cas9: Development and application in rice breeding. *Rice Sci.* 26:265–281.

Shoeb, E., Badar, U., Venkataraman, S., Hefferon, K. 2021. Chapter 10 - CRISPR/Cas9 and Cas13a systems: a promising tool for plant breeding and plant defence, Editor(s): Kamel A. Abd-Elsalam, Ki-Taek Lim, In *Nano Biotechnology for Plant Protection, CRISPR and RNAi Systems*, Elsevier.

Sánchez-León, S., Gil-Humanes, J., Ozuna, C.V., Giménez, M.J., Sousa, C., Voytas, D.F., Barro, F. 2018. Low-gluten, non transgenic wheat engineered with CRISPR/Cas9. *Plant Biotechnol. J.* 16:902–910.

Sashidhar, N., Harloff, H.J., Potgieter, L., Jung, C. 2020. Gene editing of three BnITPK genes in tetraploid oilseed rape leads to significant reduction of phytic acid in seeds. *Plant Biotechnol. J.* 18:2241–2250.

Saxer, G., Havlak, P., Fox, S.A., Quance, M.A., Gupta, S., Fofanov, Y., Strassmann, J.E., Queller, D.C. 2012. Whole genome sequencing of mutationaccumulation lines reveals a low mutationrate in the social amoeba Dictyostelium discoideum. *PLoS One* 7(10):e46759

Schaart, J.G., van de Wiel, C.C.M., Lotz, L.A.P., Smulders, M.J.M. 2016. Opportunities for products of new plant breeding techniques. *Trends Plant Sci.* 21:438–449.

Schachtsiek, J., Stehle, F. 2019. Nicotine-free, nontransgenic tobacco (Nicotiana Tabacum l.) Edited by CRISPR-Cas9. *Plant Biotechnol. J.* 17:2228–2230.

Schearer, O.D. 2013. Nucleotide excision repair in eukaryotes. *Cold Spring Harb PerspectBiol* 5:a012609

Schindele, A., Dorn, A., Puchta, H. 2020, CRISPR/Cas brings plant biology and breeding into the fast lane. *Curr. Opin. Biotechnol.* 61:7–14.

Schmidt, C., Schindele, P., Puchta, H. 2020. From gene editing to genome engineering: restructuring plant chromosomes via CRISPR/Cas. *aBiotech* 1:21–31.

Schwartz, C., Lenderts, B., Feigenbutz, L., Barone, P., Llaca, V., Fengler, K., Svitashev, S. 2020. CRISPR–Cas9-mediated 75.5-Mb inversionin maize. *Nat. Plants* 6:1427–1431.

Shan, Q., Wang, Y., Li, J., Gao, C. 2014. Genome editing in rice and wheat using theCRISPR/Cas system. *Nat. Protocols* 9:2395–2410.

Shan, Q., Wang, Y., Li, J., Zhang, Y., Chen, K., Liang, Z., et al. 2013. Targeted genomemodification of crop plants using a CRISPR-Cas system. *Nat. Biotechnol.* 31:686–688.

Shao, X., Wu, S., Dou, T., Zhu, H., Hu, C., Huo, H., He, W., Deng, G., Sheng, O., Bi, F., et al. 2020. Using CRISPR/Cas9 genome editingsystem to create MaGA20ox2 gene-modified semi-dwarf banana. *Plant Biotechnol. J.* 18:17–19.

Shen, L., Wang, C., Fu, Y., Wang, J., Liu, Q., Zhang, X., et al. 2018. QTL editing confersopposing yield performance in different rice varieties. *J. Integr. Plant Biol.* 60:89–93.

Shi, J., Gao, H., Wang, H., Lafitte, H.R., Archibald, R.L., Yang, M., et al. 2017. ARGOS8 variants generated by CRISPR-Cas9 improve maize grain yield under field droughtstress conditions. *Plant Biotechnol. J.* 15:207–216.

Shimatani, Z., Kashojiya, S., Takayama, M., Terada, R., Arazoe, T., Ishii, H., Teramura, H., Yamamoto, T., Komatsu, H., Miura, K., et al. 2017. Targeted base editing in rice and tomato using a CRISPR-Cas9 cytidine deaminase fusion. *Nat. Biotechnol.* 35:441–443.

Shlush, I., Ben Samach, A., Melamed-Bessudo, C., Ben-Tov, D., Dahan-Meir, T., Filler-Hayut, S., Levy, A.A. 2021. Crispr/Cas9 inducedsomatic recombination at the crtiso locus in tomato. *Forests* 12:9.

Si, X., Zhang, H., Wang, Y., Chen, K., Gao, C. 2020. Manipulating gene translation in plants by CRISPR–Cas9-mediated genomeediting of upstream open reading frames. *Nat. Protoc.* 15:338–363.

Sommer, A. 2008. Vitamin A deficiency and clinical disease: An historical overview. *J. Nutr.* 138:1835–1839.

Spampinato, C.P. 2017. Protecting DNA fromerrors and damage: An overview of DNA repair mechanisms in plants compared tomammals. *Cell Mol. Life Sci.* 74:1693–1709.

Sparvoli, F., Cominelli, E. 2015. Seed biofortification and phytic acid reduction: A conflict of interest for the plant? *Plants* 4:728–755.

Steinert, J., Schiml, S., Puchta, H. 2016. Homology-based double-strand break inducedgenome engineering in plants. *PlantCell Rep.* 35:1429–1438.

Sun, X., Hu, Z., Chen, R., Jiang, Q., Song, G., Zhang, H., Xi, Y. 2015. Targeted mutagenesis in soybean using the CRISPR-Cas9 system. *Sci. Rep.* 5:1–10.

Sun, Y., Jiao, G., Liu, Z., Zhang, X., Li, J., Guo, X., et al. 2017. Generation of high-amyloserice through CRISPR/Cas9-mediated targeted mutagenesis of starch branching enzymes. *Front. Plant Sci.* 8:298.

Symington, L., Gautier, J. 2011. Double-strandbreak end resection and repair pathway choice. *Annu. Rev. Genet.* 45:247–271.

Taagen, E., Bogdanove, A.J., Sorrells, M.E. 2020. Counting on crossovers: Controlled recombination for plant breeding. *Trends Plant Sci.* 25:455–465.

Tang, L., Mao, B., Li, Y., Lv, Q., Zhang, L., Chen, C., Zhao, B. 2017. Knockout ofOsNramp5 using the CRISPR/Cas9 system produces low Cd-accumulating indica ricewithout compromising yield. *Sci. Rep.* 7:14438

Tatham, A.S., Shewry, P.R. 2008. Allergens to wheat and related cereals. *Clin. Exp. Allergy.* 38:1712–1726.

Thomazella, D.P.d.T., Seong, K., Mackelprang, R., Dahlbeck, D., Geng, Y., Gill, U.S., Qi, T., Pham, J., Giuseppe, P., Lee, C.Y., et al. 2021. Loss of function of a dmr6 ortholog in tomato confers broad-spectrum disease resistance. *Proc. Natl. Acad. Sci. USA.* 118:2026152118

Tian, Y., Liu, X., Fan, C., Li, T., Qin, H., Li, X., Chen, K., Zheng, Y., Chen, F., Xu, Y. 2021. Enhancement of tobacco (Nicotiana Tabacum L.) seed lipid content for biodiesel production by CRISPR-Cas9-mediated knockout of NtAn1. *Front. Plant Sci.* 11:1–13.

Tran, M.T., Doan, D.T.H., Kim, J., Song, Y.J., Sung, Y.W., Das, S., Kim, E.J., Son, G.H., Kim, S.H., Van Vu, T., et al. 2021. CRISPR/Cas9-based precise excision of SlHyPRP1 domain(s) to obtain salt stress-tolerant tomato. *Plant Cell Rep.* 40:999–1011.

Tripathi, L., Ntui, V.O., Tripathi, J.N. 2020. CRISPR/Cas9 based genome editing of banana for disease resistance. *Curr. Opin. Plant Biol.* 56:118–126. doi:10.1016/j.pbi.2020.05.003

Ueda, T., Nakamura, C. 2011. Ultravioletdefense mechanisms in higher plants. *Biotechnol. Biotechnol. Equip.* 25:2177–2182.

Upadhyay, S.K., Kumar, J., Alok, A., Tuli, R. 2013. RNA-guided genome editing for target gene mutations in wheat. *G3 Genes Genomes Genet* 3:2233–2238.

Vats, S., Kumawat, S., Kumar, V., Patil, G.B., Joshi, T., Sonah, H., Sharma, T.R., Deshmukh, R. 2019. Genome editing in plants: Exploration of technological advancements and challenges. *Cell* 8:1386

Veley, K.M., Okwuonu, I., Jensen, G., Yoder, M., Taylor, N.J., Meyers, B.C., et al. 2021. Gene tagging via CRISPR-mediated homology-directed repair in Cassava. *Genes Genomes Genet.* 11(4):jkab028. doi:10.1093/g3journal/jkab028

Verma, P., Greeberg, R.A. 2017. Noncanonicalviews of homology-directed DNA repair. *Genes Dev* 30:1138–1154.

Voytas, D., Gao, C. 2014. Precision genomeengineering and agriculture: opportunitiesand regulatory challenges. *PLoS Biol.* 12(6):e1001877.

Wada, N., Ueta, R., Osakabe, Y., Osakabe, K. 2020. Precision genome editing in plants: State-of-the-art in CRISPR/Cas9-based genomeengineering. *BMC Plant Biol.* 20:234.

Waltz, E. 2016. Gene-edited CRISPR mushroom escapes US regulation. *Nature* 532:293.

Waltz, E. 2015. Non browning GM apple cleared for market. *Nat. Biotechnol.* 33 (4):326–327.

Wan, D.-Y., Guo, Y., Cheng, Y., Hu, Y., Xiao, S., Wang, Y., Wen, Y.-Q. 2020. CRISPR/Cas9-mediated mutagenesis of VvMLO3 results inenhanced resistance to powdery mildew in grapevine (*Vitis vinifera*). *Hortic. Res.* 7:116.

Wang, H., Russa, M., Qi, L. 2016. CRISPR/Cas9in genome editing and beyond. *Annu. Rev. Biochem.* 85:22.1–22.38.

Wang, L., Chen, L., Li, R., Zhao, R., Yang, M., Sheng, J., Shen, L. 2017. Reduced drought tolerance by CRISPR/Cas9-mediated SlMAPK3 mutagenesis in tomato plants. *J. Agric. Food Chem.* 65:8674–8682.

Wang, P., Zhang, J., Sun, L., Ma, Y., Xu, J., Liang, S., et al. 2018. High efficient multisites genome editing in allotetraploid cotton *Gossypium hirsutum* using CRISPR/Cas9system. *Plant Biotechnol. J.* 16 (1):137–150.

Wang, S., Zhang, S., Wang, W., Xiong, X., Meng, F., Cui, X. 2015. Efficient targeted mutagenesis in potato by the CRISPR/Cas9 System. *Plant Cell Rep.* 34:1473–1476.

Wang, T., Xun, H., Wang, W., Ding, X., Tian, H., Hussain, S., Dong, Q., Li, Y., Cheng, Y., Wang, C., et al. 2021. Mutation of GmAITRgenes by CRISPR/Cas9 genome editing results in enhanced salinity stress tolerance in soybean. *Front. Plant Sci.* 12:779598.

Wang, W., Pan, Q., Tian, B., He, F., Chen, Y., Bai, G., Akhunova, A., Trick, H.N., Akhunov, E. 2019. Gene editing of the wheat homologs of TONNEAU1-recruiting motif encoding gene affects grain shape and weight in wheat. *Plant J.* 100:251–264.

Week, D., Spalding, M., Yang, B. 2016. Use ofdesigner nucleases for targeted gene andgenome editing in plants. *Plant Biotechnol* J14(2):483–495.

Westra, E., Buckling, A., Fineran, P. 2014. CRISPR-Cas systems: beyond adaptive immunity. *Nat. Rev. Microbiol.* 12(5):317–326.

Xie, Z., Nolan, T.M., Jiang, H., Yin, Y. 2019. AP2/ERF transcription factor regulatory networks in hormone and abiotic stress responsesin Arabidopsis. *Front. Plant Sci.* 10:228

Xu, R., Yang, Y., Qin, R., Li, H., Qiu, C., Li, L., et al. 2016. Rapid improvement of grainweight via highly efficient CRISPR/Cas9-mediated multiplex genome editing in rice. *J. Genet. Genom.* 43 (8):529–532.

Xu, Z., Xu, X., Gong, Q., Li, Z., Li, Y., Wang, S., Yang, Y., Ma, W., Liu, L., Zhu, B., et al. 2019. Engineering broad-spectrum bacterialblight resistance by simultaneously disrupting variable tale-binding elements of multiple susceptibility genes in rice. *Mol. Plant.* 12:1434–1446.

Yang, Q., Zhong, X., Li, Q., Lan, J., Tang, H., Qi, P., Ma, J., Wang, J., Chen, G., Pu, Z., et al. 2020. Mutation of the D-hordein gene by RNA-guided Cas9 targeted editing reducing the grain size and changing grain compositions in barley. *Food Chem.* 311:125892

Yang, Y., Zhu, G., Li, R., Yan, S., Fu, D., Zhu, B., et al. 2017. The RNA editing factorSlORRM4 is required for normal fruit ripening in tomato. *Plant Physiol.* 175:1690–1702.

Yin, X., Biswal, A.K., Dionora, J., Perdigon, K.M., Balahadia, C.P., Mazumdar, S., Chater, C., Lin, H.C., Coe, R.A., Kretzschmar, T., et al. 2017. CRISPR-Cas9 and CRISPR-Cpf1 mediated targeting of a stomatal developmental gene EPFL9 in rice. *Plant Cell Rep.* 36:745–757.

Yu, Q.-h, Wang, B., Li, N., Tang, Y., Shengbao, Y., Yang, T., et al. 2017. CRISPR/Cas9-induced targeted mutagenesis and gene replacement to generate long shelf-lifetomato lines. *Sci. Rep.* 7.

Zhang, H., Zhang, J., Wei, P. et al. 2014 TheCRISPR/Cas9 system produces specific andhomozygous targeted gene editing in rice inone generation. *Plant Biotechnol. J* 12:797–807.

Zhang, A., Liu, Y., Wang, F., Li, T., Chen, Z., Kong, D., Bi, J., Zhang, F., Luo, X., Wang, J. 2019. Enhanced rice salinity tolerance viaCRISPR/Cas9-targeted mutagenesis of the OsRR22 gene. *Mol. Breed.* 39:47.

Zhang, M., Cao, Y., Wang, Z., Wang, Z.Q., Shi, J., Liang, X., Song, W., Chen, Q., Lai, J., Jiang, C. 2018. A retrotransposon in an HKT1family sodium transporter causes variation of leaf Na(+) exclusion and salt tolerance in maize. *New Phytol.* 217:1161–1176.

Zhang, Y., Bai, Y., Wu, G., Zou, S., Chen, Y., Gao, C., Tang, D. 2017. Simultaneous modification of three homoeologs of TaEDR1 bygenome editing enhances powdery mildew resistance in wheat. *Plant J.* 91:714–724.

Zhang, Y., Liang, Z., Zong, Y., Wang, Y., Liu, J., Chen, K., Qiu, J.L., Gao, C. 2016. Efficient and transgene-free genome editing in wheat through transient expression of CRISPR/Cas9 DNA or RNA. *Nat. Commun.* 7:1–8.

Zhang, Y., Wang, X., Luo, Y., Zhang, L., Yao, Y., Han, L., Chen, Z., Wang, L., Li, Y. 2020. OsABA8ox2, an ABA catabolic gene, suppresses root elongation of rice seedlings and contributes to drought response. *Crop J.* 8:480–491.

Zhao, C., Zheng, X., Qu, W., Li, G., Li, X., Miao, Y.-L., et al. 2017. CRISPR-offinder: a CRISPR guide RNA design and off-target searching tool for user-defined protospacer adjacent motif. *Int. J. Biol. Sci.* 13:1470–1478.

Zhou, H., Liu, B., Weeks, D.P., Spalding, M.H., Yang, B. 2014. Large chromosomal deletions and heritable small genetic changes induced by CRISPR/Cas9 in rice. *Nucleic Acids Res.* 42:10903–10914.

Zhou, X., Liao, H., Chern, M., Yin, J., Chen, Y., Wang, J., Zhu, X., Chen, Z., Yuan, C., Zhao, W., et al. 2018. Loss of function of a riceTPR-domain RNA-binding protein confers broad-spectrum disease resistance. *Proc. Natl. Acad. Sci. USA* 115:3174–3179.

Zong, Y., Wang, Y., Li, C., Zhang, R., Chen, K., Ran, Y., et al. 2017. Precise Base Editing In Rice, Wheat and Maize with a Cas9-Cytidine Deaminase Fusion. *Nat.Biotechnol.* 35(5):438–440. doi:10.1038/nbt.3811

8 Potential of commercialization of genome-edited crops

Shanta Karki and Govinda Rizal

8.1 Introduction

People have selected plants and animals as per their needs, preferences, and usefulness. At first, people began to select crops based on their superior phenotypes. Then the selection was based on the study of effects such as the genes and environments causing the phenotypes. This human intervention in the diversification of traits started from conventional plant breeding, vegetative propagation, distance hybridization, and recently genetic modifications. However, those technologies were time-consuming and often resulted in untargeted byproducts.

The advancement in science and technology, and the need to increase agricultural production to meet the rapidly growing population have compelled plant breeders to dive deeper into the gene level and beyond to make a formidable selection for better and higher-yielding varieties.

The discovery of DNA, transposons, mutation, and mechanisms of bacterial gene transfer have led to the genetic modification of targeted traits for desired phenotypes which have overcome the tedious manual hybridization to develop new varieties of interest which also involved screening a huge population. The genetic modification needed the help of bacteria to freight sections of DNA cloned from other organisms. It led to the concept and technology related to the development of genetically modified organisms commonly known as GMOs. The employment of bacteria as vectors of gene transfer and its residual reminiscence in the recipient genomes has been an issue of concern and debate. While immediate adverse effects of such residual exotic genes can be studied, nullified, or corrected, there are no satisfactory answers to what will happen to them when they are inherited across generations which has led to the division of countries or societies for and against GMOs or even the GE crops.

Plant breeders dedicated to genomic studies have gone further on different crop improvement methods sophisticating the science of variety development and technologies. Recent advances in genetic technology have enabled scientists and plant breeders to precisely target sections of DNA and correct them through the insertion of desirable sequences or deletion of unwanted

DOI: 10.4324/9781003382102-8

sequences. This precise technique of genetic correction for desired traits is known as gene- or genome-editing technology (GET). The GET has proven to be an important tool and gained popularity because it helps in developing crops resistant to diseases and insects which helps in reduced use of pesticides thereby helping in protecting our environment as well as providing safe food for consumption. Development of drought-tolerant varieties through GET help in minimizing risks of crop failures due to shortage of water. The development of nutrition-rich crop varieties helps in addressing nutrition deficiency so the use of GET has many such potentials that can be harnessed for a better life on earth if we rationally use these technologies with appropriate scientific regulation.

There are numerous hurdles along the production pathway from the conceptualization of genome-edited crops (GECs) to marketing. The hurdles include clarification of concepts and differences between GEC and GMO, their regulations, and awareness of the science behind the technologies so that the concerned stakeholders can make informed choices.

This chapter, based on the analysis of published research articles, aims to report the global trend of GEC production, the science behind the technologies, incentives, and challenges among the creators, and an update on the commercialization of the GEC.

8.2 Methodology of the study

We conducted a review of research, review, and opinion articles, news highlights, and other published information on GEC from across the globe. All cited information have been duly acknowledged through referencing.

8.2.1 History of crop breeding

With the domestication of crops, people started selecting for higher productivity, better taste, preferable quality traits, and faster-growing cycle that have revolutionized the phenotypes of crops. The traditional selection process had sustained population growth in the past. The period of peace and population boom following the two world wars was sustained through the green revolution of the 1960s that increased the yields of major staple grain crops mainly wheat, maize, and rice. This revolution was accomplished through coordinated approaches in the breeding of elite varieties, development of hybrid crops, increase in the use of chemical fertilizers, improvement in management practices, and substantial increase in public investment (Pingali, 2012). The impact of the green revolution was visible in most countries where despite steep population growth, there was a reduction in malnutrition and poverty. The gap between the crop yield in the field and their potential yield was reduced to a minimum in most cereals thereby substantially increasing the overall crop production.

By the 1980s, molecular and transformation technologies became new ways of increasing food production. Bioengineered genes were introduced into plant genome, making them robust crops. Crops that are resistant to insect pests or externally applied chemical herbicides were commercialized. Although the new technology reduced the need to tillage, and time of sun drying of tilled land; the seeds compatible with the technology were high yielding and available at a commercial scale, there were precautionary guidelines to follow to prevent outcrossing with wild or weedy species, natural selection of resistance in insects (Tabashnik et al. 2013), pests, and weeds (Duke, 2015). The genetic transformation technologies have had visible benefits to farmers and end-users. Application of molecular-marker-assisted breeding (Al-Khayri et al. 2015), and the combination of methods for shortening the breeding cycle (Rizal et al. 2014) led to rapid screening and analysis of novel genes in elite varieties. Identification of genes responsible for specific traits aided in the development of varieties with desired traits through genetic transformation which would be more specific than through conventional plant breeding (hybridization) methods.

Banana (*Musa acuminate*) transformed with plant ferredoxin-like protein (Pflp) gene from sweet pepper (*Capsicum annuum*) expressed resistance against banana *Xanthomonas* wilt caused by *Xanthomonas campestris* pv. *Musacearum* (Namukwaya et al. 2012). Papaya (*Carica papaya* L.) resistant to papaya ringspot virus was developed through transgenic methods (Chen et al. 2001), cultivated, and marketed in the USA.

The drought tolerant banana, maize (Castiglioni, 2008), low acrylamide potatoes (Rommens et al. 2008), low*trans*- and saturated fats containing canola (Jones et al. 2014), non-browning apples (Murata, 2001), pro-vitamin A fortified rice (Tang et al. 2009) or banana (Paul et al. 2017), virus-resistant papaya (Fitch et al. 1992), were technological boons to the consumers. However, everything was not celestial with bioengineered or GM crops. The European Union (EU) largely banned the commercial cultivation of GM crops and restricted their trade within Europe. Several countries, where the technology was exotic, also followed the EU policies of restriction on GM crops. Several pro-organic or anti-GM organizations resorted to equivocal debates and physical restrictions on commercialization. The development of cost-effective, rapid sequencing technologies increased the effectiveness of the GM system. There were regulation hurdles, public opposition, and social stigmatization against the trade and consumption of GM products. There were numerous questions regarding the potential ill consequences of the use of GM products. The fear of GM foods ranged from being harmful to the health of the consumers and environment to even being poisonous. The development and use of selective herbicides and the involvement of multinational companies, that are specific not only to GM foods, increased the concern of the consumers in doubt. There are psychological essentialism, intuitions, and strong organized opposition to GM foods that label GM products as Franken foods (Blancke, 2015).

The fears of the uncertain modification that could take place in the genome of GM crops in the future keep appearing in debates and news. All questions arising from such debates could be answered through scientific evidence and adherence to ethical guidelines as well as systematic regulatory processes in place. However, the fact that GM crops had a residual presence of foreign DNA, especially from the selectable markers in final GM products, created a need for even more advanced technology to produce foreign-DNA-free products. The advanced technology that was developed to overcome the demerits of GM system was the GE system. If we exclude this logic of residual DNA in GM products and the need for foreign-DNA-free products, the development of GE system was a technology that evolved on the foundation of research on gene manipulation. At times, disgust among consumers is severe to the extent that it impairs their moral judgments making them criticize all those involved in any aspect of GM products (Mozter, 2022).

Unlike in GM crops where foreign genes are introduced into the genome of the host crop, in GE crops, the genome of the crops can be edited within itself without incorporating foreign genes.

8.2.2 Types of GE technologies

Genome editing (GE) is a generic term for a precise and targeted approach of genetic engineering by which one or more sites are edited in the genome to obtain a desirable trait, as such GE can be referred to as a precision breeding tool. Therefore, the GE technologies or tools should have the ability to find a specific sequence of DNA within the genome, the ability to cleave the DNA at that predetermined location, and the ability of innate DNA repair mechanism. Some of the commonly used GE technologies are meganucleases, zinc finger nucleases (ZFN), transcription activator-like effector nucleases (TALENs), and the most recent one is clustered regularly interspaced short palindromic repeats (CRISPR) which have become the most dominant GE technology as it is more versatile than the others mentioned above. CRISPR has become popular because it can be programmed easily to direct the nuclease which cleaves the DNA at the desired location within the genome (Menz et al. 2020).

8.2.3 Process of genome editing

The process of genome editing from planning to producing up to marketing involves several steps and each major step is accomplished through numerous trials and improvisations. The major steps include gene cloning and expression, product development by using tissue culture technology, field testing, breeding into multiple elite varieties, product characterization and meeting the regulatory confirmations, public acceptance trials, marketing, and market feedback.

The tedious and technically long procedures make the process of developing GE crops and getting consumers' acceptance expensive. Whenever single gene targets are discovered or patents are involved, the creators get incentives for more work that adds to the input cost. The costly endeavors need huge investment which only a few public or private institutions can afford to take up the challenge. These successful private institutions often end up monopolizing the technology. Efforts are necessary to make technology a bridge between Global North and Global South countries, make the improved products accessible and affordable to needy people, and at the same time simplify the process with systematic regulation.

8.2.4 Genome-edited crops

GECs can be developed through site-directed nuclease (SDN), which are of three broad categories: i) SDN-1 or the induction of InDels or single point mutations; ii) SDN-2 or the editing of a few base pairs or short insertion using an external DNA template sequence; and SDN-3 or the insertion of longer stands of cisgenes or transgenes (Podevin et al. 2013).

Canada was the first country to commercialize gene-modified flax and canola back in the mid-1980s. The first reported GE crop is soybean which has increased oleic acid and reduced linolic acid content in seeds (Haun et al. 2014; Demorest et al. 2016). High oleic acid (a monosaturated fatty acid) is preferred over polyunsaturated fatty acids, such as linolic acid and linolenic acid for health benefits (Ascherio, 1999). Gene editing has been tried in a number of crops ranging from vegetables such as tomato and Chinese kale, tubers like potato and sweet potato, to fruit plants like banana and pome-granate (see Table 8.1).

8.2.5 Genome-edited crops in the pipeline of commercialization

Potato tubers are usually stored at cold temperatures to check untimely sprouting and extend shelf life. Cold temperature triggers the accumulation of reducing sugars which on exposure to higher temperatures react with free amino acids leading to browning, producing a bitter taste, and increasing level of acrylamide, a potential carcinogen (Mojska et al. 2007). Using RNAi technology, the vascular invertase gene (VInv) that encodes the protein involved in the breakdown of sucrose to fructose was silenced and Vlnv was knocked out using TALEN in potato variety Ranger Russet. The full Vlnv-knocked-out products were reported to have undetectable levels of sugars, reduced levels of acrylamide in processed potato chips, and the chips were light colored (Clasen et al. 2016).

Cereals with resistant starch and high amylose content offer health benefits to people. Using CRISPR/Cas 9 technology targeted mutagenesis was induced in starch branching enzyme (SBE) that determines the physical properties and structure of starch. A SBE mutant, *sbell,* was reported to have a higher proportion of long chains with debranched amylopectin, increased resistant starch, and higher amylose content in rice (Sun et al. 2017).

Genome editing tools have experimented with crops such as pomegranate, banana, Chinese kale, tomato, etc. (see Table 8.1). Such trials have been

Table 8.1 List of gene-edited crops published in research articles

Crop	Trait/gene edited	Advantage	Reference
Banana	F-carotene content	Increased F-carotene content	(Kaur et al. 2020)
Brassica rapa	Sucrose, fructose, and glucose contents	Increased sucrose content; decreased fructose and glucose contents	(Jiang, 2020)
Carrot	Anthocyanin biosynthesis	Validating the functional role of a gene; discoloration of calli used as visual marker for screening	(Klimek-Chodacka, 2018)
Chinese kale	Carotenoid isomerase (BoaCRTISO) gene	Reduction in the color-masking effect of chlorophyll on carotenoids for market appeal.	(Sun, 2020)
Grape	Tartaric acid content	Decreased tartaric acid content	(Ren, 2016)
Lettuce	Ascorbate content	Increased tolerance to oxidation stress; increased ascorbate content	(Zhang, 2018)
Mushroom	Phenolic substrates	Reduces browning by converting phenolic substrates to quinones	(Waltz, 2016)
Peanut	Oleic acid content	Increased oleic acid oil content	(Yuan, 2019); (Wen, 2018)
Pomegranate	Gallic acid 3-O- and 4- O-glucosides	Accumulation of gallic acid 3-O- and 4-O-glucosides (galloyl glucose ethers)	(Chang, Wu, and Tian, 2019)
Potato	Millard reaction; Steroidal glycoalkaloid content; acrylamide	Decreased browning; reduced level of steroidal glycoalkaloid content; decreased level of acrylamide	(González, 2020); (Nakayasu, 2018); (Clasen et al. 2016)
Rapeseed	Seed oil content; oleic acid content; linolic acid and linolenic acid contents	Increased seed oil content; increasedoleic acid content; decreased linolic acid and linolenic acid contents	(Karunarathna, 2020); (Okuzaki, 2018)
Rice	Amylose content; Cesium accumulation		

(Continued)

Table 8.1 (Continued)

Crop	Trait/gene edited	Advantage	Reference
Sorghum	α-kafirins (a storage protein)	Increased content of resistant starch; reduced cesium accumulation	(Sun et al. 2017); (Nieves-Cordones, 2017)
Soybean	Oleic acid	Increased protein quality and digestibility	(Li et al. 2018a)
		Increase the seed oleic acid content from 33% to 80% and decrease of linolic acids to less than 3%	(Demorest et al. 2016); (Haun et al. 2014)
Sweet potato	Amylose content; amylopectin content	Decreased amylose content; decreased amylopectin content	(Wang et al. 2019);
Tomato	Anthocyanin content; phenylalanine-derived volatile content; volatile organic compounds; lycopene content	Decreased anthocyanin contents; increased phenylalanine-derived volatile content; decreased volatile organic compounds; increased lycopene content	(Yan, 2020); (Tikunov, 2020); (Zhi, 2020); (Ito, 2017)
Wheat	Gluten content	Reduced gluten content	(Sánchez-León, 2018)
Wild tomato	Vitamin C content; lycopene content;	Increased vitamin C content; increased lycopene content	(Li et al. 2018b); (Zsogon, 2018)

either to increase the content such as an increase of sucrose content in *Brassica rapa* (Jiang, 2020) or decrease tartaric acid in grapes (Ren, 2016) or as research materials such as reduction of chlorophyll content in carrots to be used as markers in subsequent experiments (Klimek-Chodacka, 2018).

The success of GE crops at the laboratory level is the foundation of the process. There are at least five major stages involved in the process.

1 The discovery of genes that correct defects or add values to traits is the start of the process. Genes, or sections of the gene, or a particular position of the DNA sequence that is necessary to correct or improve traits is studied and a proposition is made for GE at this stage.
2 Proof of concept is the second step in the process. The concept of GE is tested in a small-scale experiment in a closed/contained environment. The technological challenges and plants' reactions to GE are studied, and recommendations are made.
3 Development of GE crops is done after the proof of concept gives a sound foundation of technological possibilities. GE crops are developed through extensive trials and tests. All major and minor tests are carried out to ensure that all answers to the questions of challenges, potentials, risks, and questions that may arise in the future are sought at this stage. Towards the later part, the GE crops are tested at multi-locations, and applications for legal clearance are made at this stage.
4 Commercialization of GE crops is the last and the most important stage of the GE process for agricultural purposes. Based on the results of laboratory, confined field trails and multi-location trials, legal clearances-approval, and consumers' acceptance, the success of GE crops is evaluated.

8.2.6 Types of genome-edited crops

Plant breeding techniques and crop improvement technological innovations have been undergoing continuous changes during the process of increasing yields, and improving the quality and nutritive values of the foods that we eat. From deliberate cross-breeding and wide hybridization to mutation breeding to molecular breeding showcase the technological advancements to achieve the trait of interest incorporated in the genome. This shift has been termed as advancing from conventional plant breeding methods to new breeding techniques that include genome editing. Genome editing has many advantages over conventional technologies; the most promising thing is that it enables targeted gene editing within the genome. CRISPR offers an easier, more versatile, and more accurate form of mutagenesis that facilitates the transfer or expression of the desired trait in the progeny without losing any efficacy (Georges and Ray, 2017). This technology causes mutation to a specific site within the targeted gene, making the effects on the plants more significant (Song et al. 2016) as it can be programmed to target specific segments of genetic code or edit DNA sequence with greater accuracy (Barrangou, 2015). Importantly, it holds great potential for public sector plant breeding including the developing countries, allowing for local and regional solutions to improving food security with the development of higher-yielding crop varieties. A Chinese research group (Miao et al. 2018) has already made use of CRISPR/Cas9 technology to create a rice variety that yields 25–31% more than the conventional varieties by simultaneously mutating the genes encoding abscisic acid (ABA) receptors.

8.2.7 Commercialization of genome-edited crops

Our food system has gone international adding challenges of global demand, changing consumption patterns with the increasing purchasing capacity of people, and maintaining sustainability starting from the production sites to the consumers' plates (farm to fork). Plant breeding which has been contributing to increasing production and filling the gaps between the demand and supply of the food has been undergoing tremendous and rapid sophistication in developing varieties. The most recent frontier in developing specific trait-targeted varieties and genome editing has become the plant breeders' latest toolbox. Genome editing allows targeted alteration in a few to the specific length of DNA codons for targeted change in expression with a minimum change in the existing genome atlas of the organisms. The development of Geed crops, follows a clear path, starting with gene discovery, followed by proof-of-concept (POC) studies, product development and deployment, confined field trials, consumer preference, safety check trials, and finally commercialization.

Crop breeding technologies developed over time could help in meeting the quality and quantity of food demanded by the rapidly growing population. GE has been the most recent and the most advanced method added to the previous skills. Through GE precise, targeted, and desired cut and/ or paste of DNA sequences within the existing genome of an organism is possible. The GE method is built upon the best knowledge of the earlier methods and practices. CRISPR Cas 9 is one of the widely used GE tools and is rapidly used in crop research and development wherever the technology has reached.

Consumers across the world are accepting, knowingly or ignorantly, GE *Brassica oleracea*, *Camelina sativa*, cucumber, flax, grapefruits, maize, mushroom, potato, rice, soybean, sugarcane, tomato, wheat crops and the products made from them.

8.2.8 Advantages of genome-edited crops

Climate change has been causing frequent fluctuations in the growing environments and imbalances in crops' responses to abiotic and biotic stresses. Therefore, there is an urgent need for the development of disease and pest-resistant varieties, flood, drought, and extreme temperature-tolerant crops to adapt to the changing climate and minimize the potential risks of crop failures. The products of GE produced through SDN1, and SDN2 are similar to the products from natural or conventional breeding in terms of environmental risks and food safety as they are strictly regulated and rigorously tested prior to their release. They have added advantage for the traits of interest as GET enables targeted precise chances to the genomes thereby helping to improve only the trait of concern. GETs have benefits such as nutritional enhancement, improved food safety, and reduced food waste by improving certain traits that help in post-harvest life extension (Pixley et al. 2022). Other advantages include the quick delivery of improved varieties as genes can be edited directly in the elite breeding lines or popular commercial varieties which eliminates the need for backcrossing, hence reducing the time needed for eliminating the linkage drag.

8.2.9 Controversies related to genome-edited crops

The researchers involved in GET think its regulation should be flexible, but the consumers are concerned about the health and environment. There are debates on whether GE crops regulations should be less stringent than that of GMO regulation.

While the addition of diversity in preferred traits has been successful, the taking of gene-modified organisms to cultivate commercially under field conditions and farms and adoption by the producers as well as acceptance by the consumers have faced more hurdles than creating it. Technology has now advanced to the extent that genetic modification for improvements has become necessary in plants, animals, and humans. The revolution in crop improvement, its fear of unseen consequences has provoked public outcry and protests in different corners of the world that has compelled governments to impose strict regulations. Newer technologies are built for fixing short-comings of the older methods because of which there is an increase in both the quality and quantity of modifications and spiraling of investments. Genetic modification will provide unimaginable benefits by correcting natural shortcomings, curing genetic diseases, or healing new and old bruises, con-ferring tolerance to adverse environmental conditions and resistance to bio-logical factors.

Although GET such as CRISPR Cas is designed for a high precision sequence editing, off-target mutations are expected during improvisation pro-cesses and trials. Such unexpected mutations can be removed during molecular characterization. Finally, only those plants that have the desired genetic changes are selected and advanced further. GECs are considered to be as safe as nat-urally evolving crop species as these are rigorously tested prior to their release.

8.2.10 Regulatory mechanisms for gene-edited crops in different countries

There are no uniform regulations for GE crops among different countries. Most countries do not have a separate regulation guideline specific to GE crops. Many countries regard GE crops as GMOs with certain flexibilities and GE crops are exempted from certain clauses of the guidelines related to the legislation on GMOs. Some of the country-specific regulatory frameworks are discussed in brief below.

8.2.10.1 *Argentina*

Argentina is one of the leading countries in terms of formulating regulations related to GMOs and growing GM or biotech crops commercially. The Argentine regulators discussed rigorously about the products from new plant breeding techniques (NBTs) including GET and designed the regulatory processes to balance the developers' need for early certainty and the requirement of end product data for their assessment. The Argentina Government was the first to issue the regulation of NBTs in the world (Whelan and Lema, 2015). Argentina's regulatory framework is a product-by-product or a case-by-case consideration of genetic modifications in light of the concept of "novel combination of genetic material." This procedure considers if a product derived from NBTs is a GMO or not by fully following the Cartagena Protocol. Argentina's regulatory system has been proactive and scientific which with continuous updating will allow it to effectively address future challenges. Following Argentina, other Latin American countries such as Brazil, Bolivia, Colombia, Paraguay practiced similar regulatory frameworks which has allowed them to be the "Biotech mega countries" declared by International Service for the Acquisition of Agri-biotech Applications, in short ISAAA (Rozas et al. 2022).

8.2.10.2 *Australia*

The Australian Government in 2019 announced that it will not regulate the use of gene-editing techniques in plants, animals, and human cell lines that do not introduce new genetic material (Mallapaty, 2019). This was done after reviewing the country's gene technology regulations as the Australian regulators said that genetic edits made without templates are no different from changes that occur in nature, and therefore do not pose an additional risk to the environment and human health. Gene-editing technologies that do use a template, or that insert other genetic material into the cell, will continue to be regulated by the Office of the Gene Technology Regulator (OGTR).

8.2.10.3 *Canada*

Canada has developed a product-based risk assessment framework which is known as "plants with novel traits (PNTs)" that regulates varieties regardless of the technologies used to develop it. The novelty of phenotype/traits resulting from genome editing would be regulated but not what genome-editing techniques were employed to develop the new variety. Therefore, Canada regulates on the basis of traits expressed and not on the basis of the method used to introduce the traits (Food Directorate, 2022). If the Food

Directorate of Canada after safety assessments concludes that the novel food is safe for consumption, it is allowed to enter the market in the same way as traditional food products and hence is subject to the same regulatory requirements as applicable to all other foods.

8.2.10.4 Japan

The Government of Japan has recognized the importance of GE agricultural organisms as GE technologies accelerate the plant breeding process with precision and shorter time to develop a variety of interests. In February 2019, the Japanese Government defined genome-edited end products derived by modifications of SDN-1 type (directed mutation without using a DNA sequence template) as not representing "living modified organisms (LMOs)" according to the Japanese Cartagena Act (Tsuda et al. 2019) which means crops developed by SDN-1 are not subjected to regulation as they are considered similar to those produced by conventional breeding methods. However, products of SDN-2 will be considered on a case-by-case basis for regulation whereas, SDN-3 products are considered as LMOs.

Unlike the above-mentioned countries, the EU puts all the products developed using any new generation technique (NGT) into one basket. The GM crop growers in Europe face challenges obtaining authorization for the cultivation and marketing of NGT products. The EU supports its researchers by promoting patents on plants produced through patented inventions. Even if the edit could have been natural, crops can be patented in Europe if such crops were developed using NGT.

8.2.11 *Future of GE Crops*

The recently developed GE technologies such as CRISPR are simpler, faster, and more accurate than the earlier GM technologies. The site specificity is not always fault-free. There are reports of gene editing tools cutting in wrong spots, thus creating off-targets (Klein et al. 2018). The GE technologies have been used in all biological fields and agriculture is one of those. The use of GE technologies in human studies adds to the controversies. The use of GE in human gene therapies is criticized for being unethical practice because there are no answers as to what will happen in cases of off-targets (National Academies of Sciences; National Academy of Medicine, 2017). In plants, such off-targets are removed through rigorous selections during the process.

GE technologies are expensive and only the rich countries and multinationals can take the challenge of high investment for the technology. Countries with low economic status may remain disadvantaged and vulnerable in competition during production and trade. While multinational companies take up the GE research in Global North countries, the governments of the Global South countries must carry out the research themselves or provide incentives to enhance the capacity of their private sectors.

Regardless of the cost and benefits, debates, and criticism revolving around GE technologies, the pioneers of GE users will reap a greater advantage. The GE technology used in agriculture is mostly the trickle-down from human and pharmaceutical research. Thus, the countries adopting GE technologies will excel in different fields, not just in agriculture.

Although the use of GE shortens the time of research, the research time in breeding fruit crops and tree species using the new technology remains longer than in other crops. Only fewer private companies have the resources to invest in such research. The research on fruit sectors will fall behind other sectors; therefore, special attention has to be given as fruits are a rich source of various vitamins and nutrients required for a balanced healthy diet.

8.3 Conclusion and recommendations

We propose that the crops contributing to food and nutrition security without any adverse effect on the health of humans, animals, and the environment can be recommended for commercialization and policies must be framed encouraging stakeholders' investment in advanced technologies and promotion of their ethical uses. As such the regulatory mechanisms should be designed to follow a product-based approach that focuses on regulating the actual source of risk-the characteristics of the final product rather than on the method used to create the product. The regulatory frameworks followed by Argentina and Canada could be examples to follow and developing a globally harmonized regulatory system would benefit the world in the long run for using GET to meet the food and nutrition security for the growing population.

The determination of the safety of a crop, food, or food item is determined through analysis of phenotypic or genotypic characters, analysis of the chemical contents, and molecular interactions. For a crop to be potentially risky, it must contain ingredients with active functional properties with adverse effects or allergens. Any method, whether conventional, natural, or biotechnological, should not make a difference. If the product is found to be risky based on rigorous testing, it should be terminated, and safe products should be promoted. The regulations should be based on the properties of the components in the products, and the methods used to develop them should be standardized with regular monitoring.

Although there is no single recommendation to increase the crop yield to meet the calorie and nourishment of the world's population that is expected to reach the peak by 2050, GE is one of the recent and reliable methods to face the challenge. The rapid advancement in gene manipulation will lead to much more promising technologies and innovations that will change the system of plant breeding and biotechnology. To take advantage of such advancements, it is equally important for all countries to keep up with the trend.

While the genes and traits are easier to modify than the climate and environment where such crops are grown, multi-faceted research is needed to tackle these changes. Although consumers' demands and preferences are being globalized, the production of food crops is becoming niche-specific or ecologically governed. Therefore, they must be either grown in limited areas where the environment permits them or under protected structures. In all cases, these technological advancements, unless adopted rationally, shall be a divider between the countries. Those countries that adopt the technology will reap its benefit and others that are slower to adopt will be left behind.

Recently genome editing has been widely used to improve agronomical traits as well as eating quality traits. Although only a few GECs have been commercialized, many crops are in the pipeline that have the better yielding capacity or improved nutritional contents. Therefore, GECs have immense potential to be commercialized and hence appropriate regulatory compliances should be in place in each country for their smooth production as well as marketing.

References

Al-Khayri, J., Jain, S., and Johnson, D. (eds.) (2015). *Advances in Plant Breeding Strategies: Breeding, Biotechnology and Molecular Tools.* Springer, Cham. doi:10.1007/978-3-319-22521-0_15

Ascherio, A.K. (1999). Trans fatty acids and coronary heart disease. *The New England Journal of Medicine*, 340, pp. 1994–1998. doi:10.1056/NEJM199906243402511

Barrangou, R. (2015). Diversity of CRISPR–Cas immune systems and molecular machines. *Genome Biology*, 16.

Blancke, S. (2015, 08 15). *Why People Oppose GMOs Even Though Science Says They Are Safe.* Retrieved from Scientific American: https://www.scientificamerican.com/article/why-people-oppose-gmos-even-though-science-says-they-are-safe/

Castiglioni, P.W. (2008). Bacterial RNA chaperones confer abiotic stress tolerance in plants and improved grain yield in maize under water-limited conditions. *Plant Physiology*, 147(2), pp. 446–455. doi:10.1104/pp.108.118828

Chang, L., Wu, S., and Tian, L. (2019, 11 08). Effective genome editing and identification of a regiospecific gallic acid 4-O-glycosyltransferase in pomegranate (Punica granatum L.). *Horticulture Research*, 6, p. 123. doi:10.1038/s41438-019-0206-7

Chen, G., Ye, C., Huang, J., Yu, M., and Li, B. (2001, 02 16). Cloning of the papaya ringspot virus (PRSV) replicase gene and generation of PRSV-resistant papayas through the introduction of the PRSV replicase gene. *Plant Cell Reports*, 20, pp. 272–277. doi:10.1007/s002990000311

Clasen, B., Stoddard, T., Luo, S., et al. (2016, 01). Improving cold storage and processing traits in potato through targeted gene knockout. *Plant Biotechnology Journal*, 14(1), pp. 169–176. doi:10.1111/pbi.12370

Demorest, Z., Coffman, A., Baltes, N., et al. (2016). Direct stacking of sequence-specific nuclease-induced mutations to produce high oleic and low linolenic soybean oil. *BMC Plant Biology*, 16, pp. 1–8. doi:10.1186/S12870-016-0906-1/FIGURES/3

Duke, S. (2015). Perspectives on transgenic, herbicide-resistant crops in the United States almost 20 years after introductoion. *Pest Management Science*, 71, pp. 652–657.

Fitch, M., Mansshardt, R., Gonsalves, D., Slightom, J., and Sanford, J. (1992). Virus resistant papaya plants derived from tissues bombarded with the coat protein gene of papaya ringspot virus. Biotechnology, 10, p. 146601472.

Food Directorate. (2022, 07). *Guidelines for the Safety Assessment of Novel Foods.* Retrieved from Government of Canada: https://www.canada.ca/en/health-canada/services/food-nutrition/legislation-guidelines/guidance-documents/guidelines-safety-assessment-novel-foods-derived-plants-microorganisms/guidelines-safety-assessment-novel-foods-2006.html#a3.2

Georges, F., and Ray, H. (2017). Genome editing of crops: A renewed opportunity for food security. *GM Crops Food, 8*, pp. 1–12.

González, M.N., M. G. (2020, 01 09). Reduced enzymatic browning in potato tubers by specific editing of a polyphenol oxidase gene via ribonucleoprotein complexes delivery of the CRISPR/Cas9 system. *Fronteir in Plant Science, 10*, p. 1649. doi: 10.3389/fpls.2019.01649

Haun, W., Coffman, A., Clasen, B., et al. (2014). Improved soybean oil quality by targeted mutagenesis of the fatty acid desaturase 2 gene family. *Plant Biotechnology Journal*, pp. 934–940. doi:10.1111/pbi.12201

Ito, Y., Nishizawa-Yokoi, A., Endo, M., et al. (2017, 11). Re-evaluation of the RIN mutation and the role of RIN in the induction of tomato ripening. *Nature Plants, 3*(11), pp. 866–874. doi:10.1038/s41477-017-0041-5

Jiang, M.Z. (2020). Brassica rapa orphan genes largely affect soluble sugar metabolism. *Horticulture Research*, p. 181. doi:10.1038/s41438-020-00403-z

Jones, P., Senanayake, V., Pu, S., et al. (2014, 07). DHA-enriched high–oleic acid canola oil improves lipid profile and lowers predicted cardiovascular disease risk in the canola oil multicenter randomized controlled trial. *The American Journal of Clinical Nutrition, 100*(1), pp. 88–97. doi:10.3945/ajcn.113.081133

Karunarathna, N.L. (2020). Elevating seed oil content in a polyploid crop by induced mutations in SEED FATTY ACID REDUCER genes. *Plant Biotechnology Journal, 18*(11), pp. 2251–2266. doi:10.1111/pbi.13381

Kaur, N., Alok, A., Shivani, K.P., et al. (2020, 05). CRISPR/Cas9 directed editing of lycopene epsilon-cyclase modulates metabolic flux for β-carotene biosynthesis in banana fruit. *Metabolic Engineering, 59*, pp. 76–86. doi:10.1016/j.ymben.2020.01.008

Klein, M., Eslami-Mossallam, B., Arroyo, D.G., and Depken, M. (2018, 02 06). Hybridization kinetics explains CRISPR-Cas off-targeting rules. *Cell Reports, 22*(6), pp. 1413–1423. doi:10.1016/j.celrep.2018.01.045

Klimek-Chodacka, M.O. (2018, 04). Efficient CRISPR/Cas9-based genome editing in carrot cells. *Plant Cell Reports, 37*(4), pp. 575–586. doi:10.1007/s00299-018-2252-2

Li, A., Jia, S., Yobi, A., et al. (2018a). Editing of an alpha-kafirin gene family increases, digestibility and protein quality in sorghum. *Plant Physiology, 177*, pp. 1425–1438. doi:10.1104/PP.18.00200

Li, R., Li, R., Li, X., et al. (2018b, 02). Multiplexed CRISPR/Cas9-mediated metabolic engineering of γ-aminobutyric acid levels in solanum lycopersicum. *Plant Biotechnology Journal, 16*(2), pp. 415–427. doi:10.1111/pbi.12781

Mallapaty, S. (2019). Australian gene-editing rules adopt 'middle ground'. *Nature.* Retrieved from https://www.nature.com/articles/d41586-019-01282-8#content

Menz, J., Modrzejewski, D., Hartung, F., Wilhelm, R., and Sprink, T. (2020, 10 09). Genome edited crops touch the market: A view on the global development and regulatory environment. *Frontiers in Plant Science*, p. 586027. doi:10.3389/fpls.202 0.586027

Miao, C., Xiao, L., Hua, K., Zou, C., Zhao, Y., Bressan, R., and Zhu, J. (2018, 05 21). Mutations in a subfamily of abscisic acid receptor genes promote rice growth and productivity. *Proceedings of the Natonal Academy of Sciences (USA)*, 115(23), pp. 6058–6063. doi:10.1073/pnas.180477411

Mojska, H., Gielecińska, I., and Szponar, L. (2007). Acrylamide content in heat-treated carbohydrate-rich foods in Poland. *Annals of the National Institute of Hygiene*, 58(1), pp. 345–349.

Mozter, P. (2022, 09 21). *Gene-edited Crops Helps the World To Ensure People's Food Security.* Retrieved from Nature World News: https://www.natureworldnews.com/articles/53186/20220921/gene-edited-crops-helps-world-ensure-people-s-food-security.htm

Murata, M.N. (2001). A transgenic apple callus showing reduced polyphenol oxidase activity and lower browning potential. *Bioscience Biotechnology and Biochemistry*, 65(2), pp. 383–388. doi:10.1271/bbb.65.383

Nakayasu, M. A. R. (2018, 10). Generation of α-solanine-free hairy roots of potato by CRISPR/Cas9 mediated genome editing of the St16DOX gene. *Plant Physiology and Biochemistry*, pp. 70–77. doi:10.1016/j.plaphy.2018.04.026

Namukwaya, B., Tripathi, L., Tripathi, J., Arinaitwe, G., Mukasa, S., and Tushemereirwe, W. (2012, 08). Transgenic banana expressing Pflp gene confers enhanced resistance to Xanthomonas wilt disease. *Transgenic Research*, 21, pp. 855–865. doi:10.1007/s11248-011-9574-y

National Academies of Sciences; National Academy of Medicine. (2017). *Human Genome Editing: Science, Ethics, and Governance.* National Academies Press. doi:10.17226/24623

Nieves-Cordones, M.M. (2017). Production of low-Cs+ rice plants by inactivation of the K+ transporter OsHAK1 with the CRISPR-Cas system. *Plant Journal*, 92(43). doi:10.1111/TPJ.13632

Okuzaki, A.O. (2018, 10). CRISPR/Cas9-mediated genome editing of the fatty acid desaturase 2 gene in Brassica napus. *Plant Physiology and Biochemistry*, 131, pp. 63–69. doi:10.1016/j.plaphy.2018.04.025

Paul, J.Y., Khanna, H., Kleidon, J., et al. (2017). Golden bananas in the field: Elevated fruit pro-vitamin A from the expression of a single banana transgene. *Plant Biotechnology Journal*, 15(4), pp. 520–532. doi:10.1111/pbi.12650

Pingali, P. (2012). Green revolution: Impacts, limits, and the path ahead. *Proceedings of National Academy of Science USA*, pp. 12302–12308.

Pixley, K.V., Falck-Zepeda, J.B., Paarlberg, R.L. et al. (2022). Genome-edited crops for improved food security of smallholder farmers. *Nature Genetics*, 54, 364–367.

Podevin, N., Davies, H., Hartung, F., Nogue, F., and Casacuberta, J. (2013, 06 01). Site-directed nucleases: A paradigm shift in predictable, knowledge-based plant breeding. *Trends in Biotechnology*, 31(6), pp. 375–383. doi:10.1016/j.tibtech.2013.03.004

Ren, C.L. (2016, 08 31). CRISPR/Cas9-mediated efficient targeted mutagenesis in Chardonnay (Vitis vinifera L.). *Scientific Reports*, p. 32289. doi:10.1038/srep32289

Rizal, G., Karki, S., Alcasid, M., et al. (2014, 03 01). Shortening the breeding cycle of sorghum, a model crop for research. *Crop Breeding & Genetics*, pp. 520–529. doi:10.2135/cropsci2013.07.0471

Rommens, C., Yans, H., Swords, K., Richael, C., and Ye, J. (2008). Low-acrylamide French fries and potato chips. *Plant Biotechnology Journal*, 6, pp. 843–853.

Rozas, P., Kessi-Perez, E., and Martinez, C. (2022, 10 20). Genetically modified organisms: Adapting regulatory frameworks for evolving genome editing technologies. *Biological Research.* doi:10.1186/s40659-022-00399-x

Sánchez-León, S., Gil-Humanes, J., Ozuna, C.V., Giménez, M.J., Sousa, C., Voytas, D.F., and Barro, F. (2018, 04). Low-gluten, nontransgenic wheat engineered with CRISPR/Cas9. *Plant Biotechnology Journal, 16*(4), pp. 902–910. doi: 10.1111/pbi.12837

Song, G., Jia, M., Chen, K., et al. (2016). (2016) CRISPR/Cas9: A powerful tool for crop genome editing. *Crop Journal, 4*, pp. 75–82.

Sun, B.J. (2020). Color-related chlorophyll and carotenoid concentrations of Chinese kale can be altered through CRISPR/Cas9 targeted editing of the carotenoid isomerase gene BoaCRTISO. Horticulture Research, 7(1), pp. 161. doi: 10.1038/s41438-020-00379-w

Sun, Y., Jiao, G., Liu, Z., et al. (2017, 03 07). Generation of high-amylose rice through CRISPR/Cas9-mediated targeted mutagenesis of starch branching enzymes. *Frontiers in Plant Science*, p. 298. doi: 10.3389/fpls.2017.00298

Tabashnik, B., Brevault, T., and Carriere, Y. (2013). Insect resistant to Btcrops: Lessons from the first billion acres. *National Biotechnology, 31*, pp. 510–521.

Tang, G., Qin, J., Dolnikowski, G.G., Russell, R.M., and Grusak, M.A. (2009, 04 15). Golden Rice is an effective source of vitamin A. *The American Journal of Clinical Nutrition, 89*(6), pp. 1776–1783. doi: 10.3945/ajcn.2008.27119

Tikunov, Y.M., Roohanitaziani, R., Meijer-Dekens, F., et al. (2020, 08). The genetic and functional analysis of flavor in commercial tomato: The FLORAL4 gene underlies a QTL for floral aroma volatiles in tomato fruit. *103*(3), pp. 1189–1204. doi: 10.1111/tpj.14795

Tsuda, M., Watanabe, K., and Ohsawa, R. (2019, 12 06). Regulatory status of genome-edited organisms under the Japanese Cartagena Act. *Frontiers in Bioengineering and Biotechnology*. doi: 10.3389/fbioe.2019.00387

Waltz, E. (2016, 04 21). Gene-edited CRISPR mushroom escapes US regulation. *Nature, 532*(7599), p. 293. doi: 10.1038/nature.2016.19754

Wang, H., Wu, Y., Zhang, Y., et al. (2019, 09 23). CRISPR/Cas9-based mutagenesis of starch biosynthetic genes in sweet potato (Ipomoea Batatas) for the improvement of starch quality. *International Journal of Molecular Sciences, 20*(19), p. 4702. doi: 10.3390/ijms20194702

Wen, S., Liu, H., Li, X., et al. (2018, 05). TALEN-mediated targeted mutagenesis of fatty acid desaturase 2 (FAD2) in peanut (Arachis hypogaea L.) promotes the accumulation of oleic acid. *Plant Molecular Biology*, pp. 177–185. doi: 10.1007/s11103-018-0731-z

Whelan, A., and Lema, M. (2015). Regulatory framework for gene editing and other new breeding techniques (NBTs) in Argentina. *GM Crops Food, 6*(4), pp. 253–265. doi: 10.1080/21645698.2015.1114698

Yan, S.C. (2020, 03). Anthocyanin Fruit encodes an R2R3-MYB transcription factor, SlAN2-like, activating the transcription of SlMYBATV to fine-tune anthocyanin content in tomato fruit. *The New Phytologist*, pp. 2048–2063. doi: 10.1111/nph.16272

Yuan, M., Zhu, J., Gong, L., et al. (2019, 04). Mutagenesis of FAD2 genes in peanut with CRISPR/Cas9 based gene editing. *BMC Biotechnology, 19*(1), p. 24. doi: 10.1186/s12896-019-0516-8

Zhang, H.S. (2018, 10 01). Genome editing of upstream open reading frames enables translational control in plants. *Nature Biotechnology, 36*, p. 894–898. doi: 10.1038/nbt.4202

Zhi, J.L. (2020). CRISPR/Cas9-mediated SlAN2 mutants reveal various regulatory models of anthocyanin biosynthesis in tomato plant. *Plant Cell Reports*, *39*(6), pp. 799–809. doi:10.1007/s00299-020-02531-1

Zsogon, A.C. (2018, 10 01). De novo domestication of wild tomato using genome editing. *Nature Biotechnology*, pp. 1211–1216. doi:10.1038/nbt.4272

Abbreviations

BoaCRTISO	*Brassica oleracea* var. *alboglabra* carotenoid isomerase
Cas	CRISPR-associated protein
CRISPR	Clustered regularly interspaced short palindromic repeats
DNA	Deoxyribonucleic acid
EU	European Union
GE	Genome edited
GEC	Genome-edited crop
GET	Gene- or genome-editing technology
GM	Genetic modification
GMO	Genetically modified organism
InDels	Insertion deletions
ISAAA	International Service for the Acquisition of Agri-biotech Applications
LMO	Living modified organism
NBT	New (plant) breeding techniques
NGT	New generation technique
OGTR	Office of the Gene Technology Regulator
PNT	Plant with novel trait
POC	Proof-of-concept
RNAi	Ribonucleic acid interference
SDN	Site-directed-nucleases
TALEN	Transcription activator like effector nuclease
VInv	Vascular invertase (gene)
ZFN	Zinc-finger-nuclease

9 Crop genome editing – regulations and policies

Elena V. Mikhaylova

9.1 The history of GMO regulation

The first GMO plants were developed much later than the first GM bacteria (1973) and animal (1974). Tobacco resistant to antibiotics was obtained by *Agrobacterium*-mediated transformation only in 1983. Insect-resistant tobacco was created soon after, in 1987 (Vaeck et al. 1987). Tobacco, resistant to Cucumber Mosaic Virus, was the first transgenic crop commercialized in China in 1992 (James, 1997). In 1994 "Flavr and Savr" tomato with delayed ripening was approved in the United States, and bromoxynil-resistant tobacco was approved in European Union (EU) (Gupta et al. 2021). In 1996 several crops came to the US market: *Bacillus thuringiensis* (Bt) cotton, potato and maize, herbicide-tolerant soybean, and canola. International Service for the Acquisition of Agri-biotech Applications (ISAAA) has begun collecting, analyzing, and publishing data on the adoption of GM crops around the world in annual reports (James, 1997). Foreign Agriculture Service of the US Department of Agriculture (USDA) also issues an annual Biotechnology Report on different countries, containing an in-depth analysis of biotech regulatory policy (https://www.fas.usda.gov/).

At that time, regulation policies for GM crops were just beginning to emerge. The question has been raised by the Organization for Economic Co-operation and Development (OECD), World Health Organization (WHO), and Food and Agriculture Organization (FAO). Finally, the report, "The safety evaluation of foods derived by modern technology – concepts and principles" was published in 1993. This document only dealt with the use of GMOs for food and recommended conducting safety assessments on a case-by-case basis through comparison to existing products that are considered safe. But until 1995 no research was conducted in order to evaluate the safety of GM products. Therefore, "Flavr and Savr" was considered "as safe as tomatoes bred by conventional means" by FDA and approved for food use without special labeling.

Things changed in 1998 when Dr. Arpad Pusztai, who was involved in one of the first safety studies, publicly announced negative health effects in lab rats fed with lectin-rich GM potatoes (Ewen and Pusztai, 1999). Soon after research

DOI: 10.4324/9781003382102-9

on the potential toxic effects of insect-resistant maize on monarch butterflies was published in Nature (Rosi-Marshall, 2009). In 2007, Irina Ermakova argued with the Nature Biotechnology journal upon her declined publication about the negative impact of GM soy on the growth and productivity of rats (Ermakova, 2007). In 2012Gilles-Eric Seralini reported an increase in tumors in rats caused by herbicide-resistant GM corn (Seralini et al. 2014).

Further studies have demonstrated that risks for human health appear to be negligible. In 2013 a review of 1783 papers found no evidence of the unsafety of GM crops (Nicolia et al. 2014), as well as a study made by the US National Academies of Sciences, Engineering, and Medicine (Gould et al. 2017).

However GMO controversial studies were criticized by the scientific community, they didn't help reduce public concern over GM food. Therefore, GM crops undergo a case-by-case assessment of health and environmental risks. It is considered that the cultivation of GM crops can carry some environmental risks, such as the spread of resistance genes or the plants themselves into the environment. Herbicide tolerance in weeds has been documented several times, as well as an increase in the use of herbicides (Myskja and Myhr, 2020). Glufosinate-ammonium has been considered toxic to mammals, therefore this herbicide along with resistant GM crops is prohibited in several countries such as Norway (Bawa et al. 2013). Other risks, such as the introduction of an allergen and horizontal gene transfer, are expected to be lower than background rates.

An international agreement on biosafety – The Cartagena Protocol – have been accepted in May 2003 (Gupta et al. 2021; Gatica-Arias, 2020). The document was supposed to ensure the safe handling, transport, and use of "living modified organisms" via risk assessment and risk management. Any living organism that possesses a novel combination of genetic material obtained through the use of modern biotechnology is considered a GMO according to the Cartagena Protocol. At the moment it has 173 parties. Each party is required to ensure that the intentional transboundary movement of such organisms is accompanied by proper documentation. The Biosafety Clearing-House database was established under the Cartagena Protocol (https://bch.cbd. int/). It contains information that Parties to the Protocol must provide, such as decisions to release or import GMOs, their characteristics, detection methods, results of the risk assessment, etc.

Currently, most countries (88%) have no GMO in agriculture, and among 72 countries that approved GM crops, only 29 are planting them, and this number did not increase since 2010 (ISAAA, 2019) (Figure 9.1). Nevertheless, the global GM seed market ($20.1 billion in 2018) keeps growing and will probably reach $30.2 billion by 2026.

Just four countries (the United States, Brazil, Argentina, and Canada) produce 85% of GM crops (Paull and Hennig, 2019). Except for Brazil, these countries are not parties to the Cartagena Protocol. But even in the United States, more than 90% of GM soybean and maize are grown for feed and biofuel production (Redden, 2021).

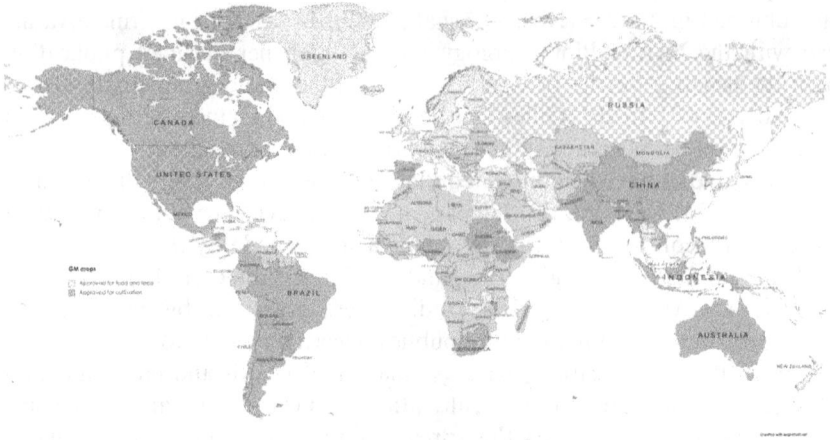

Figure 9.1 Countries with approved GM Crop Events.

Three more countries (Malawi, Nigeria, and Ethiopia) adopted GM crops in 2019, however, Burkina Faso and Egypt suspended the approvals for cultivation of GM cotton, according to Global gene editing regulation tracker. There was also a 0.7% decline in the area occupied by GM crops to 190.4 million hectares in 2019.

In general, there have been 4,485 approvals of 403 GM events from 29 biotech crops, mostly herbicide and insect resistant. The most popular GM crops, accounting for 99% of GMO agriculture hectares, are soybean, maize, cotton, and canola. Some countries also grow the less common sugar beets, papaya, alfalfa, potatoes, safflower, eggplant, squash, sugarcane, apples, and pineapple, but their contribution does not exceed 1% (Redden, 2021). The only plants authorized in EU are carnations, modified to express a new color, and insect-resistant maize cultivated in Spain and Portugal due to the need to control the pest *Ostrinia nubilalis* (Myskja and Myhr, 2020).

Certainly, these 72 countries which already approved GM crops have great potential to introduce new plant breeding technologies (NPBTs) such as genome editing, which can significantly increase the number of traits that can be modified in crops. But there are still more than 100 countries without proven regulations and policies regarding genetic modification. Some of them are known for growing GMOs without state registration or illegally (Nazar et al. 2019; Turnbull et al. 2021). However, all of these countries need to decide whether genome-edited plants will be regulated as GMOs or not.

9.2 Genome edited plants: GMOs or not?

Experience in regulating GM crops simplifies the introduction of genome-edited plants. However, in many countries, especially in the EU, the cultivation of GM crops is prohibited. They still have to enact regulatory criteria to assess if genome-edited plants are to be considered genetically modified or not.

While traditional genetic engineering involves the introduction of foreign (exogenous) genes, genome editing allows performing a greater variety of manipulations, namely a knock-out of genes, base pair changes, transcriptional programming, and a knock-in of a donor DNA in a precise location.

However basic knock-ins and knock-outs can be done with natural site-directed nucleases (SDNs), and nickases (nCas) achieved by point mutations in a catalytic domain allow for base editing and prime editing. dCas, deactivated by mutations in both catalytic domains and fused to an effector, can be used for transcriptional activation and repression of a target gene (Mikhaylova et al. 2022).

In most studies, CRISPR elements, as well as the selective gene, are cloned inside T-DNA and inserted into the plant genome (Xing et al. 2014; Ma et al. 2015; Čermák et al. 2017; Lowder et al. 2018; Hahn et al. 2020). Integration occurs at a random site, which may adversely affect the editing efficiency (Kuluev et al. 2019). Moreover, such plants are actually transgenic because they carry foreign DNA. Backcrossing and cross-pollination can be used to get rid of unwanted insertions. Visible selection markers and RNA interference-based strategy (CRISPR-S) can be used to identify transgene-free progeny (Gu et al. 2021). RNAi expression cassette, aimed to silence a native herbicide-resistance gene, is added to a CRISPR/Cas9 construct in order to select transgene-free plants by spraying the herbicide (Lu et al. 2017). Transgene Killer CRISPR system was proposed to induce automatic self-elimination of the transgene. Such genetic constructs contain two cassettes expressing the suicide genes controlled by the embryo-specific promoter, which ensures that plants carrying transgenes are killed in early embryonic stages (Yubing et al. 2019). However, these approaches are time-consuming and cannot be applied to several species such as potatoes, grapevine, and apples due to their long juvenile stages or vegetative propagation (Metje-Sprink et al. 2019).

DNA-free gene editing requires the separate delivery of a nuclease and guide RNA (gRNA) or preassembled Cas9 protein-gRNA ribonucleoproteins (RNPs) into the protoplasts or cells instead of the plasmids (Toda et al. 2019; Park and Choe, 2019). RNPs immediately cleave DNA at the target sites before being degraded by cell proteases. Despite greater prospects for practical use, this method is rare, probably because it is incompatible with *Agrobacterium*-mediated transformation (Kuluev et al. 2019). Selection of successful events is also complicated due to the absence of selective genes (Shan et al. 2013; Piatek et al. 2015).

The limitation of both approaches can be avoided by using transient expression vectors based on viral replicons (Zhang et al. 2022; Čermák et al. 2015; Butt et al. 2017; Gil-Humanes et al. 2017). Modified viruses with large cargo capacities and high gene stability, carrying genome editing instruments, replication-associated proteins, and replication points, can be cloned inside transfer DNA (T-DNA) and undergo a rolling circle replication upon arrival of T-DNA in the nucleus (Kuluev et al. 2019). *Geminiviruses, Cytorhabdoviruses, Nucleorhabdoviruses,* and *Potexviruses* have been used in genome editing (Zhang et al. 2022). Expression of SDNs, gRNAs, and donor DNAs from viral vectors can be up to 80 times higher than from the plant genome. The replicon can remain in the cell for up to 8 weeks before being eliminated without integration into plant DNA (Piatek et al. 2015). Therefore, obtained plants do not become transgenic.

Transient Expression of CRISPR/Cas9 DNA and RNA (TECCDNA and TECCRNA) system was developed by Zhang et al. (2016). Constructs expressing gRNA and Cas9 are introduced to plant cells via particle bombardment. RNA vectors allow to avoid the integration of small degraded vector fragments, as far as RNA molecules are unlikely to integrate into the plant genome. However, this method requires sequencing of all regenerated seedlings.

There is another plant-only approach to producing transgene-free plants through the graft-mobile gene editing system (Yang et al. 2023). The grafting of shoots onto roots that contain mobile Cas9 and guide RNA transcripts fused to tRNA-like sequence motifs, which convert into the corresponding protein allows editing of the seeds without integration into the genome. Roots and shoots may belong to different species (for example, Arabidopsis and oilseed rape). The authors declare the surprisingly high efficiency of the method. However, it is new and has yet to be tested.

It should be noted that transgene-free plants can emerge occasionally during traditional *Agrobacterium*-mediated transformation. For example, 17.2% of seedlings regenerated from tobacco leaf-disc explants carried the desired mutation in the PHYTOENE desaturase gene, but no signs of integration of CRISPR/Cas system (Chen et al. 2018).

Based on the adopted approach and introduced changes, three types of genome editing can be distinguished (Figure 9.2) (Redden, 2021; Friedrichs et al. 2019; Gupta et al. 2021; Buchholzer and Frommer, 2022). Many countries use this system to legally categorize gene editing events.

SDN1 type involves small base substitutions or indels without integration of foreign DNA due to the nonhomologous end-joining repair of a targeted double-strand break. SDN1 does not use any templates to guide genome repair. It is considered a very low risk for human health and the environment because similar mutations can be achieved by conventional breeding.

The SDN2 procedure involves homologous recombination using a DNA repair template. The template carries one or several predefined mutations flanked by two sequences matching both ends of the double-strand break, called homology arms. When using the prime editing method, a sort donor sequence is carried by pegRNA.

The SDN-3 plants are considered transgenic because an entire gene or an even longer genetic element is integrated in a precise location using the same technique. Being the most difficult to perform, SDN-3 is quite rare. Knock-ins

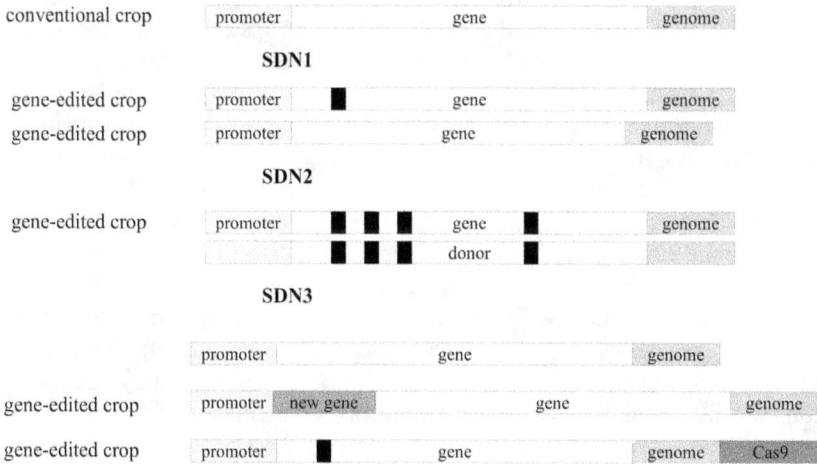

Figure 9.2 Three types of genome editing.

require a significant amount of donor DNA in the cell (Kuluev et al. 2019). There have been several approaches to increase the efficiency of the programmable insertion of DNA, including the use of transient viral vectors (Čermák et al. 2015) and tools such as PASTE (Programmable Addition via Site-specific Targeting Elements). PASTE involves the use of integrase simultaneously with Cas enzyme and guides RNA and allows to deliver genes as long as 36 kb (Yarnall et al. 2022).

Thus, some relaxation can be made in assessing the risks of gene flow and the health safety of SDN1-type varieties, as compared to traditional GMOs. This makes it possible to speed up the commercialization of genome-edited crops. But the emergence of new types of NPBTs requires reconsideration of regulations and policies which can serve as a guarantee of their sustainable application.

9.3 New challenges in regulations and policies

Genome editing policies have already been developed in several countries. Others need to clarify if genome-edited plants are covered by existing regulations. The choice of regulation method can be based on two approaches (Ishii and Araki, 2017; Entine et al. 2021; Gould et al. 2022).

Product-based regulation assumes that a new combination of genetic material was obtained via modern biotechnology. Traditional varieties are taken as the basis for the risk assessment of novel crops. Plants without new traits, that could be developed via conventional breeding, are not supposed to be regulated. For example, a change of one or several base pairs or a deletion of any size could occur naturally. This approach was chosen in the United States, Canada, Argentina, Uruguay, Colombia, and Japan.

Process-based regulation focuses on the method of genetic modification, which can be differently assessed for risk analysis. Crops created via the SDN1 procedure, which works similarly to natural mechanisms, can avoid regulation in Australia, China, and India. If donor DNA of transgenic origin has been used to induce even the slightest mutation (SDN2), plants may fall under regulations in some countries. In the EU, Mexico, and New Zealand all plants that were not obtained naturally by mating or natural recombination are considered GMOs. In 2014, the Environmental Protection Authority of New Zealand was going to deregulate gene-edited plants without foreign DNA. However, this decision was overturned by the High Court (Jones et al. 2022). The European Court of Justice decision to regulate gene-edited crops as GMOs, made on 25 July 2018, provoked a debate over this new technology (Wasmer, 2019). Soon after Argentina, Australia, Brazil, Canada, the United States, Paraguay, Uruguay, Colombia, the Dominican Republic, Honduras, Guatemala, Jordan, Vietnam, and the Secretariat of the Economic Community of West African States issued a joint statement to the World Trade Organization supporting relaxed regulations for gene editing (Schiemann et al. 2020). Applications for the use of gene-edited crops can also be submitted in other countries such as China, Nigeria, Kenya, and Japan (Whelan et al. 2020; Buchholzer and Frommer, 2022). In Nigeria, products without recombinant DNA fall under non-GMO regulatory classification. In Kenya, even processed products that do not contain DNA (such as oils) were excluded from regulation (Metje-Sprink et al. 2020).

Most of the countries, such as Mexico, India, China, Brazil, and EU, have a specific Biosafety Law, regulating GMOs and NPBTs. In other countries, such as Argentina and United States, such crops are regulated by general laws for environmental and food protection and individual Resolutions. Agricultural Ministries are usually implementing the laws, however, some countries have specific responsible agencies, such as National Technical Commission on Biosafety (CTNBio), or The Genetic Engineering Appraisal Committee (GEAC).

The countries that have already developed policies on genome-edited crops mostly agree that SDN-1 events do not require field trials to test the safety for humans and the environment (Table 9.1). Food produced from these plants also is not considered a novel food.

Some countries still have ongoing discussions over the regulation of SDN-1 plants (Gupta et al. 2021). Some restrict the cultivation of any kind of biotechnological crops, such as Norway and Switzerland. Others did not announce the regulatory status of gene-edited crops yet.

In 2015, Argentina was the first country to declare a policy for the agricultural use of genome-edited crops. Plants without any foreign DNA were not supposed to be regulated under biosafety legislation (Eriksson et al. 2019; Schmidt et al. 2020). Now, the regulatory system in Argentina is among the most experienced ones, being based on the breeding methodology, introduced trait, phenotype, the accomplished genetic changes, and removal of the temporary transgene.

Table 9.1 Regulation of gene-edited crops established in different countries (Eriksson et al. 2019; Gatica-Arias, 2020; Schmidt et al. 2020)

Country	Approved GM events	Genome editing regulation status	Regulation document	Responsible agency	Regulation approach	Party to the Cartagena Protocol	Cultivation of GM crops
Argentina	80	plants without foreign DNA are regulated as conventional plants (SDN-1 and SDN-2)	Resolution 173/15	MAGYP, CONABIA	Case-by-case Product-based	no	yes
Australia (Redden, 2021)	144	SDN-1 de-regulated as conventional plants	2019 Amendments to The Gene Technology Act 2001	OGTR, FSANZ	Process-based	no	yes
Brazil	112	plants without foreign DNA are regulated as conventional plants (SDN-1 and SDN-2)	Normative Resolution No. 16	CTNBio, CNBS, CIBios	Case-by-case Product-based	yes	yes
Canada	192	SDN-1 and SDN-2 are deregulated	Guidelines for the Safety Assessment of Novel Foods	CFIA	Product-based	no	yes
Chile	3	plants without foreign DNA are regulated as conventional plants (SDN-1 and SDN-2)	Resolution 1523/ 2001	SAG	Case-by-case Product-based	no	yes
China	77	plants without foreign DNA are regulated as conventional plants (SDN-1 and SDN-2). Only the commercialization of	2021: Biosecurity Law of China 2022: Guidelines for Safety Evaluation of	MARA	Process-based	yes	yes

(Continued)

Table 9.1 (Continued)

Country	Approved GM events	Genome editing regulation status	Regulation document	Responsible agency	Regulation approach	Party to the Cartagena Protocol	Cultivation of GM crops
Colombia	109	China-produced GM crops is allowed plants without foreign DNA are regulated as conventional plants	Agricultural Gene Editing Plants Resolution 29299/ 2018	ICA	Case-by-case Product-based	yes	yes
Costa Rica	20	no specific legislation is required for the approval of GM and genome-edited crops	Regulation #29782-MAG, Law #26921	NTBC	Case-by-case Product-based	yes	yes
Ecuador (Santos et al. 2014)	0	plants without foreign DNA are regulated as conventional plants. Cultivation of GM crops is prohibited	National Agrarian Authority. 2019: Decree 752, Organic Code of the Environment	Ministry of Agriculture and Livestock	Product-based	yes	no
European Union (Bruetschy, 2019; Turnbull et al. 2021)	116	All types of gene editing regulated as GMOs	2001 GMO Directive	EFSA	Process-based	yes	Spain and Portugal
Guatemala	0	plants without foreign DNA are regulated as conventional plants. GMO cultivation is allowed only for research purposes	Resolution No. 60–2019	CONAP	Product-based	yes	no

Country		Description	Regulation/Act	Agency	Approach		
Honduras	8	simplified procedure to approve genetically edited products	Resolution No. 60–2019 Agreement for genome editing C.D.-008-2019	SENASA CTNBio	Product-based	yes	yes
India (Metje-Sprink et al. 2020)	11	relaxed regulations for SDN-1 and SDN2 that do not contain foreign DNA	Rules 1989, Guidelines for the Safety Assessment of Genome Edited Plants, 2022	FSSAI, GEAC	Process-based	yes	yes
Indonesia (Metje-Sprink et al. 2020; Jones et al. 2022)	22	plants without foreign DNA are regulated as conventional plants	Government Regulation No. 21/2005 on the Biosafety of Genetically Modified Products	KKH-PRG	Product-based	yes	yes
Israel (Schmidt et al. 2020; Duensing et al. 2018; Justo-Hanani, 2016)	1	plants without foreign DNA are regulated as conventional plants	Seed Regulation Act of 2005	NCTP	Product-based	no	yes
Japan (Wasmer, 2019)	196	SDN-1 and SDN-2 type organisms are not GMOs if the genetic change is no more than a few base pairs	Japanese Cartagena Act, the Food Sanitation Law, the Feed Safety Law	MOE	Product-based	yes	no
Kenya (Schmidt et al. 2020)	1	gene-edited crops will be regulated as conventional plants unless they contain	Kenya's Biosafety Act of 2009, Guidelines for determining the	NBA	Case-by-case Product-based	yes	yes

(Continued)

Table 9.1 (Continued)

Country	Approved GM events	Genome editing regulation status	Regulation document	Responsible agency	Regulation approach	Party to the Cartagena Protocol	Cultivation of GM crops
		DNA from sexually incompatible species	regulatory process of genome-edited organisms and products in Kenya				
Russia (Dobrovidova, 2019; Turnbull et al. 2021)	28	gene editing is allowed only for research purposes	Federal Law 358-FZ 2020 Food Security Doctrine	Agricultural Ministry	Process-based	no	no
Mexico	188	All types of gene editing regulated as GMOs	Biosafety Law	SAGARPA	Process-based	yes	yes
New Zealand	114	All types of gene editing regulated as GMOs	Hazardous Substances and New Organisms (HSNO) Act 1996	FSANZ, EPA	Process-based	yes	no
Nigeria	29	relaxed regulations for plants without the new combination of genetic material	National Biosafety Guideline on Gene Editing	NBMA	Product-based	yes	yes
Pakistan	6	plants without foreign DNA are regulated as conventional plants	Pakistan Biosafety Rules, 2005	NBC, IBC, TAC	Product-based	yes	yes
Paraguay	22	plants without foreign DNA are regulated as conventional plants	Resolution No. 565	CTNBio	Case-by-case Product-based	yes	yes

Country	No.	Description	Regulation/Act	Agency	Approach		
Philippines (Metje-Sprink et al. 2020; Jones et al. 2022)	129	only plants with novel combinations of genetic material, which is not possible to achieve through conventional breeding are considered GMOs	Joint Department Circular No. 1, Series of 2016, revised on February 15, 2022 Memorandum Circular No. 8, Series of 2022	DOST, DA, JAG	Product-based	yes	yes
South Korea (Metje-Sprink et al. 2020; Jones et al. 2022)	170	plants without foreign DNA are regulated as conventional plants	Living Modified Organism (LMO) Act.	MOTIE, MAFRA	Product-based	yes	no
United States (Piatek et al. 2015)	216	SDN-1 and SDN-2 are deregulated	"Federal Insecticides, Fungicides, and Rodenticides act," "Federal Food, Drug, and Cosmetic Act"	APHIS, USDA, FDA and EPA	Product-based	no	yes
Uruguay	17	plants without foreign DNA are regulated as conventional plants	2008: Decree No. 353/008	GNBio	Product-based	yes	yes

Biotechnology develops very fast, therefore some of new techniques are not captured by the existing regulations (Metje-Sprink et al. 2020). For example, in 2018 Philippines started reviewing the status of eight new types of NPBTs (SDNs, oligonucleotide-directed mutagenesis, cisgenesis, intragenesis, RNA-dependent DNA methylation, grafting with GM material, reverse breeding, agroinfiltration and synthetic genomics). As a result, the original document – Joint Department Circular No. 1 (2016) was revised in 2021 and 2022, and a new document (Memorandum Circular No. 8) was issued in 2022 (Jones et al. 2022). The new policies reduce the timeline for approvals and costs for commercializing NPBTs and GMOs. Plants without any novel combinations of genetic material will be considered conventional products.

Other countries also start revising their policies on the use of GMOs and NPBTs. India was one of the first countries to issue a regulation document called "Rules for the Manufacture, Use, Import, Export, and Storage of Hazardous micro-organisms/Genetically engineered organisms or cells" in 1989, and several guidelines and memoranda under these Rules. To regulate NPBTs, India released 'Guidelines for the Safety Assessment of Genome Edited Plants, 2022' (Buchholzer and Frommer, 2022).

Thailand evaluated the use of GM crops under the Plant Quarantine Act (1964). According to the document, import and transit of all transgenic plants were prohibited, and GM food required obligatory labeling. Later new documents such as Biosafety Guidelines for R&D, Biosafety Guidelines for Plants Carrying Stacked Genes and Their Derivatives, and Guideline on Plant Biosafety in Research Greenhouses, were adopted. At the moment, Thailand considers several new documents: The Draft Genetically Modified Foods Regulation, The Draft Biodiversity Law, and The Draft Biosafety Assessment Guidelines for Genome Editing Technology. NPBTs will be still regulated as GM crops, however, risk assessment requirements will be relaxed for SDN-1 plants (Buchholzer and Frommer, 2022).

Canada regulates plants with novel traits, regardless of the applied method. To clarify the regulatory status of gene-edited plants, this country drafted guidance to Part V of the Seeds Regulations. New seeds must be authorized before releasing into the environment only if they are not substantially equivalent to the seeds that are already present in Canada (Metje-Sprink et al. 2020). In such cases, plants produced by chemical or radiation mutagenesis are regulated in the same way as genome-edited plants.

In Australia, The Gene Technology Act 2000 underwent a series of revisions. As a result, SDN-1 products were excluded from the regulations, however, methods that require a template to guide genome repair (SDN-2 and SDN-3) were supposed to be regulated as GMOs. If the introduced traits do not change product characteristics compared to conventional food, gene-edited products are to be considered equivalent to conventional food in Australia (Jones et al. 2022).

After leaving the EU, UK introduced a Genetic Technology (Precision Breeding) Bill that would allow the commercial cultivation and sale of gene-edited crops when passed by Parliament.

It is highly important to harmonize the regulations and policies on NPBTs across the planet. Studies of the existing regulatory regimes are required to discover the most flexible policy which would ensure the legalization of all products developed via new gene editing techniques. Therefore, the product-based approach will probably prevail over the process-based approach in the next few years.

There are several organizations studying policies in the field of gene editing. Global gene editing regulation tracker, developed by The Genetic Literacy Project, is an interactive tool that sums up gene editing regulations worldwide (Ewa et al. 2022; Hundleby and Harwood, 2022).

Another organization studying the potential and impact of plant genome editing, including regulatory frameworks, is COST (European Cooperation in Science and Technology). Five working groups are involved in the action "Genome editing in plants - a technology with transformative potential" (PlantEd) since 2018. Group No. 3, working on Regulations and policies of gene editing, includes 236 scientists from Turkey, Serbia, Spain, Sweden, Finland, France, New Zealand, Italy, Belgium, Germany, Netherlands, United Kingdom, Slovenia, Armenia, Denmark, Ireland, Latvia, Bulgaria, Albania, Hungary, Tunisia, Moldova, Greece, Cyprus, Croatia, Lithuania, Portugal, Austria, Belarus, Switzerland, Azerbaijan, Israel, Estonia, Australia, United States, Mexico, Poland, India, Norway, Iceland, Jordan, and Romania. This group evaluates attitudes toward editing techniques and the resulting products in the legislation of COST countries and compares them with that of other countries that have already announced the regulatory status of gene-edited crops. COST members also monitor the emergence of patents related to genome editing tools. The project is supposed to end in October 2023 (https://plantgenomeediting.eu/).

GeneBEcon is a new Horizon project that conducts research on the potential of gene editing, as well as its social, economic, and regulatory issues in EU (https://www.plantetp.eu/new-horizon-europe-project-on-gene-editing-in-plants/). It aims to represent a powerful toolbox for gene editing in potatoes and microalgae, which is supposed to be used for case studies, including the evaluation of the benefits and risks of NGT-derived products. The project started in September 2022 and will last for 3 years.

9.4 Approaches to risk assessment of gene-edited crops

Regulations and policies on GMOs and NPBTs are being developed because of the risks of their cultivation. In countries with more relaxed regulations these risks are considered to be small, however, in others, the use of biotech plants can be completely prohibited. It may depend on the native flora and fauna of the country. For example, Mexico is the center of origin and diversity of maize, therefore cultivation of GM or gene-edited maize in this country poses an additional risk, as compared to the cultivation of this crop in Europe. But in most countries, stricter rules for the commercialization of biotech plants are associated with putative risks to human health (Jones et al. 2022). Genome editing provides a level of precision and predictability that was previously unavailable, which means improved safety of gene-edited seeds and products (Pixley et al. 2022). In most cases, risk assessment uses comparison with varieties obtained through conventional breeding. If NPBT plants do not substantially differ from traditional crops, they can be deregulated (Buchholzer and Frommer, 2022). Nevertheless, common plants can be quite dangerous due to the accumulation of pesticides, toxins, and allergens.

There are several risks, specific for gene-edited plants, which can be easily reduced:

• Off-target effects, which may occur at sites similar to the target site. The frequency of off-target effects is close to those of spontaneous natural mutagenesis. At the stage of planning an experiment, recently developed, more specific editing systems can be used. For example, Cas12b is more precise than Cas9 (Mikhaylova et al. 2022). Alternative guide RNA sequences without off-target sites might be chosen if unspecific editing is detected (Jones et al. 2022). Usually, proper design of gRNAs allows to solve the problem. The genomic sequence of the plant can be done before commercialization. Even if off-target edits arise, they may bring no harm to the consumers and the plant itself.
• The introduction of undesired DNA sequences from the plasmid backbone. Even with the use of technologies that do not require the integration of genome-editing tools in the genome (such as viral vectors or TECCDNA), spontaneous insertions may occur. The use of RNA instead of DNA (TECCRNA and grafting methods) and full-genome sequencing of the designed variety will fully guarantee the absence of undesired insertions.
• Synthesis of new peptides that might be toxic or allergenic. Editing might result in the emergence of a new ORF or a change in protein function instead of its knockout. To avoid this problem, the DNA and protein sequences surrounding the edit site, as well as changes in gene expression, can be checked.
• Problematic detection. Traditional GMOs and plants with integrated genome-editing tools (such as Cas gene) have a unique genetic signature that can be detected via a simple PCR. However, only the developer knows what changes in the genome have occurred in transgene-free NPBTs. Complicated regulations may urge the developers to register gene-edited plants as conventional varieties or cultivate them illegally. In such cases, there will be no guarantees that the products do not contain new allergens or toxins because varieties obtained through traditional breeding do not undergo safety evaluation.

Despite relaxed regulations of SDN-1 and SDN-2 in many countries, the developer is faced with the task of proving that the new variety is safe, contains no foreign DNA or novel combination of genetic material, or they are possible to achieve through conventional breeding. The required evidence may differ depending on the country where the plant will be commercialized. For example, China and India have very strict safety precautions for experiments with genome-edited plants until it can be ensured that they are transgene-free (Buchholzer and Frommer, 2022). When registering an NPBT, the application must include detailed documentation on the editing technique, introduced mutations, stability of the trait, absence of off-target effects, and vector sequences. NPBTs are certified only after providing sufficient evidence of compliance with local policies. Applications are being reviewed to determine whether the new varieties are GMOs or they fall under relaxed regulations. Non-GMOs can be released from further safety trials.

According to calculations, developing and bringing a genome-edited crop to market could take 5 years and cost $10.5M, however, a usual GM event is twice as expensive, and its creation and commercialization takes three times longer (Friedrichs et al. 2019). After just 10 years since the invention of CRISPR, the gene editing market value reached USD 5.4 billion in 2021 and will continue its growth up to 19.9 billion in 2030, according to Global Market Insights Inc. Therefore, it can be assumed that gene-edited plants will become more and more common.

9.5 Application of genome editing

Deregulation of genome-edited crops in several countries complicates data collection. As far as SDN-1 events are indistinguishable from conventional crops, we may never know the extent of their adoption. It is suggested that only six genome-edited crops—soybean, rice, maize, canola, mushroom, and camelina—have been approved for commercial use (Pixley et al. 2022). It has also been reported that non-browning Arctic Gala apples, obtained by reduction of polyphenol oxidase production, reached the market (Waltz, 2015). Biotech cotton with reduced gossypol content (event TAM66274) achieved a non-regulated status in United States. So did non-Browning GreenVenus™ Romaine lettuce (ISAAA, 2019).

According to the Biosafety Clearing-House database, the only genome-edited event registered in the Parties of the Cartagena Protocol is borer-resistant, herbicide-tolerant maize BCH-CON-SCBD-14931-2 bred by Pioneer (Corteva) (https://bch.cbd.int/). But CRISPR/Cas9 was used only for a site-specific integration of exogenous genes of insecticidal protein IPD079Ea and phosphinothricin N-acetyltransferase into the maize genome. Therefore, this plant is actually transgenic. The company is also working on the adoption of CRISPR-Cas edited waxy corn with increased content of amylopectin. It has already been considered a non-GMO in the United States, Canada, Brazil, Argentina, and Chile. According to another source,

Kenya, being a party to Cartagena Protocol, has approved five gene-edited events (Metje-Sprink et al. 2020).

A lot of biotech companies are involved in the development of gene-edited crops (Metje-Sprink et al. 2020; Waltz, 2019). CRISPR-edited maize developed by Dupont was exempted from regulation in the United States (Waltz, 2016). Bayer is exploring the application of gene editing in the improvement of disease resistance, stress tolerance, plant growth, and development. The company is experimenting with traditional and new plant breeding technologies to generate short-stature corn.

CIBUS claims it has six different traits that will be implemented in five different crops using Rapid trait development system gene-editing platform. One such product is herbicide-tolerant SU Canola™, which is already approved in the United States and Canada.

CALYXT performs gene editing using TALEN. The company launched its first commercial high-oleic soybean variety in 2018 as the "first commercially approved gene-edited crop in the U.S." However, the browning-resistant mushroom was deregulated in the United States as early as in 2015; Waltz., 2016). Calyxt's Cold-storable "PPO_KO" potato has been considered a non-GMO in the United States and Canada. The company began proof-of-concept trials of reduced lignin IQ™ Alfalfa in 2021. This variety is supposed to enter the market in 2023.

An Ireland-based CRISPR startup company PLANTeDIT produced a soybean with high oleic oil content named SOlive by mutating FAD2 gene via CRISPR. The company is specializing in non-transgenic, regulatory-free, sustainable, consumer-oriented genome-edited plant products. However, no information is available on the application of SOlive variety or other varieties developed by PLANTeDIT.

A US company Pairwise launched the brand Conscious™ Foods. The first product, a salad with reduced pungency called Conscious Greens, is supposed to hit the market in 2023. The company is also developing seedless blackberries, black raspberries, and pitless cherries. Pairwise currently uses Cas12 and works on developing novel editing tools.

Inari is a Cambridge-based company, performing multiplex gene editing through its SEEDesign™ platform. It has been granted two U.S. patents related to gene-edited insect-resistant INIR6 corn and herbicide-tolerant MON-89788 soybean. However, both varieties are transgenic (SDN-3) due to the insertion of donor DNA. The company also conducts genome editing of wheat.

TreeCo is a company founded in 2019 and focused on developing gene-edited trees with increased quality of timber and fiber and pest resistance. First of all, they aim to create trees with desired commercial traits such as lower lignin or syringyl/guaiacyl ratio.

Simplot and Plant Sciences Inc. expect to commercialize gene-edited strawberries with improved shelf life. The Agriculture Department has already determined that the use of gene editing on strawberries replicates a natural process and doesn't require regulatory approval. Previously Simplot released Innate®—branded Z6 and W8 potatoes with low acrylamide potential, black spot bruise, and late blight resistance traits. Being a GMO, it is unlikely to pose a plant pest risk, in 2017 it was deregulated in the United States but considered as a novel food in Canada (Halterman et al. 2016; Nadakuduti et al. 2018).

Neoplants is a Paris-based company that was started in 2018. They are trying to contribute to solving the problem of climate change by designing a houseplant *Epipremnum aureum* to fight air pollution by efficiently removing volatile organic compounds. The Neo P1 variety, created via gene editing, will soon be available on the market.

The Dutch company Hudson River Biotechnology employs a CRISPR-based non-transgenic approach to gene editing using a molecular breeding workflow called TiGER (Target identification, Guide selection, Entry into the cell, and Regeneration) based on MAD7 nuclease and machine learning technology AccelATrait™. At first, they tried to apply CRISPR to marigold plants to increase their lutein content, but the work was not successful. It is not clear what exactly the company is developing now.

An Australian research company Nexgen Plants created salt-tolerant rice and virus-resistant plants, like Tomato or Sorghum, through INTtraitTM technology that inserts into plants only parts of their own genome. In 2019 virus resistant tomato was considered non-GM by APHIS in the United States. Nevertheless, it is not clear if this technology uses SDNs.

Unlike GMOs which were primarily produced by large companies such as Monsanto to benefit farmers (via herbicide resistance and insect resistance), NPBTs are supposed to benefit the consumer. Most of the gene editing events are notable for increased product quality and agronomic value. New traits, visible to consumers, are likely to win their trust. For example, British scientists are about to bring to market a gene-edited tomato, rich in provitamin D3, as soon as the Genetic Technology Bill successfully gets through Parliament (Li et al. 2022).

Thus, there are many young and ambitious companies in the gene editing market that are trying to release new products with varying degrees of success. Companies may face copyright issues that can hinder the release of gene-edited crops, allowing scientists to achieve greater editing success. Patents play an important role in the regulation and commercialization of NPBTs. Only the US Patent Office has more than 6,000 CRISPR patents and patent applications. But most of them are owned by universities and research organizations, and not companies. This is one of the key factors in the availability of technology for scientists.

The debate over who invented CRISPR has been going on for years. The emergence of new CRISPR-related technologies makes the problem even more complex. Jennifer Doudna and Emmanuelle Charpentier filed for the US patent just a few months earlier than the team led by Feng Zhang, who claimed to be the first to invent the technology. The two teams have been arguing since then. The US patent office has repeatedly ruled in favor of the Feng Zhang team, but the patent office in Europe has reached a different decision (Ledford, 2022).

For many companies that want to sell products developed via CRISPR technology requesting a license will remain quite problematic until the right holder is clearly identified. But the ongoing dispute will not keep gene-edited crops from reaching the market. For academic and non-profit research use, the license is usually not necessary. The use of alternative approaches also provides the option to avoid these patents. Among CRISPR patents, 45% are directed to technical improvements in the field, including modified Cas9 and guide RNAs and multiplex editing (Scheinerman and Sherkow, 2021).

The genome of at least 40 plant species was modified in the last few years (Metje-Sprink et al. 2020). For example, 27 gene-edited events of rice, maize, tomato, rapeseed, camelina, peanut, sugarcane, and wheat, developed in universities and research institutes, currently undergo field studies in 5 countries (Metje-Sprink et al. 2020). Out of these events, 19 are being tested in China, which has recently published trial rules for the approval of gene-edited plants.

EU-SAGE is a database for peer-reviewed scientific publications about genome-edited crop plants (https://www.eu-sage.eu/). There are 648 search results at the moment, 213 of them refer to rice. There are not many reports on field trials because SDN-1 events do not require safety testing. The invention of new gene editing tools, such as new nucleases, provides new opportunities for the application of the resulting new varieties of crops.

9.6 Conclusion

The absence of ethical concerns allows the development and application of genome-edited plants; however, animal experiments and gene therapy can face more complicated regulatory issues. A large number of traits have been introduced into economically valuable crops to increase productivity and overcome hunger and lack of nutrients. Due to the relaxation of policies, they may soon enter the market and start a new era in agriculture.

Declaration of interest statement: The authors have no conflicts of interest to declare.

Funding: No funding was received for this work.

References

Bawa, A.S., & Anilakumar, K.R. (2013). Genetically modified foods: safety, risks and public concerns—a review. *Journal of food science and technology, 50*(6), 1035–1046. 10.1007/s13197-012-0899-1

Bruetschy, C. (2019, August). The EU regulatory framework on genetically modified organisms (GMOs). In *Transgenic research* (Vol. 28, No. 2, pp. 169–174). Springer International Publishing. 10.1007/s11248-019-00149-y

Buchholzer, M., & Frommer, W.B. (2022). An increasing number of countries regulate genome editing in crops. *New Phytologist*. 10.1111/nph.18333

Butt, H., Eid, A., Ali, Z., Atia, M.A., Mokhtar, M.M., Hassan, N., Lee, C., Bao, G., & Mahfouz, M.M. (2017). Efficient CRISPR/Cas9-mediated genome editing using a chimeric single-guide RNA molecule. *Frontiers in Plant Science, 8*, 1441. 10.3389/fpls.2017.01441

Čermák, T., Baltes, N.J., Čegan, R., Zhang, Y., & Voytas, D.F. (2015). High-frequency, precise modification of the tomato genome. *Genome Biology, 16*(1), 1–15. 10.1186/s13059-015-0796-9

Čermák, T., Curtin, S.J., Gil-Humanes, J., Čegan, R., Kono, T.J., Konečná, E., Belanto, J., Starker, C., Mathre, J., Greenstein, R.,& Voytas, D.F. (2017). A multipurpose toolkit to enable advanced genome engineering in plants. *The Plant Cell, 29*(6), 1196–1217. 10.1105/tpc.16.00922

Chen, L., Li, W., Katin-Grazzini, L., Ding, J., Gu, X., Li, Y., Gu, T., Wang, R., Lin, X., Deng, Z., McAvoy, R., Gmitter, F., Deng, Z., Zhao, Y., & Li, Y. (2018). A method for the production and expedient screening of CRISPR/Cas9-mediated non-transgenic mutant plants. *Horticulture Research, 5*. 10.1038/s41438-018-0023-4

Dobrovidova, O. (2019). Russia joins in global gene-editing bonanza. *Nature, 569*(7756), 319–320. 10.1038/d41586-019-01519-6

Duensing, N., Sprink, T., Parrott, W.A., Fedorova, M., Lema, M.A., Wolt, J.D., & Bartsch, D. (2018). Novel features and considerations for ERA and regulation of crops produced by genome editing. *Frontiers in Bioengineering and Biotechnology, 6*, 79. 10.3389/fbioe.2018.00079

Entine, J., Felipe, M.S.S., Groenewald, J.H., Kershen, D.L., Lema, M., McHughen, A., Nepomuceno, A.L., Ohsawa, R., Ordonio, R.L., Parrott, W.A., Quemada, H., Ramage, C., Slamet-Loedin, I., Smyth, S.J., & Wray-Cahen, D. (2021). Regulatory approaches for genome edited agricultural plants in select countries and jurisdictions around the world. *Transgenic Research, 30*(4), 551–584. 10.1007/s11248-021-00257-8

Eriksson, D., Kershen, D., Nepomuceno, A., Pogson, B.J., Prieto, H., Purnhagen, K., Smyth, S., Wesseler, J., & Whelan, A. (2019). A comparison of the EU regulatory approach to directed mutagenesis with that of other jurisdictions, consequences for international trade and potential steps forward. *New Phytologist, 222*(4), 1673–1684. 10.1111/nph.15627

Ermakova, I.V. (2007). GM soybeans—revisiting a controversial format. *Nature Biotechnology, 25*(12), 1351–1354.

Ewa, W.G., Agata, T., Milica, P., Anna, B., Dennis, E., Nick, V., Godelieve, G., Selim, C., Naghmeh, A., & Tomasz, T. (2022). Public perception of plant gene technologies worldwide in the light of food security. *GM Crops & Food, 13*(1), 218–241. 10.1080/21645698.2022.2111946

Ewen, S.W., & Pusztai, A. (1999). Effect of diets containing genetically modified potatoes expressing Galanthus nivalis lectin on rat small intestine. *The Lancet, 354*(9187), 1353–1354. 10.1016/S0140-6736(98)05860-7

Friedrichs, S., Takasu, Y., Kearns, P., Dagallier, B., Oshima, R., Schofield, J., & Moreddu, C. (2019). An overview of regulatory approaches to genome editing in agriculture. *Biotechnology Research and Innovation, 3*(2), 208–220. 10.1016/j.biori.2019.07.001

232   *Elena V. Mikhaylova*

Gatica-Arias, A. (2020). The regulatory current status of plant breeding technologies in some Latin American and the Caribbean countries. *Plant Cell Tissue and Organ Culture (PCTOC)*, *141*(2), 229–242. 10.1007/s11240-020-01799-1

Gil-Humanes, J., Wang, Y., Liang, Z., Shan, Q., Ozuna, C.V., Sánchez-León, S., Baltes, N., Starker, C., Barro, F., Gao, C., & Voytas, D.F. (2017). High-efficiency gene targeting in hexaploid wheat using DNA replicons and CRISPR/Cas9. *The Plant Journal*, *89*(6), 1251–1262. 10.1111/tpj.13446

Gould, F., Amasino, R.M., Brossard, D., Buell, C.R., Dixon, R.A., Falck-Zepeda, J.B., Gallo, M., Giller, K., Glenna, L., Griffin, T., Hamaker, B., Kareiva, P., Magraw, D., Mallory-Smith, C., Pixley, K., Ransom, E., Rodemeyer, M., Stelly, D., Stewart, N., & Whitaker, R.J. (2017). Elevating the conversation about GE crops. *Nature Biotechnology*, *35*(4), 302–304. 10.1038/nbt.3841

Gould, F., Amasino, R.M., Brossard, D., Buell, C.R., Dixon, R.A., Falck-Zepeda, J.B., Gallo, M.A., Giller, K.E., Glenna, L.L., Griffin, T., Magraw, T., Mallory-Smith, C., Pixley, K.V., Ransom, E.P., Stelly, D.M., & Stewart Jr, C.N. (2022). Toward product-based regulation of crops. *Science*, *377*(6610), 1051–1053. 10.1126/science.abo3034

Gu, X., Liu, L., & Zhang, H. (2021). Transgene-free genome editing in plants. *Frontiers in Genome Editing*, 3.

Gupta, S., Kumar, A., Patel, R., & Kumar, V. (2021). Genetically modified crop regulations: scope and opportunity using the CRISPR-Cas9 genome editing approach. *Molecular Biology Reports*, *48*(5), 4851–4863. 10.1007/s11033-021-06477-9

Hahn, F., Korolev, A., Sanjurjo Loures, L., & Nekrasov, V. (2020). A modular cloning toolkit for genome editing in plants. *BMC Plant Biology*, *20*(1), 1–10. 10.1186/s12870-020-02388-2

Halterman, D., Guenthner, J., Collinge, S., Butler, N., & Douches, D. (2016). Biotech potatoes in the 21st century: 20 years since the first biotech potato. *American Journal of Potato Research*, *93*(1), 1–20. 10.1007/s12230-015-9485-1

Hundleby, P., & Harwood, W. (2022). Regulatory Constraints and Differences of Genome-Edited Crops Around the Globe. In *Genome Editing* (pp. 319–341). Springer, Cham. 10.1007/978-3-031-08072-2_17

ISAAA. (2019). Global Status of Commercialized Biotech/GM Crops in 2019: Biotech Crops Drive Socio-Economic Development and Sustainable Environment in the New Frontier. ISAAA Brief No. 55. ISAAA: Ithaca, NY.

Ishii, T., & Araki, M. (2017). A future scenario of the global regulatory landscape regarding genome-edited crops. *GM crops & food*, *8*(1), 44–56. 10.1080/21645698.2016.1261787

James, C. (1997). Global Status of Transgenic Crops in 1997. ISAAA Briefs No. 5. ISAAA: Ithaca, NY. pp. 31.

Jones, M.G., Fosu-Nyarko, J., Iqbal, S., Adeel, M., Romero-Aldemita, R., Arujanan, M., Kasai, M., Wei, X., Prasetya, B., Nugroho, S., Mewett, O., Mansoor, S., Awan, M.J.A., Ordonio, R.L., Rao, S.R., Poddar, A., Hundleby, P., Iamsupasit, N., & Khoo, K. (2022). Enabling Trade in Gene-Edited Produce in Asia and Australasia: the Developing Regulatory Landscape and Future Perspectives. *Plants*, *11*(19), 2538. 10.3390/plants11192538

Justo-Hanani, R. (2016). Israeli Regulation and Policy of GM Food and Crops. In *International Food Law and Policy* (pp. 1409–1425). Springer, Cham. 10.1007/978-3-319-07542-6_54

Kuluev, B.R., Gumerova, G.R., Mikhaylova, E.V., Gerashchenkov, G.A., Rozhnova, N.A., Vershinina, Z.R., Khyazev, A.V., Matniyazov, R.T., Baymiev, An. Kh., Baymiev, Al. Kh., & Chemeris, A.V. (2019). Delivery of CRISPR/Cas components into higher plant cells for genome editing. *Russian Journal of Plant Physiology, 66*(5), 694–706. 10.1134/S102144371905011X

Ledford, H. (2022). Major CRISPR patent decision won't end tangled dispute. *Nature, 603*(7901), 373–374. 10.1038/d41586-022-00629-y

Li, J., Scarano, A., Gonzalez, N.M., D'Orso, F., Yue, Y., Nemeth, K., ... & Martin, C. (2022). Biofortified tomatoes provide a new route to vitamin D sufficiency. *Nature Plants*, 1–6. 10.1038/s41477-022-01154-6

Lowder, L.G., Zhou, J., Zhang, Y., Malzahn, A., Zhong, Z., Hsieh, T.-F., Voytas, D.F., Zhang, Y., & Qi, Y. (2018). Robust transcriptional activation in plants using multiplexed CRISPR-Act2.0 and mTALE-Act systems. *Molecular Plant, 11*(2), 245–256.

Lu, H.P., Liu, S.M., Xu, S.L., Chen, W.Y., Zhou, X., Tan, Y.Y., Huang, J.Z., & Shu, Q.Y. (2017). CRISPR-S: an active interference element for a rapid and inexpensive selection of genome-edited, transgene-free rice plants. *Plant Biotechnology Journal, 15*(11), 1371. 10.1111/pbi.12788

Ma, X., Zhang, Q., Zhu, Q., Liu, W., Chen, Y., Qiu, R., Wang, B., Yang, Z., Li, H., Lin, Y., Xie, Y., Shen, R., Chen, S., Wang, Z., Chen, Y., Guo, J., Chen, L., Zhao, X., Dong, Z., & Liu, Y.G. (2015). A robust CRISPR/Cas9 system for convenient, high-efficiency multiplex genome editing in monocot and dicot plants. *Molecular plant, 8*(8), 1274–1284. 10.1016/j.molp.2015.04.007

Metje-Sprink, J., Menz, J., Modrzejewski, D., & Sprink, T. (2019). DNA-free genome editing: past, present and future. *Frontiers in Plant Science*, 1957. 10.3389/fpls.2018. 01957

Metje-Sprink, J., Sprink, T., & Hartung, F. (2020). Genome-edited plants in the field. *Current Opinion in Biotechnology, 61*, 1–6. 10.1016/j.copbio.2019.08.007.

Mikhaylova, E.V., Khusnutdinov, E.A., Chemeris, A.V., & Kuluev, B.R. (2022). Available Toolkits for CRISPR/CAS Genome Editing in Plants. *Russian Journal of Plant Physiology, 69*(1), 1–14. 10.1134/S1021443722010137

Myskja, B.K., & Myhr, A.I. (2020). Non-safety assessments of genome-edited organisms: should they be included in regulation? *Science and Engineering Ethics, 26*(5), 2601–2627. 10.1007/s11948-020-00222-4

Nadakuduti, S.S., Buell, C.R., Voytas, D.F., Starker, C.G., & Douches, D.S. (2018). Genome editing for crop improvement–applications in clonally propagated polyploids with a focus on potato (Solanum tuberosum L.). *Frontiers in Plant Science, 9*, 1607. 10.3389/fpls.2018.01607

Nazar, B.I., Kushnir, H.V., Boiko, H.I., & Murska, S.D. (2019). A necessity of introduction of the system of registration of GMO is for Ukraine. *Scientific Messenger of LNU of Veterinary Medicine and Biotechnologies. Series: Veterinary Sciences, 21*(94), 152–156. 10.32718/nvlvet9428

Nicolia, A., Manzo, A., Veronesi, F., & Rosellini, D. (2014). An overview of the last 10 years of genetically engineered crop safety research. *Critical Reviews in Biotechnology, 34*(1), 77–88. 10.3109/07388551.2013.823595

Park, J., & Choe, S. (2019, August). DNA-free genome editing with preassembled CRISPR/Cas9 ribonucleoproteins in plants. In *Transgenic Research* (Vol. 28, No. 2, pp. 61–64). Springer International Publishing. 10.1007/s11248-019-00136-3

Paull, J., & Hennig, B. (2019). New world map of genetically modified organism (GMO) agriculture: North and South America is 85%. *Acres Australia, 101*, 59–60.

Piatek, A., Ali, Z., Baazim, H., Li, L., Abulfaraj, A., Al-Shareef, S., Aouida, M., & Mahfouz, M.M. (2015). RNA-guided transcriptional regulation in planta via synthetic dC as9-based transcription factors. *Plant Biotechnology Journal, 13*(4), 578–589. 10.1111/pbi.12284

Pixley, K.V., Falck-Zepeda, J.B., Paarlberg, R.L., Phillips, P.W., Slamet-Loedin, I.H., Dhugga, K.S., Campos, H., & Gutterson, N. (2022). Genome-edited crops for improved food security of smallholder farmers. *Nature Genetics, 54*(4), 364–367. 10.1038/s41588-022-01046-7

Redden, R. (2021). Genetic modification for agriculture—Proposed revision of GMO regulation in Australia. *Plants, 10*(4), 747. 10.3390/plants10040747

Rosi-Marshall, E. (2009). GM crops: battlefield. *Nature, 461*, 27–32. 10.1038/461027a

Santos, E., Sánchez, E., Hidalgo, L., Chávez, T., Villao, L., Pacheco, R., & Navarrete, O. (2014, August). Status and challenges of genetically modified crops and food in Ecuador. In *XXIX International Horticultural Congress on Horticulture: Sustaining Lives, Livelihoods and Landscapes (IHC2014): 1110* (pp. 229–235). 10.17660/ActaHortic.2016.1110.33

Scheinerman, N., & Sherkow, J.S. (2021). Governance choices of genome editing patents. *Frontiers in Political Science, 3*, 745898. 10.3389/fpos.2021.745898

Schiemann, J., Robienski, J., Schleissing, S., Spök, A., Sprink, T., & Wilhelm, R.A. (2020). Plant genome editing–Policies and governance. *Frontiers in Plant Science, 11*, 284. 10.3389/fpls.2020.00284.

Schmidt, S.M., Belisle, M., & Frommer, W.B. (2020). The evolving landscape around genome editing in agriculture: many countries have exempted or move to exempt forms of genome editing from GMO regulation of crop plants. *EMBO reports, 21*(6), e50680. 10.15252/embr.202050680

Seralini, G.E., Clair, E., Mesnage, R., Gress, S., Defarge, N., Malatesta, M., Hennequin, D., & de Vendômois, J.S. (2014). Republished study: long-term toxicity of a Roundup herbicide and a Roundup-tolerant genetically modified maize. *Environmental Sciences Europe, 26*(1), 1–17. 10.1186/s12302-014-0014-5

Shan, Q., Wang, Y., Li, J., Zhang, Y., Chen, K., Liang, Z., Zhang, K., Liu, J., & Gao, C. (2013). Targeted genome modification of crop plants using a CRISPR-Cas system. *Nature Biotechnology, 31*(8), 686–688. 10.1038/nbt.2650

Toda, E., Koiso, N., Takebayashi, A., Ichikawa, M., Kiba, T., Osakabe, K., Osakabe, Y., Sakakibara, H., Kato, N., & Okamoto, T. (2019). An efficient DNA-and selectable-marker-free genome-editing system using zygotes in rice. *Nature Plants, 5*(4), 363–368. 10.1038/s41477-019-0386-z

Turnbull, C., Lillemo, M., & Hvoslef-Eide, T.A. (2021). Global regulation of genetically modified crops amid the gene edited crop boom–a review. *Frontiers in Plant Science, 12*, 630396.doi: 10.3389/fpls.2021.630396.

Vaeck, M., Reynaerts, A., Höfte, H., Jansens, S., De Beuckeleer, M., Dean, C., Zabeau, M., Van Montagu, M., & Leemans, J. (1987). Transgenic plants protected from insect attack. *Nature, 328*(6125), 33–37. 10.1038/328033a0

Waltz, E. (2015). Nonbrowning GM apple cleared for market. *Nature Biotechnology, 33*(4), 326–328. 10.1038/nbt0415-326c

Waltz, E. (2016). Gene-edited CRISPR mushroom escapes US regulation. *Nature, 532*(7599), 293. 10.1038/nature.2016.19754

Waltz, E. (2019). With CRISPR and machine learning, startups fast-track crops to consume less, produce more. *Nature Biotechnology*, 1251–1253. 10.1038/d41587-019-00027-2

Wasmer, M. (2019). Roads forward for European GMO Policy—Uncertainties in wake of ECJ judgment have to be mitigated by regulatory reform. *Frontiers in Bioengineering and Biotechnology*, 7, 132. 10.3389/fbioe.2019.00132

Whelan, A.I., Gutti, P., & Lema, M.A. (2020). Gene editing regulation and innovation economics. *Frontiers in Bioengineering and Biotechnology*, 303. 10.15252/embr.202050680

Xing, H.L., Dong, L., Wang, Z.P., Zhang, H.Y., Han, C.Y., Liu, B., & Chen, Q.J. (2014). A CRISPR/Cas9 toolkit for multiplex genome editing in plants. *BMC Plant biology*, *14*(1), 1–12. 10.1186/s12870-014-0327-y

Yang, L., Machin, F., Wang, S., Saplaoura, E., & Kragler, F. (2023). Heritable transgene-free genome editing in plants by grafting of wild-type shoots to transgenic donor rootstocks. *Nature Biotechnology*, 1–10.

Yarnall, M.T., Ioannidi, E.I., Schmitt-Ulms, C., Krajeski, R.N., Lim, J., Villiger, L., Zhou, W., Jiang, K., Garushyants, S., Roberts, N., Zhang, L., Vakulskas, C., Walker, J., Kadina, A., Zepeda, A., Holden, K., Ma, H., Xie, J., Gao, G., Foquet, L., Bial, G., Donnelly, S., Miyata, Y., Radiloff, D., Henderson. J., Ujita, A., Abudayyeh, O., & Gootenberg, J.S. (2022). Drag-and-drop genome insertion of large sequences without double-strand DNA cleavage using CRISPR-directed integrases. *Nature Biotechnology*, 1–13. 10.1038/s41587-022-01527-4

Yubing, H.E., Min, Z.H.U., Lihao, W., Junhua, W.U., Qiaoyan, W., Rongchen, W., & Yunde, Z. (2019). Improvements of TKC technology accelerate isolation of transgene-free CRISPR/Cas9-edited rice plants. *Rice Science*, *26*(2), 109–117. 10.1016/j.rsci.2018.11.001

Zhang, C., Liu, S., Li, X., Zhang, R., & Li, J. (2022). Virus-Induced Gene Editing and Its Applications in Plants. *International Journal of Molecular Sciences*, *23*(18), 10202. 10.3390/ijms231810202

Zhang, Y., Liang, Z., Zong, Y., Wang, Y., Liu, J., Chen, K., Qiu, J., & Gao, C. (2016). Efficient and transgene-free genome editing in wheat through transient expression of CRISPR/Cas9 DNA or RNA. Nature Communications, *7*(1), 12617, 10.1038/ncomms12617

List of abbreviations

CRISPR clustered regularly interspaced short palindromic repeats
GMO genetically modified organism
SDN site-directed nucleases
NPBT new plant breeding technologies
T-DNA transfer DNA
gRNA guide RNA

10 Biosafety and biosecurity concerns associated with plant genome editing

Rama Krishna Satyaraj Guru,
Ashutosh Sawarkar Ganpatrao,
Atul Pradhan Madhao, Rojalin Pradhan,
Ayesha Mohanty, Kaushik Kumar Panigrahi,
Ranjan Kumar Tarai, and Bushra Khatoon

10.1 Introduction

There are several strategies for introducing desired traits in plants. These strategies span from nearly century-old techniques such as random selection of agricultural important features to purposeful selection and planned improvement of traits via breeding, and finally to functional genomic modification techniques like genetic engineering and genome editing (GE). Nonetheless, due to the specific benefits associated with genome modification, all genetic improvement methodologies, whether modern or traditional, will remain viable (Schiemann et al. 2019). Traditional breeding procedures have contributed to increasing food yields enormously, but they are time-consuming, unproductive, and inefficient. This led to the creation of new breeding technologies (NBTs), which in turn influenced agriculture in the 21st century. These methodologies transformed and changed from unregulated, arbitrary mutagenesis, whether a physical or chemical and biological modification to remarkably high precision as well as restricted gene muzzling and genetic manipulation stimulate meticulous changes in the plant genetic material at a particular place to enhance crop traits (Horsch, 1993; Europejska, 2017).

Scientists are always reluctant to proclaim something incredibly safe. Nevertheless, if we acknowledge that non-genetically modified (non-GM) crop species are safe enough, we could indeed evaluate the safety of genetically engineered crops. The primary horticultural and agricultural crops that are grown and used around the world seem to have transformed over centuries if not millennia. As a result, they believe they are healthy and safe. However, the majority of these crops, if not all, contain toxic chemical or allergenic compounds (Liener, 1980). Since the majority of these compounds evolved to protect against animal grazers or microbial pathogens, it's not astounding that these substances are also hazardous to human beings. Glycoalkaloids from potatoes, linseed's cyanogenic glycosides, cruciferous oilseed crop's glucosinolates, proteinase inhibitors from soybean, and other leguminous plant seeds are one illustration (Aletor 1993). Genetic modification could have caused toxins to be introduced, it is unimaginable that all

DOI: 10.4324/9781003382102-10

these or numerous other widely consumed foods would be approved for consumption (Simee, 2011). Nonetheless, with exceptions including the introduction of new food crop types and varieties resulting from traditional breeding does not necessarily require any distinctive validation for the existence of irritants or toxicants, despite the fact that genes came from exotic cultivars or closely related wild relatives through making use of the wide crossing.

As evidenced by the successful implementation of genetically modified organisms (GMOs) in authorized cultivation regions of the world, the recent genetically modified variety seems to meet farmers' requirements. Furthermore, despite the reality that agricultural biotechnology continues to hold many commitments, a few remain shortfalls among farmers' requirements as well as biotechnological investigations (Ricroch et al. 2015). There is now a huge multitude of scientific publications that seem to be accessible on the fiercely contested potential hazards associated with the commercial application of genetically modified cultivars. Nonetheless, some organizations remain vehemently against the application of genetically modified organisms. Clearly, despite the accumulation of scientific understanding, beliefs the media and the internet do not converge on agricultural biotechnology, and so on.

Gene suppression by utilizing RNA interference (RNAi) is among the most recent and widely used approaches in plant science. There are a number of approaches to inculcating genetically modified (GM) crops. MicroRNA (miRNA) and double-stranded RNA (dsRNA) components are also both programmed into RNAi-based GM crops. Dicer proteins break these precursor compounds into the small interfering RNAs (siRNAs) or microRNAs that are 20–30 nucleotides long. RNAi-based transgenic crops seem to be reportedly focused on the inactivation or reduced expression of plant-specific mRNA of pathogenic strains in pests and pathogens. The binding of siRNAs or miRNAs to messenger RNA (mRNA) with a high degree of complementarity and leads to gene suppression. RNA interference (RNAi) using siRNAs and miRNAs is widely used to suppress unfavorable genes in plants, thereby enhancing their resistance to both biotic and abiotic stress factors (Papadopoulou et al. 2020). Accordingly, the creation of innovative methods for genetic manipulation, (such as GE using the CRISPR technique) laid the groundwork for more accurate editing with delicacy (base insertion, elimination, transposition) of the specific genes by not having an impact on the entire genome.

As a result, the CRISPR system has a wide range of possible uses, including the immense opportunity to enhance human food and nutrition security via means of crop breeding programs. The CRISPR/CAS9 method demonstrated to remain a major leading scientific research narrative of the past decade, and it has significantly contributed to crop improvement globally. The system's limitations, as well as the regulatory characteristics of its final product, inhibit it from being extensively utilized in the improvement of plant cultivars (Bogdanove et al. 2018; Zhang et al. 2020). Off-target effects, which can cause additional phenotypes, are a serious limitation alike RNA interference and CRISPR systems. As a consequence, scientists think that CRISPR-associated methods and by-products will necessarily require regulatory systems as well as a comprehensive analysis of just about any potential threats to the surroundings and health hazards

Modern biotech products should pass definite regulation testing developed by each country in accordance with their government regulations and policies before they are commercialized. These regulatory checks necessarily require experimental support that the products are bio-safe for humans, other living creatures, and the surroundings. Because of concerns in terms of its suitability for human consumption and environmental impact, the use of GMOs and their by-products is constantly debated. Different countries have tried to approach GMO research, development, and commercialization in different ways. For instance, the European Union (EU) keeps a close eye on GM crop products for biosafety concerns (including GM crops produced using the CRISPR system) and required extensive labeling of GM crops/by-products prior they may well be popularized (Tagliabue, 2017).In contrast, legal perceptions of biotechnology supervisory oversights issued by the United States, Israel, Argentina, Canada, Australia, Chile, and Brazil exclude GMO law exempting Transgenic plants. Nevertheless, this exclusion is largely determined by the lack of template DNA (Australia), pest-related factors (USA), phenotypic traits (the United Kingdom), and modification (Canada and other countries) (Schiemann et al. 2019). The World Health Organization has pronounced that genetically modified crops and their by-products are safe for human consumption in a study.

10.2 Problems with genetically manipulated crops

For more than two decades, the discussion over genetically engineered crops has spanned the world and shows no signs of subsiding. It is arguably one of our time's most interesting topics, as well as a test of science's and researchers' capacity to overcome magical thinking as well as fear-mongering. There is no such thing in Europe[1].

10.2.1 Concerns raised by conflicting biosafety study results

Herbicide-resistant soybeans, for example, were endorsed to be used in food in the late 1980s. They have since been commonly consumed and analyzed without causing concern. Dr. Irina Ermakova of the Institute of Higher Nervous Activity and Neurophysiology of the Russian Academy of Sciences revealed in October 2005 that she had conducted an experiment in which young whose mothers have been subjected to a diet consisting of herbicide-tolerant genetically engineered soybeans exceeded rodents supplied non-GM soybeans in terms of mortality and body weight (Ermakova and Barskov, 2006; Ermakova, 2006a). The results were published without the benefit of scientific peer review (Ermakova, 2006b). Toxicology experts, including members of key high-level expert bodies in the United Kingdom and the European Food Safety Authority, made complaints they weren't able to conduct a comprehensive assessment of the research due to a lack of information regarding the experiments' methodology.

South Dakota State University's Denise Brake and Donald Evenson undertook analogous feeding trials on rodents in 2004 and reported their findings in a peer-reviewed journal Food and Chemical Toxicology. They discovered that the GM soybeans had not resulted in any harm. This did not stop the Soviet daily Pravda from forecasting that wide-spread consumption of GM soybeans would decrease average lifespan, nor did the Daily Mail in the United Kingdom warns of the hazards to unborn infants. Reconfigured maize has come under fire notwithstanding the fact that endorsed and utilized infeed and food ever since the early 1990s. Gilles-Eric Séralini along with his team at France's University of Caen conducted a new analysis and challenged the way animal feeding was interpreted in the 2007 experimentations by the GM insect-resistant maize variety MON863. A Bt gene in MON863 acquired resistance to the maize rootworm. Greenpeace funded Séralini's 1999 establishment of the Committee of Research and Independent Information on Genetic Engineering (CRIIGEN) which also promotes against genetically modified products and crops.

He attempted to claim that rats fed MON863 grain sustained multiple organ damage (i.e. liver, kidney, adrenal gland, heart, and blood-forming system damage) as well as impacts on body mass, triglyceride levels in female rats, and urine composition in male rats. In spite of that, there was a wide range of growing MON810 rather than MON863. This research was considered a precedent by the French government as a justification for prohibiting the increasing production of genetically engineered maize in France in violation of the European Commission regulatory frameworks. In 2006 and 2007, genetically modified maize cultivation gained prominence among farmers in France, although it had become somewhat increasingly dubious. Although The European Food Safety Authority (EFSA) examined Séralini's research work and gave confirmation about all the hematological parameters, the French authorities have upheld this prohibition because the organ weight readings of the rats fed on genetically modified maize were within standard ranges and inappropriate statistical methodologies were utilized. Séralini's research had also been heavily criticized by the Commission du Génie Biomoléculaire (CGB), France's expert committee.

In 2008, at the University of Vienna, Austria's authorities announced the findings of Jürgen Zentek's study (Velimirov et al. 2008), attempting to claim that rodents fed a genetically modified corn hybrid obtained after a cross of MON810 and glyphosate-tolerant line NK603 (the cross thereby holding arrayed attributes of insect resistance and herbicide tolerance) experienced fertility issues. The report's release stimulated considerable discussions and was used to justify genetically-modified crop and food bans throughout

Austria and other European states. The Austrian government, on the other hand, declared in October 2009 that the investigation would indeed be revoked. They asserted that Zentek and his colleagues had been unable to provide a credible report about it, particularly in terms of the statistical data analysis and the authorities did not expect to receive it anymore. As a consequence, the government is no more power intended about writing any such report. Naturally, it's quite unfortunate that the Austrian government didn't confirm the factual information or join the queue for just a peer assessment prior to actually making it public.

Séralini reverted in 2009, releasing a new study based on toxicological data re-evaluation, this time on the genetically-modified corn lines NK603, MON810, and MON863. As per Séralini, the analysis suggests digestive system, kidney, and cardiovascular problems in rats nourished with GM maize (Séralini et al. 2014). The study was re-evaluated by EFSA, which further decided that the data was insufficient and did not agree with Séralini's claims, as well as the statistical analysis was misleading. The study was also reviewed by the French Haut Conseil des Biotechnologies (HCB), which conclusively proved that there was "no legally valid scientific component likely to attribute any hematological, hepatic, or nephrotoxicity toxicity to the three re-analyzed GMOs." Séralini's independence was also questioned by the Haut Conseil des Biotechnologies (HCB).

Séralini persisted, publishing another review of animal feeding studies in 2011 (Séralini et al. 2011) and summarizing that genetically modified food produced sex- and dose-dependent effects on the kidney and liver. Séralini published the findings of a 2-year investigation on rats fed genetically - modified corn and glyphosate in the journal Food and Chemical Toxicology in 2012. He tried to claim that rodents nourished with GM corn and herbicide were at a higher risk of developing cancer cells compared to rats fed a normal diet Researchers and regulatory agencies were outraged by the study, particularly the use of the Sprague Dawley rat. Sprague Dawley rats are bred for scientific science and, regardless of diet, develop tumor cells on a routine basis throughout their lives. According to the data provided by the suppliers, the tumor cells developed by study rodents seemed to be typical of those observed in that breed of rat. As a consequence, the French Society of Toxicological Pathology mentioned about Séralini's paper to be ambiguous because it displayed images of tumor cells from rats that were fed genetically-modified corn and glyphosate without showcasing tumor cells from healthy rats with normal diets. The paper (Séralini et al. 2012) had been withdrawn in November 2013 by Food and Chemical Toxicology, quoting its lack of logical coherence and validity. In June 2014, Environmental Sciences Europe republished the article in a revised form (Séralini et al. 2014). Meanwhile, in April 2010, EFSA released a report on its re-evaluation of genetically engineered maize accepted to be used in European food and animal feed. The agency confirmed in that report that it has seen no scientifically based reason to have any concern. However, these instances and the subsequent hue and cry serve to show once again that the European public, in addition to European politicians and policymakers if they so choose, primarily simply ignore the beliefs of experts from Europe scientific advisory boards, as well as those of individual Member States.

10.2.2 *Antibiotic resistance marker genes*

A number of international scientific organizations, including the World Health Organization and regulatory organizations committee members established by the European Union and a number of state governments, examined the potential risks associated with antibiotic resistance genes in food. These groups ascertained that the antibiotic-resistance alleles being utilized do not endanger one's health risk. However, The British Medical Association has highlighted its concerns, and The Advisory Committee on Novel Foods and Processes (ACNFP) in Great Britain had also requested the creation as substitute indicators. In the mid-2000s, the biotechnology sector ascertained that anti-microbial resistance genetic markers were safe, but they also continue to pose a public relations problem and should not be utilized in basic research. The presumed threat of using antimicrobial resistance marker genes is the fact that they may access the gastrointestinal tract or soil and then disperse to microbes that cause infections. In reality, the antimicrobial resistance marker genes are employed in agricultural improvement bestow resistant antibiotics, for example, kanamycin and neomycin, which are rarely used in oral antibiotics (Manyi-Loh et al. 2018).

Antimicrobial resistance marker genes are also employed only for cloning, modifying, and bulking up genetic material for transient expression. The gene in this case is designed to work in bacteria. When the whole viral vectors are employed to modify plant cells through the use of particle bombardment, the above gene, together with the gene of interest, may be integrated into the host chromosome. One of the antibiotic-resistance genetic markers used only for bacterial genetic modification provides resistance to the frequently used antibiotic ampicillin. This gene was discovered in the diverse array of genetically modified insect-resistant corn, however, it's not found in current varieties. Is it dangerous to possess any one of these genotypes in a genetically modified plant? First, the gene insertion in the DNA of a Genetically Modified plant is done, and the possibility for the gene to be horizontally transferred from the genetic material of the plant to soil or intestinal microbiota is incredibly low, if not non-existent. Moreover, which really do these elements are derived from in the very first place? The ampicillin resistance gene has been provided by *E. coli,* a human intestinal bacterium (Li et al. 2019).

Merely A brief analysis of the available research materials indicates that numerous species of bacteria possess naturally an ampicillin-resistant gene including *Kluyvera ascorbata, Pyrococcus furiosus, Proteus mirabilis, Bacillus subtilis, Klebsiella pneumoniae strain H18, K. pneumoniae strain G122,*

Pseudomonas aeruginosa, Pseudomonasfuriosus, Staphylococcus aureus, and *Synechocystis sp. PCC 680.* To put it another way, the "concern" with antibiotic-resistance genes utilized for plant biotechnology is – these genes may penetrate already-existing bacterial populations. Only antimicrobial resistance genetic markers that provide a selective advantage survive in bacterial populations. Antibiotic-resistant strains of bacterial pathogens do present a health risk, but they emerge innately and flourish as a result of poor antibacterial drug administration in animal and human medicine, not from crop biotechnology's use of antibiotic resistance marker genes (Sultan et al. 2018).

10.2.3 Insect resistance to Bacillus thuringiensis (Bt)crops

Weeds' probable consequences of developing resistance to control measures, along with insect infestation conferring resistance to pesticide residues, is constantly on the thoughts of agricultural producers and those involved in the manufacturing of agrochemicals. This occurred prior to genetic manip-ulation, but the beginning of transgenic plants equipped with their own pesticide introduced a new twist to the old problem. In the United States, the Environmental Protection Agency (EPA) was assigned the task to regulate and monitoring the plants that have been genetically engineered and implemented to be insect-resistant because of the inclusion of a Cry gene (Abbas, 2018). Since the same insecticidal protein is incorporated into GM plants and the pesticides, the EPA argued that they should fall under the purview of the same regulatory agency. In some ways, this makes sense, but other transgenic crops are currently under the control of the United States Department of Agriculture's Animal and Plant Health Information Service (APHIS), and this division of duties for genetically engineered crops may have contributed to the Star Link debacle.

The Bt pesticide is frequently used by vegetable and organic farmers in the United States because most organic regulatory authorities allow the use of the Bt insecticide which is believed to be safe to use till harvest. The occurrence of insecticide resistance, on the other hand, was thought to be a major risk, and the EPA had long been genuinely worried about all this. The EPA continued to insist that growers who are using GM crops supposed to contain the Cry1A gene also plant some non-GM crops in reaction to the introduction of Bt Genetically engineered crops in 1996. This would be a safe haven for insects that had gained immunity to the Bt protein's consequences and would no longer have a survival advantage (in fact they would be at a selective disadvantage). According to the EPA's forecasts of how the insect population will react, the percentage of the non-GM crop would have to vary greatly depending on what other insect-resistant GM crops were cultivated in a given location. In response to farmer representatives' opprobrium, the EPA issued a report forecasting pervasive Bt resistance throughout insect populations within 3–5 years only if recommendations were not followed. Insect-resistant

GMO and Bt pesticides would be rendered obsolete as a direct consequence. The EPA managed to win, and the refuge legislation was put into effect. Thus far, almost two decades later, it has seemed to have occurred huge success. Nevertheless, activists in Europe raised the absolute catastrophic nightmare projection out of the EPA's (Abbas, 2018) study and used it as one of the explanations for their opposition to genetically engineered crops, anticipating the advent of "superbugs."

10.2.4 *"Superweeds"*

The concept "superweed" (Bain et al. 2017) is an excellent illustration despite the subjective ambiguity that has captivated Europe's debate over genetically engineered crops. It is based on the notion that an herbicide tolerance allele in a GM crop might "circulate" into the related wild populations using cross-pollination, culminating in varieties that might turn into unmanageable weeds. Cross-pollination of GE crops with wild species must always be assessed on a case-by-case basis, taking the species and genetic polymor-phisms into account. Maize and potato do not cross with any wild species in Great Britain (even though forced crosses between potato and black night-shade can indeed be made in the research lab), and wheat does not cross with any native plant species to generate fertile recombinants. Wheat is almost entirely self-pollinating. Chinese cabbage, Brussels sprouts, Indian mustard, hoary mustard, wild radish, and charlock are among the other cultivated and wild Brassicas that will cross with rapeseed. The extent of such crossings in agricultural systems is still being evaluated, but this doesn't necessarily indicate that genetically – modified rapeseed is a concern; instead, the potential danger symbolized by genetically – modified rapeseed differs from that signified by wheat. In Canada, where millions of hectares of GM herbicide-tolerant rape-seed are produced, no hybridization between the GM crop and natural species has been discovered. Even if a hybrid between a GM plant and a wild relative did occur, the gene linked with it would succeed if it provided a competitive advantage, therefore herbicide-tolerance genes are extremely unlikely to exist in nature.

A whole other way herbicide tolerance in weed species might spring up is if the elevated evolutionary pressure levied by the widespread use of only one herbicide creates weed species to develop tolerance to that herbicide by themselves. Monsanto created (Duke and Powles, 2008) its "Roundup Ready Xtend Crop System," which is centered on genetically-modified varieties with stacked glyphosate and dicamba tolerance characteristics, in reaction to the invention of glyphosate-tolerant weeds in the United States. Stacking herbi-cide attributes in this way allows farmers to manage glyphosate-tolerant weeds while reducing the risk of weed resistance to the herbicide policy. If the problem of herbicide-tolerant weed populations becomes too severe, the herbicide and the genetically engineered crop with which it was packaged will be rendered ineffective. Farmers need to make use of new herbicides. Farmers have been aware of these problems ever since the pervasive use of herbicides, long before the genetically engineered crop was developed. Farmers should also be extremely careful that transgenic crops with different herbicide acceptance criteria need not cross to deliver hybrids with innumerable tolerance levels.

10.2.5 The 35S RNA gene promoter of the cauliflower mosaic virus

The use of the Cauliflower mosaic virus 35S RNA gene promoter, also known as the CaMV35S promoter, in plant biotechnology continues to pose some sort of threat to human health, pertaining to yet another terrifying story that has surfaced in Europe's genetically-modified debate. The virus is recognized as the "cauliflower mosaic virus" because that infiltrates cauliflowers and other Brassica (cabbage) crops. Although it doesn't spread the disease in humans or animals, it infests and has probably always infested the majority of the UK's Cruciferous crops. As a result, it's not the best candidate for a food scare.

The CaMV35S gene promoter is the only part of the viral genome that is present. It is used in plant biotechnology when a gene introduced through genetic manipulation must be active throughout the plant. The alleged threat is that this promoter and an animal virus will recombine, likely to result in the creation of a new "supervirus." It is ambiguous how this is expected to happen, why it is more likely to occur when the CaMV35S promoter is incorporated into a plant genome rather than a viral genome, or why the virus that ultimately resulted would be particularly dangerous.

10.2.6 "Terminator" technology

Terminator technology is the development of crop varieties that produce unfertilized seeds, also recognized as "suicide seeds" in some contexts. Anti-GM activist groups have long attempted to interconnect genetically modified plants to "terminator" technology; in fact, the anti-GM campaign has utilized the term "terminator" and the likelihood that biotech companies will use this innovation to compel farmers, particularly those in developing countries, to purchase seed from them on a yearly basis. This is worth mentioning because none of the genetically-modified cultivars used in commercial agricultural production produce infertile seeds (Yousuf et al. 2017).

Making plant seeds sterile through genetic manipulation is certainly possible; it is also feasible and has already been done in a non-GM manner. Correspondingly, pollen can be rendered infertile using both GM and non-GM techniques; in fact, it is a well-established technique in plant breeding for producing hybrid seeds. However, the fertility of hybrid seeds produced from these plants and marketed to farmers has indeed been completely restored. Sterile plants can be employed commercially in highly specialized, small-scale applications such as vaccine and pharma industrial production, where cross-contamination with food crops is critical.

10.3 Genetically-modified (GM) crops and food legislation

The ACNFP (Advisory Committee on Novel Foods and Processes) assessment basically adheres to World Health Organization (WHO) guidelines, and nearly all regulatory agencies globally adhere to the same guidelines. The principle of substantial equivalence is core to the procedure. Anti-GM activists quite often argue that asserting that somehow a GM plant or food is nearly similar to its non-GM counterpart is satisfactory. Undoubtedly, substantial equivalence presupposes a thorough comparison of a GM food's biochemical and molecular properties to those of its conventional counterpart, as well as a thorough examination of any differences. Despite being considered acceptable as safe to consume, a small percentage of the foods we eat presently has become the main topic of toxicological investigation. When conventional risk analysis and toxicological screening methods are applied to whole foods, whether GM or not, it is well almost impossible to ensure complete safety. As a direct consequence, the considerable equilibrium method aims to determine whether GM food is as safe as its conventional counterpart if one exists. To begin, GM food is compared to its closest traditional counterpart to see if there are any intended or unintended differences. The following are the results of a survey conducted by the National Institute of Standards and Technology (NIST). The following elements are considered in the safety assessment:

- The identity of novel genes and their source (particularly, is the origin a very well source of food or is it totally new to a food web?).
- The content of the plant and/or food is in contrast to its traditional counterparts.
- The consequences of cooking and refining
- The procedures used to construct the GM plant.
- The consistency and replicability of the unique genes or genes.
- The characteristics of the protein produced by the unique genes or genes.
- Modifications in how recently found genes and proteins perform.
- The novel proteins' potential toxin potential.
- The capability of unique proteins to cause allergies.
- Potential knock-on effects on metabolic processes and modifications in the production of nutrients, anti-nutrients, toxicants, allergens, and physiologically bioactive molecules as a result of the unique gene or genes' expression, such as gene disruption in the host plant.
- The prospective intake and nutritional consequences of the emergence of GM foods.

In recent years, significant advances have been made there in innovation that can be used to perform these types of studies. The activity of thousands, if not tens of thousands of genes can be ascertained in a single experiment. Correspondingly, the vast numbers of proteins and metabolic products found in plants can be directly measured. These transcriptomics, proteomics, and biomolecular methods are trying to transform safety checks. New crop varieties formed using any of the plant breeding techniques, mutagenesis, genetic manipulation, genome editing, or anything else—may exhibit alterations in gene expression, protein synthesis, and metabolite profiles. This is both possible and likely. The challenge will be trying to identify substantial modifications and making rational decisions about them.

10.3.1 Genetically modified (GM) plants' biosafety in containment

Any physical, chemical, or biological barriers when employed to restrict an organism's interaction with its surroundings or other organisms, is known as containment. In order to control bacterial containment, it is common to use genetically modified/disabled strains that are unable to survive outside of a controlled environment. Similarly, in the case of plants, containment measures can be taken by using species that are unable to survive in the local environment and do not crossbreed with wild plants or by taking preventive measures such as removing flowers before pollen is dispersed. An example of this is wheat, which does not thrive outside of agriculture. The use of herbicides is an example of a chemical barrier that can be employed to prevent the spread of a GM plant, although this method is usually applied during post-harvest treatment of a field test site rather than in a strictly controlled containment environment. To ensure the safety and control of genetically modified organisms, specialized laboratory facilities, plant growth rooms, and greenhouses are required as part of physical containment measures. These measures are typically subject to legal regulations that apply specifically to the use of genetically modified plants in research laboratories and greenhouses.

Even if genetically engineered organisms are assumed to not pose any risk to humans or to their surroundings, but still the research facilities where such organisms are stored or utilized must fulfill basic standards. For example, the laboratory should be organized to be easily cleaned. Sealing of benchtops and floors is necessary, and if mechanical ventilation is installed, it should ensure the inward flow of air. To maintain proper hygiene, hand-washing stations should be placed near the laboratory exit, and standard protective equipment like lab coats and disposable gloves should be worn and discarded before exiting the lab. The research facility should only be accessible to authorized personnel and access should be strictly controlled. Aerosol-generating processes must be decided to carry out in a laminar flow cabinet, effective disinfectants have to be readily accessible beside every sink, spills need to be treated promptly and cataloged, Benchtops must be scrubbed after use, and a high level of sanitation must be practiced. It is required to store all contaminated glassware securely and maintain sterility after use. Additionally, all contaminated waste should undergo autoclaving (a process that uses temperature and pressure to decontaminate) before proper disposal. Expert risk assessment is central to GMO safety. All GMO-related projects should undergo a risk assessment undertaken by a qualified individual such as the project supervisor, and the risk assessment (or risk analysis) must be taken into account by the organization's internal Genetic Modification Safety Committee or Biological Safety Committee. The project team members are expected to provide details about their training and experience. While the assessment of any risk to human health, the leader of the project should evaluate the likelihood of inducing or intensifying toxic effects and/or allergies in comparison to the parent plant, and also consider the risk of accumulating toxicity through food chains and its impact on human health. Furthermore, the project manager must evaluate any environmental threat posed by the suggested transgenic plants, especially the possibility that they will be more "weedy" than the parent organism. This assessment includes the capacity to colonize, seed dispersal pathways, resistance to control measures such as herbicides, increased toxicity to insects and other grazers, and any other significant move in how the plants interact with their environment. Particularly if the plants have the ability to transmit unique genetic material to the plant species, the potential for and consequences of sexual nucleic acid transfer between transgenic plants and other plants of the same species or a species compatible with them must be considered.

For example, a virus, bacterium, or other vector can horizontally transfer genes to unrelated species, with risks and consequences. Last but not least, the likelihood that GM plants will harm animals or beneficial microorganisms must be assessed. Regardless of whether the plants will be kept in specially designed greenhouses or climate-controlled rooms, this risk evaluation must be completed before the experiment started. These have screened deleterious air pressure ventilation, sealed drains, and a chlorination treatment system for drainage water to make sure that no viable plant material escapes into the environment.

10.3.2 Safety of field releases of GM plants and GM foods

Laws regulating the field release of transgenic plants will generally include the risk assessments listed below:

- Information about host organism:
  - The full species record of the plant in addition to the breeding line used.
  - The plant's sexual reproduction, generation time, and sexual compatibility with other cultivated or wild plant species.
  - Knowledge on the geographic range of the species, as well as the viability and spread of the plant.
- The nature and source of the plant-modifying vector, as well as an explanation of the genetic manipulation processes.
- Comprehensive information on genetic manipulation

- Information about the novel genes implemented into the plant, such as their size, intended function, and the organisms from which they originated.
- Information on the unique gene or genes' activity in the plant, as well as the methods used to ascertain this.
- Information regarding the number of copies of the unique gene or genes and the position of the inserted novel DNA in plant cells.
- The size and function of any genetically engineered host plant genome were omitted.
- A glance at how stable genetically unique gene or genes are.

- An examination of GM plants:

  - A description of the genetically engineered plants' altered or new characteristics and traits.
  - Investigate any differences in how rapidly the genetically –modified plant reproduces, disperses, and can thrive when compared to its parent.
  - Characterization of the techniques used to detect and recognize genetically- modified plants.

- An appraisal of prospective health and/or environmental repercussions:

  - A glance at any potential negative consequences on any other living beings
  - A risk assessment of the transgenic plant's likelihood to become invasive in natural settings or more persistent in agricultural habitats compared to the original or parent plants.
  - Characterization of the framework by which the genetically modified plants interact with target species (for example, insects) as well as any possible significant influences on unintended species.
  - An inspection of how the interaction of the GM plant with either target or non-target organisms may influence the environment.
  - A look at the probability of genetic information from GM plants spreading to other organisms.
  - Selective pressure resulting from the transfer of genetic material from the GM plant may have positive or negative effects on a separate species.
  - Data of past releases of GM plants

- Records of the site where GM plants are released.

  - site of release/location of release
  - Enlisting the major information related to the climate, plant, and animal life of the release site ecosystem.
  - It is important to identify and document any cultivated or wild plant species that share sexual compatibility with the animals inhabiting the release sites.
  - The location of the release sites in relation to legally recognized protected areas may have a potential effect that should be considered.

- Classify the confinement provisions that will be used throughout the duration of the field trial and following its completion.

  - It is important to provide an account of any actions taken to prevent crossbreeding between the genetically modified plant and other sexually compatible plants, as well as any measures taken to limit or prevent the dissemination of pollen or seeds.
  - A description of the treatment procedures for the site after it has been released. The site will most likely be plowed, and irrigated to encourage seed germination and sprouting plants will be removed using an appropriate total herbicide, and the site will be left fallow for one to two years before being monitored.
  - An explanation of the strategy for disposing of genetically modified plant material.
  - Details of the process of monitoring the site after the trial
  - The emergency plans will be described in case of plant proliferation or an adverse impact during the experiment.
  - An assessment of the potential for theft of genetically modified material from the trial site, the impact of any vandalism that may occur during the trial, and the possibility of accidental transfer off-site through means such as field machinery.

10.3.3 Genome editing legislation

Because of the advancement of genome editing, many government regulators have actually realized that regulations written in the nineties must be updated to reflect technological developments. It is already known that the European Commission finds itself in a challenging position, due to the rigorous restrictions already in place, and the potential for significant political fallout and criticism from Non-governmental organizations (NGOs) despite the existence of scientific evidence.

Based on scientific evidence, there is no justification for treating plants developed using genome-editing techniques differently from those created through chemical or radiation mutagenesis. Therefore, regulations that apply to genetically modified plants should not be extended to cover genome-edited plants.

The United States Department of Agriculture (USDA) has declared an unrestricted mushroom that has been genetically altered using the gene-editing tool CRISPR-Cas9, denoting that the USDA agrees with that viewpoint. Amidst this, the States of America is as of now redrafting its agricultural biotechnology regulatory requirements in light of breakthroughs in genome editing. The Office of Science and Technology Policy released the National Strategy for Modernizing the Regulatory System for Biotechnology Products in September 2016. It highlighted a "vision" for reshaping the national regulatory structure so that prospective bioengineering products could be assessed. It issued a "Notification to the Collaborative Guideline for the Legislation of Biotechnology" in January 2017, making clear the roles and responsibilities of the Food and drug administration, Environmental protection agency, and United States Department of Agriculture. In addition, the Food and drug administration has initiated a public counseling session on the legislation of foods derived from plants whose genomes have indeed been edited.

Argentina and Canada also have legislations and policies for genome-edited plants, implicitly saying that new strains will be assessed on a case-by-case basis, but that the product is non-GM when no novel gene material is combined, and no transgenes are used. This includes plants in which a transgene was employed to create a new variety but was removed prior to commercial exploitation (as is the case with most plants produced using specific target nucleases techniques like TALENs and CRISPR/Cas 9).

The European Commission, in contrast, has avoided the matter by postponing a decision on genome-edited crops to an undefined future date. Regardless of the fact that the European food safety authority genetically-modified panelist outlined that "there are a few applications of these [genome editing] tools (i.e. where no recombinant DNA remains in the commercialized plant and where the "edit" is impossible to distinguish from a natural mutation) that seem so not to be a GMO as currently defined," Sweden, an EU member, has gone it alone. In response to officially submitted questions from Swedish researchers, the Swedish Board of Agriculture released a statement in 2015 (Eriksson, 2018) asserting that plants with genomes edited utilizing CRISPR/Cas9 technology do not fulfill the European Genetically Modified definition.

10.4 The regulatory components of RNA interference and CRISPR methods

To be able to be sold, RNAi products based on genetically modified technology had to undergo a multi-step method for determining inherent hazards and detrimental consequences that were known all over the world. Formulation of the problem is the first step in the risk assessment process, and it helps to ensure that the data supplied is beneficial for decision-making and offers a perfectly rational and trying to trace farming strategy to downward threat assessment stages. The first step in problem formulation is typically the recognition of detrimental reactions of genetically engineered crops in contrast to their non-GM counterparts.

This method of comparative risk analysis emphasizes the possible ways that genetically engineered crop products could harm people, animals, and the surroundings. In the research framework, policy objectives and legislative action are employed to define assessment nodes, test hypotheses, and navigate resultant record-keeping and evaluation steps. A strategy of assessment and inclusionary measures for future research are then developed using one or more conceptual frameworks (Wolt et al. 2010; Schiemann et al. 2019). Regardless of the existence of international doctrines for controlling genetically altered crops, their implementation obviously varies across borders. The judicial system, for instance, involves determining the necessity of a framework of methods and techniques based on the methodology used in addition to the uniqueness of the product.

One possible reason for the discrepancies in Transgenic crop regulatory requirements all over jurisdictions is that only a few nations (Canada, the United States, and Argentina) abide by the Cartagena Protocol for Biosafety, a supplementary international treaty to Convention on Biological Diversity that went into force on September 11, 2003. The Cartagena Protocol contributed to the creation of national biosafety guidelines (NBG) and regulation systems to guarantee sufficient protection during the transfer, management, and utilization of GMOs derived from biotechnologies (Diversity, 2000; Devos et al. 2012).

There are a few questions regarding the regulation of GE crops: (1) to what extent does GMO legislation apply to GE crops (Araki and Ishii, 2015)? Can certain GE crop categories be exempted from regulations based on techniques used (to develop GE crops) or product properties (exclusion of low-risk products from regulatory oversight)? What safety data must be provided to regulatory agencies in nations where GE plants will be marketed? Zannoni, 2019. There seem to be currently stringent regulatory and sociocultural structures in place that govern the regulation and adoption of genetically engineered crops (Davison and Ammann, 2017), with two types of regulatory frameworks in use to regulate GE crops, process- or product-based (Ishii, 2017). The regulatory process for genetically engineered crops (RNAi-based) and GE crops (CRISPR-based) in various countries is described in detail below.

10.4.1 *European union*

In contrast to the United States and Canada, regulation of GM and GE crops focuses on the innovation or process used rather than the trait launched. Bioengineered plants intended for release into the environment are distinguished by EU legislation from those intended for feed or food (Schiemann et al. 2019). The European Union (EU) policy on the use of genetically modified organisms (GMOs) strives to provide a substantial amount of safeguards for human, animal, and environmental health even while guaranteeing a well-functioning EU internal market. The framework governs the release of genetically modified organisms into the environment as well as their usage as or in food and feed. It is built on three pillars: pre-market authorization based on a risk assessment, traceability, and labeling. Thus far, the EU has authorized the commercialization of 118 GMOs under this regulatory framework (Bruetschy, 2019). With the implementation of the GMO law, new genetic engineering techniques, including novel mutagenesis methodologies, have been developed, raising concerns about the application of the GMO legislation and attracting a great deal of attention from stakeholders and the general public.

10.4.2 Canada

The standards for biotech crops in the United States are solely based on "the product and its novel trait," with no due consideration for the process. As a result, Canada's GM crop regulatory oversight operation is significant compared to that of the United States. The Canadian Food Inspection Agency (CFIA) is in charge of overseeing new plants and feed for livestock in Canada. Crops developed through traditional breeding, mutagenesis, transgenic technology, and genetic manipulation are governed by the Canadian food inspection agency and the Ministry of Health. These plants must undergo the exact same approval procedures as other plants for commercial cultivation (Gao et al. 2018).

Department of Health is in charge of upholding legislative changes, especially with regard to the investigation of novel foods that are meant for human consumption. Uniqueness is a major trigger for regulation and oversight in Canada, according to Smyth and Mchughen (2008), and it is interpreted differently for completely new traits in food, feed, and plants. Under the Seed Act, Feed Act, and their respective regulations, the CFIA characterizes the possible effects of food and feed products intended for consumption by both humans and animals.

Both acts are concerned with the regulatory frameworks of a unique trait rather than its initiation. Department Of Health validates how unique features influence health upon consumption under the Food and Drug Act (Prince, 2000; Kochhar et al. 2005). Ministry Of Health first must label foods obtained from novel plant traits (Wolt et al. 2016). Herbicide-tolerant canola was released as the first commercial GE agricultural product for commercial production in 2014 by the Canadian food inspection agency and the Department Of Health (Jones, 2015).

10.4.3 USA

The regulation of bioengineered plants produced through GM, GE, or any other novel breeding systems is controlled by a Collaborative Framework consisting of three Federal Departments. The legislation is centered on the product's quality instead of the process used to generate the plants. The Collaborative Framework for the Assessment of Genetically altered Crops was established by the US Department of Agriculture (USDA), the Food and Drug Administration of the United States (FDA), and the United States Environmental Protection Agency (EPA). These organizations evaluate the impact of genetically engineered crops on agricultural production, human and animal health, and the surroundings. The Department of Agriculture is controlled by the Animal and Plant Health Inspect Services (APHIS) division of Biotechnology Regulation Services. APHIS has been assigned to safeguard agriculture production from diseases and pests by employing the Plant Protection Act to evaluate the possible dangers connected with GM or GE crops (Mchughen, 2006; Gostek, 2015). Under the Environmental protection agency mandate, which covers the protection of both humans and the environment from pesticide use, the Insecticide, Fungicide, and Rodenticide Act (Vogt et al. 2001) helps to regulate GM plants with altered pesticide features. The FDA regulates the safety of human and animal food and feeds through the Federal Food, Drug, and Cosmetic Act. All of these organizations are free to meet; moreover, the Food and drug administration has just evaluated those food and feed products made from genetically modified crops that are now available to consumers (Mchughen, 2006). The US government recently relaxed its genetically modified crop regulations. The new regulations explicitly excluded from regulatory oversight GE crops that might have been created through traditional breeding and have minor alterations, such as a change in two amino acid bases or deletion of a component of DNA.

10.4.4 *Approval for purposeful release and food and feed purposes*

The deliberate release of genetically altered crops requires approval under Regulations 2001/18/EC, which was later revised by Regulations (EU) 2018/ 350. The process includes the European Commission (EC), each European Union member state, and, on occasion, the European Food Safety Authority (EFSA). The EC representative for the member state looks at the request for the release of genetically altered crops to the market for possible risk and issues a fact sheet on the plight of the GM crop, which includes whether it is appropriate to be sold. If the EC representative of the member state has made a conclusive result, the notification will be sent to all other EC member states via the EC. The assessment report is then re-examined by member countries and the EC, who could object. (Sprink et al. 2016; Globus and Qimron, 2018; Schiemann et al. 2019).

   If both the EC and the member states comply that there's no risk involved with the crop or if any objections brought up by the applicant seem to be justified, the member states that at first performed the threat assessment will endorse the genetically- modified crop. If any of the member countries and the European Commission do not arrive at a satisfying resolution and reject the genetically altered crop in question, EFSA is compelled to offer an alternative scientific consensus that considers the member states' opposition to the risk assessment findings. The EFSA's opinion is then used to draw up the EC's decision, which is subsequently conveyed to the regulatory agencies. If most eligible bodies abandon the genetically altered crop, the matter will go to the ministerial council for consideration. Eventually, if the suggested decision does not obtain a qualified majority of approval or opposition from the Council of Ministers, the EC has the power to determine its fate (Schiemann et al. 2019).

The approval of genetically altered crops for use in food and feed is completely controlled by Rules (EC) 1829/2003 and trying to implement Regulation (EU) 503/2013. The applicant presents its marketing case in front of a prospective representative member state of the European Union. The representative of EC member countries can forward this application to the European food safety authority for risk evaluation surveys. When undertaking risk evaluation analyses, the EFSA considers the scientific consensus of the member states. If the application includes cultivation, EFSA uses the Guidelines 2001/18/EC and (EU) 2018/350 to direct the competent authority to carry out ERA. The European Commission rewrites a decision according to the scientific opinions of the European food safety authority and submits it to the Central Committee on Food Chain and Animal Health. If none of these agencies can attain an eligible bias in favor of or against the application, the claim goes to the Ministry Council for a course of action. However, if the Council of Ministers fails to approve or decline the application by a qualified majority, the EC has the power to determine its fate (Schiemann et al. 2019).

Following genetically altered crop authorization, EU regulations require referential integrity and labeling prerequisites. These requirements apply to all food and agricultural products (including oils) obtained or derived from genetically altered crops. The EC Regulation No. 1830/2003 allows for a 0.9% labeling cut-off point for authorization product lines if these traces seem to be technically inevitable and unintentionally, whereas another EC Standing order 2001/18/EC requires post-marketing surveillance of genetically altered crops and by-products after they are released into the marketplace to confirm for instantaneous or deferred, either anticipated or unpredicted impacts on human, animal, and environmental safety. The EC Mandate 2001/18/ and Legislation 1829/2003 have established processes to eliminate a genetically altered crop or product from the market of a member country if a potential hazard linked with the crop is ascertained or noticed, therefore the EC can restrict or forbids the marketing of that commodity, as happened with the case of MON-810 (Devos et al. 2019).

10.5 Health effects of RNAi and CRISPR and threat appraisal

The assessment of associated risks with living organisms, society, and the surroundings because of new technologies and their implications is known as risk assessment. Toxicology is the study of toxic substances and their effects on plants, animals, and other organisms. It is a scientific discipline that is developed through case-by-case testing. To improve public and environmental health, toxicology research should be conducted using cutting-edge technologies. RNAi specificity is determined by the sequence homology between small silencing RNAs and their mRNA targets. Consequently, one of the most toxic aspects of RNAi is uncovered when the tiny silencing RNAs pair up with transcripts besides the intended complementary base pairing of small silencing

RNAs with transcripts other than the target molecule, resulting in off-target silencing (Ramon et al. 2014; Casacuberta et al. 2015).

There are possibilities of immediate unintended consequences in the genetically modified plant or indirect effects on the consumers of GM products. Identifying the unintended impacts can assist in risk evaluation. The GMO panel based on the EFSA devised *in silico* expression analysis, a standard for assessing the risks linked with RNAi products; however, this is inadequate to predict risk in humans as well as animals (Pinzo'n et al. 2017). The computational method (Lück et al. 2019), can also be used to detect an off-target pairing of small silencing RNA with a transcript. The abundance of small RNA produced is another factor that contributes to off-target effects.

The EU GMO panel established a risk assessment method for RNAi-based gene silencing based on bioinformatics. The GMO panel's parameters are appropriately applied to both miRNA and siRNA. Based on miRNA target specificity knowledge, complementary incompatibilities among the small RNA and other off-target genes are usually observed. The vulnerabilities of the maize events MON87427MON89034MIR162MON87411 and MON87411 were evaluated using this risk assessment method. Since the risk evaluation divulged no probable off-target impacts, the events underwent further analysis (EFSA Panel on Genetically Modified Organisms et al. 2018; EFSA Panel on Genetically Modified Organisms et al. 2019). To ensure a comprehensive risk assessment, it is crucial to consider prior field test findings which include anticipated as well as incidental changes, particularly in terms of phenotype, agronomy, and composition gathered for commercial reasons. Subsequent testing will be required to validate the findings of the molecular approach if a probable off-target is identified (Papadopoulou et al. 2020).

As per external scientific reports, ncRNAs, also widely recognized as silencing RNAs, are consumed by humans as well as animals as part of their diets. These types of RNAs are readily absorbed and deteriorated by enzymes, gastrointestinal factors like pH, intracellular impediments, such as the lysosomal system, and impediments present outside the cells, such as the intestinal mucosa, preventing ncRNA ingesting.

As a result, dietary silencing RNA ingested by humans, birds, mammals, and fishers as their regular dietary intake is non-toxic and easily decomposed by the digestive tract until and unless their concentration levels and consistency are modified chemically to increase their concentration. Detection of foreign RNA in bodily fluids, even at low concentrations, must be investigated to identify whether it is a result of technical artifacts or a source of contamination. As a result, even at low concentrations, the existence of foreign RNA in bodily fluids of animals and humans needs to be investigated the presence of exogenous RNA in biological fluids of humans and animals must be investigated to determine whether it is the result of technical artifacts or a source of contamination. Thus far, RNA silencing has shown no adverse influences in diets of humans and animals (Petrick et al. 2016).

The environmental intoxication of RNAi-based dsRNA expressing pathogen and insect-resistant crops is well documented due to their potential to harm valuable NTOs (non-target organisms), particularly arthropods, as well as the ecosystem (Taning et al. 2019). The plants with RNAi-assisted modifications are categorized as hazardous in case adequate consumption of dsRNA by NTO causes adverse effects (Christiaens et al. 2018). NTO may be exposed to dsRNA when they consume associated products, such as pollens or preys-consuming GM products, or through below-ground exudates in the soil (Dubelman et al. 2014).

The probable unintended implications of RNAi-based GM plants and gene silencing on non-target organisms (NTOs) are evaluated through a risk assessment process (Lundgren and Duan, 2013). NTOs demonstrated susceptibility to dsRNA from RNAi-based GM plants receive particular attention. Bioinformatics analysis is utilized to identify NTOs with genes that are similar to the dsRNA intended to inhibit the expression of a specific gene in the pathogen. This analysis also takes into account the sequence complementarity between the NTO transcripts and the siRNA derived from dsRNA (Devos et al. 2019). Such findings will guide future risk assessment testing of RNAi-based GM plants before they are released into natural settings.

No subsequent tests are required if computational analysis shows no sequence homology of dsRNA with NTO. Nonetheless, the use of bioinformatics to assess the risk of dsRNA for NTOs has some constraints:

1 Limited availability of genome sequences for all NTOs;
2 Differences in the RNAi-based machinery of NTOs, especially regarding mismatches; and
3 A lack of scientific credibility regarding generalized principles that dictate how transcripts and siRNA interact.

A more accurate bioinformatics-based risk assessment could be achieved by:

1 Conducting genome sequencing for all NTOs,
2 Creating efficient and trustworthy algorithms for precise estimations, and
3 Risk assessment capacity development (Christiaens et al. 2018; Devos et al. 2019).

10.5.1 CRISPR contamination and threat assessment

Despite the fact that CRISPR-Cas9 has achieved major advancements in the creation of plants that are resistant to biotic (Ahmad et al. 2020) and abiotic stress, as well as crop improvement (Biswal et al. 2019), toxicity and risk-related issues require further investigation (Ahmad et al. 2021).

Even while Cas9 nucleases and programmable gRNA have extremely specific functions, they considerably simplified the genetic engineering in plants. Yet, off-target effects can occur due to the tolerance of mismatching as many as five base pairs in the 5' region of the gRNA sequence (Wolt, 2017). These off-target effects can be observable or unobservable and may result in potential toxic effects. Thus, CRISPR-Cas9-based GE products must undergo extensive risk assessment for probable hazards (Liu et al. 2017). Plants have a very high specificity in CRISPR-based GE compared to mammalian systems, where early investigations indicated as much as 50% off-target results (Endo et al. 2015; Peterson et al. 2016).

Some researchers, nevertheless, presume that the off-target effects of CRISPR in plants are directly analogous to those in animals and humans, and that designed to detect those off-target effects in plants have procedural constraints. In plants, off-target detection may be limited by screening a small subset of the genome that includes target events and homologous regions, although off-targeting may eventuate in some non-homologous regions with adequate sequence homology with gRNA (Peterson et al. 2016). Whole-genome sequence studies are performed to determine off-target effects, but they are limited to the disclosed reference genomes. Another possible reason for the recognition of minor off-target effects is the use of a small sample size to analyze on- and off-target effects. A small number of genotypes cannot be used to convincingly represent the entire genome of a plant. As a result, because they only cover a portion of the genome, current screening tests have a narrow scope (Brooks et al. 2014).

One CRISPR-based GE study, for example, targeted the Arabidopsis JASMONATE-ZIM-DOMAIN Protein 1 locus. The mutations in the developed lines vary from flower to flower. These various mutations were not only present in the same generation but they were also passed down to subsequent generations. These findings suggested that toxicity and risk-related activities should not be limited to a single plant stage or generation, but that these prospective off-target consequences could last for generations (Feng et al. 2016). For CRISPR toxicity and risk assessment studies, the surveyor nuclease and T7 endonuclease I (T7EI) assay, loss of restriction site assay, high-resolution melting analysis, and PAGE-based genotyping can all be used (Shan et al. 2014). Each technique, however, has advantages and disadvantages for detecting off-target mutations. T7EI and surveyor nuclease assays, as well as PAGE-based genotyping, are ineffective at detecting homozygous mutants. Similarly, the presence of restriction sites in the target region (which is a transgene) is required for the restriction site loss assay so that the mutated plant can withstand digestion analysis. HRMA-based risk and toxicity studies necessitate the use of expensive equipment, and their sensitivity is compromised if the melting temperatures of the mutant and wild-type versions differ only slightly (Bao et al. 2019).

Another potential risk associated with CRISPR is the activity of the Cas9 protein in subsequent generations (Wang et al. 2018, Ahmad et al. 2021). These Cas9 molecules have the potential to cause an unintended mutation in a genetically stable line, which could be toxic. The persistence of Cas9 activity in subsequent generations was explained using an Arabidopsis-based study.

Four *Arabidopsis* genes were modified using the CRISPR-Cas9 tool: AtJAZ, AtBRI1, AtAP1, and AtGAI. After three generations of research, Cas9 activity was detected in the mutants. The T0 contained a variety of mutations, the majority of which were somatic and passed down to the next generation. In contrast, the inherited germline mutation followed the Mendelian genetic model. The intriguing findings were made in the majority of T1 and T2 lines that carry Cas9 but do not mutate at the target site. When planted in T3, however, that line demonstrated mutagenesis at the target location. These surprising findings suggest that the occurrence of CRISPR machinery by itself is insufficient for GE; rather, the machinery should be present in a physiologically active state (Feng et al. 2016).

To probe further, researchers used a reciprocal crossing mechanism to cross a Cas9/gRNA positive corn line with its wild type to observe the transmission of CRISPR machinery in the wild type. Cas9 had high mutagenic activity at the target site in all Cas9-containing plants, and this correlated positively with mRNA transcript abundance (Char et al. 2017). These findings confirmed that Cas9 activity persists across generations, implying that Cas9-free plants should be obtained prior to genotyping. Because the presence of Cas9 will result in the continued production of undesirable chimeras. Because the majority of T0 mutants are somatic, CRISPR edits should be tested using next-generation progeny. Given Cas9 activity in subsequent generations, this trait could be used to use gene drive technology to transfer a gene in a population to control weeds or pests. Because gene drive is beyond the scope of this chapter, readers interested in the subject should read scientific studies on the subject (Courtier-Orgogozo and Morizot, 2017).

To eliminate the persistence of Cas9 activity, researchers propose using plasmid-free gRNA integration and CRISPR editing. Viral-based expression vectors that do not require integration into the host nuclear genome are another option. The use of a viral-based expression vector would reduce off-target effects while also allowing for a quick fix to obtain Cas9-free plants as early as T0 plants. However, the editing efficiency of viral-based expression vectors is 100 times lower than that of stable delivery systems. Another constraint of viral-based expression systems is their relatively small cargo capacity (around 3 kb), with only the Cas9-coding DNA region exceeding 4 kb. As a result, using a viral delivery system to deliver gRNA sequences into plant genomes with a stable Cas9 background is preferable. However, in subsequent generations, the edits in viral-based expression vectors are poorly inherited, and the edited gene reverts to its wild form. Only germline mutations are passed down to future generations, which are extremely rare in comparison to the population of entire plant cells (Glass et al. 2018).

The DNA-free method of editing, which involves delivering preassembled gRNA: Cas9 complexes to plant protoplasts, is another technique used by scientists to create Cas9-free plants (Woo et al. 2015; Zhang et al. 2016). This method has been used to genetically engineer rice, tobacco, lettuce, and Arabidopsis, with a high (46%) heritable mutation rate at the T0 stage

(Woo et al. 2015). Because it is a DNA-free technology, it eliminates the need for transgene outcrossing and is an equally effective strategy for asexually propagated plants. Although it is a suitable method for gene editing, it cannot be used to insert a foreign gene because it lacks support for an additional/foreign DNA template. Furthermore, protoplast-based gene editing protocols in plants are not widely available.

10.6 Regulatory framework for CRISPR-based regulations in different countries

The use of transgenic crops to enhance crop production has been a topic of discussion among EU regulatory authorities (Pauwels et al. 2014). The legal and scientific experts do not seem to be in agreement on a strategic plan, with just some arguments that GM foods should indeed be strictly controlled using a process-based framework, whereas others make the argument that it should be policed based on the product's nature and threat assessment (Sprink et al. 2016). The EU must generate scientifically valid and pragmatic rules and regulations. They would more than presumably be determined by the kind of finished product rather than the methodology utilized (Wolt et al. 2016). In light of the present scenario, EU regulatory authorities appear to be remarkably quiet on the destiny of genetically engineered crops (Wolt, 2017).

Currently, the United Kingdom and the European Union categorize genetically engineered plants, microorganisms, animals, and fungi as GMOs (Spicer and Molnar, 2018). As a direct consequence, companies should first abide by EFSA's rigorous GMO regulations before promoting products for commercial cultivation (Globus and Qimron, 2018). Only a few nations, Finland and Sweden, have created standard operating processes and protocols developed on the opt-in law as well as prefer non-GMO labeling of genetically engineered crops (Wolt, 2017). Other countries, such as the Netherlands and Germany, are in the phase of establishing an opt-in process for genetically engineered crops and prioritizing non-GMO labeling of GM crops (Spicer and Molnar, 2018). Several more representatives from the European Academies Science Advisory Council, the EU, EFSA, as well as the former Chief Scientific Advisor to the President of the EC have indeed asserted that products derived from GMOs shouldn't be applicable to GMO regulatory requirements as they do not contain any inserted foreign DNA. Furthermore, the Swedish Agricultural Agency has ascertained that CRISPR-Cas9-based GM crops seem to be excluded from the clauses of transgenic crop statutory provisions (Zhang et al. 2020).

The EU GMO provisions are incredibly strict, as the approximate expenditure for creating and reviewing GMO products is US$35 million, with a 6-year hold for final authorization (Jouanin et al. 2018). However, the repercussions are still unclear because final approval is completely reliant on several political and social considerations and isn't exclusively based on empirical facts. Chosen to give the time and expense involved in GM rules and regs, GE's fundamental advantages of creating value-added goods that fulfill the needs of the community might be ended up losing. As a result, EU-based technology companies are relocating to the United States to experiment with and popularize GM crops, culminating in a mass migration to the EU (Burger and Evans, 2018). There is a widespread concern that controlling GM crops will stifle innovation, competitiveness, as well as the EU's access to nutritious products (Jouanin et al. 2018).

10.6.1 China

In China, plant products based on RNAi silencing have been regulated as GMOs. The authorization for genetically modified crops (based on RNAi and CRISPR) is complex and difficult and directly controlled by the Ministry of Agriculture and Rural Affairs (MARA). The process starts whenever a biotech crop developer appears to apply for a bio-safety certificate. The request will be denied if the identical commodity has not yet been marketed in the country of origin. To fulfill the first federal requirement, the GM crop is subjugated to rigorous examination to assess food hygiene and safety, non-target organism (NTO) effects, genetic drift, as well as other possible risk variables. Following the completion of each of these assessments, a three-phase evaluation process that involves field tests (as a small, completed trial in the United States), environmental release tests (field trials for farmers in the United States), and pre-production trials begin. Pre-production trials are not needed for the genetically- modified product shipped for processing (Huang et al. 2008).

Parallel to these three studies, MARA-designated research organizations and academic institutions are undertaking bio-safety investigations. Upon accepting MARA's biosafety accreditations, imported crops can be commercialized as processing or raw material, but cultivation needs three additional documents: (1) a variety registration certificate, (2) a production license, and (3) a marketing license for commercialization. The data show that the approval period in China is comparatively brief, averaged nearly approximately 34 months for 50 cases, with a minimum of 18 months for the MIR604 maize activity and a maximum of 71 months for MIR162, another maize activity (Jin et al. 2019). China could expedite the commercial viability of genetically engineered crops by further authorizing their commercialization status fairly shortly after they've been endorsed in their nation of origin. This will enhance the Chinese economy's competitive advantage with all these crop varieties (Jin et al. 2019).

10.6.2 Japan

In Japan, GMO crops are regulated by the Ministry of Environment. In Japan, GMOs are characterized by the end product instead of the procedure. Furthermore, no matter what method is used for bioengineering, each living organism comprising foreign nucleotides will be tightly controlled. This definition specifically excludes RNAi and GMO crops from GMO regulations. This interpretation of GMOs was agreed upon at the second meeting of the Ministry of Environment's Advisory Board on GMOs in August 2018 (Zannoni, 2019).

10.6.3 Australia

The progress of genetically altered crops appears to require prior permission from the Biosafety Advisory Board and the Office of the Genetic Engineering Regulatory Agency (OGTR). RNAi-based crops and genetically modified crops seem to be liable to the exact same regulatory framework as traditional genetic manipulation. The OGTR comprehensively updated the regulations for genetically engineered crops in April 2019 and unearthed that now the product of NBTs, which really is transgene-free, is controlled correspondingly to traditional breeding products. As a direct consequence, it must have consented that instances in which prefabs weren't utilized and modifications that took place were exactly equivalent compared to those that naturally occur would be regarded in a manner that did not pose an added risk. However, technology to transform the use of DNA templates and the introduction of unique genetic material into the host chromosome would be considered part of GMOs and would therefore be governed by the genetically modified regulatory framework (Mallapaty, 2019).

10.6.4 Brazil

The Brazilian National Biosafety Technical Commission (CTNBio) was formed in 2014 as a group of experts of scientific professionals tasked with creating a regulatory regime for NBT products. In Resolution No. 16 (RN16), the 2018 certified CTNBios professionals assessed cases of NBTs that ought to be strictly controlled as GMO or non-GMO. According to CTNBios scientists, the process is insignificant if the progeny lacks transgenes and is non-GMO if somehow the organism evolved is comparable to that developed through traditional breeding. Such genetic variations are assumed to be comparable to those triggered by older mutagenesis methodologies, which can also happen naturally. As a consequence, the professionals asserted that regulations would pertain on a case-by-case basis (Eriksson et al. 2019).

10.6.5 New Zealand

The Hazardous Substances and New Organisms (HSNO) Act of 1996 authorizes genetically engineered crops through a process-based framework. This is coherent with the HSNO Act's definition of GMO as an organism with whom the genetic composition has indeed been altered by in vitro methods, or compounds obtained or acquired from the organism whose genetic composition has been reconfigured by in vitro methods. In simple terms, it involves the modification of genetic material in organisms using in vitro methods. As a result, RNAi-based GMO, transgenic crops, and genetically engineered crops are all governed by the GMO Law (Fritsche et al. 2018). Section 26 of the HSNO Act equips the Competent Authority to decide if a certain kind of NBT is a GMO or not. Moreover, it emboldens the law enforcement body to decide if the item of NBTs is harmful to humans and to share its observations with the public (Kershen, 2015). The GMO regulatory system applies to all GMO crops and RNAi-based GMO crops (Fritsche et al. 2018).

10.6.6 Argentina

Argentina was the initial nation worldwide to commercialize NBT commodities (Sánchez, 2020). The Argentine Secretariat of Agriculture, Livestock, Fisheries, and Food (SAGPyA) is in charge of overseeing genetically engineered crops in conformity with Argentine policy initiatives, declarations, and regulations. To evaluate the effect of genetically engineered crops on the agro-ecosystem, the National Direction of Agricultural Food Markets (DNMA), the National Institute of Seeds (INASE), the National Service of Agricultural and Food Health and Quality (SENASA), and the National Advisory Commission on Agricultural Biotechnology (CONABIA) are key regulatory bodies under the administrative control of SAGPyA. TACs are used by SENASA to determine the safety of genetically modified plant feed and food for animal and human consumption. The DNMA regulates the commercial aspects of genetically engineered crops, while INASE controls the licensing and supervising of commercially available seedlings (Burachik and Traynor, 2002). SAGPyA Resolution No. 173/15 involves determining the question of whether GE's product should really be categorized as a GMO. The CONABIA regulates genetically engineered plant products. The legislation is concerned with determining if the genetically engineered plant encloses transgenic organisms. According to Argentina's assessment guidelines (Whelan and Lema, 2015), if indeed the commercial final product of a genetically altered crop is free of a transgene, the crop is not regulated as a GMO; otherwise, the end product is subject to the genetically modified regulatory framework (Gao et al. 2018).

10.6.7 Chile

Chile is the second state, after Argentina, to start introducing NBT-based products. It has commercially produced eight NBT products after proclaiming them non-GMO under its regulatory regime, including two varieties of *Brassica napus* (for silique splinter resistance), three maize cultivars (for fatty acid component change, drought resistance, and yield increase), two from *Glycine max* (for fatty acid composition change), and one from *Camelina sativa* (for fatty acid composition change) (Sanchez, 2020). To regulate genetically altered crops, the Chilean Ministry of Agriculture has established an operational group of government entities under its administrative supervision.

Chilean Resolution 1523, approved in 2001, establishes guidelines for the regulatory oversight of genetically altered crops, from field propagation to harvest and export. Before genetically engineered crops could be propagated and cultivated, the Chilean Ministry of Environment must evaluate their risk. They generally follow guidelines to Argentina as long as the end product of GM crops is free of transgenes, it will be regulated as a traditional breeding product, otherwise as a GMO. Besides that, the Agricultural and Livestock Service helps to regulate genetically altered crop products through its own Forestry and Agricultural Protection department (Eriksson et al. 2019, Ahmad et al. 2021).

10.7 Different international conventions and protocols on biosafety

Biosafety is the protection of substantial loss of biological integrity, with just an emphasis on both ecological systems and public health. Precautionary mechanisms involve frequent biosafety evaluations in lab conditions in addition to clear regulations that need to be followed. The primary objective of biosecurity is to prevent potentially dangerous incidents. A constant bio-safety risk assessment and enforcement procedures are employed in numerous blood-borne pathogen laboratories. Failure to adhere to such protocols increases the risk of being exposed to health hazards or pathogens. Human error and poor engineering contribute to unnecessary exposure and put the most effective protection mechanisms in jeopardy.

The Biosafety Protocol emerged from the Convention on Biological Diversity (CBD); an intergovernmental pact aimed at bio-diversity preservation (BSP). The CBD's laudable objectives are to safeguard biodiversity, focus on ensuring its sustainable use, and promote an equal and fair distribution of the advantages in terms of genetic assets located in biodiversity. In adhering to the Convention for the Protection of Biological Diversity, a Biosafety Working Group (BSWG) was formed to draw up a Proper procedure, particularly regarding the "transfer, managing, and utilization of living modified organisms (LMOs) deduced from biotechnology that may have a negative impact on the preservation and sustainable utilization of biodiversity." The BSP's scope, advanced briefed arrangements for LMO shipments, threat assessment parameters, obligation, labeling, confidentiality, quasi-status, and connection to other international treaties are all being managed to negotiate by the BSWG.

10.7.1 Convention on Biological Diversity (CBD)

The Convention on Biological Diversity, also known as the Agreement on Biological Diversity, is a multilateral treaty. The Convention's primary objectives are threefold: the safeguarding of species diversity, sometimes referred to as biological diversity; appropriate utilization of its constituents; and the equal sharing of genetic resource benefits. It is widely regarded as a crucial memorandum for sustainable growth, with the purpose of creating national plans for the preservation and long-term use of species diversity. The Rio de Janeiro Earth Summit kicked off the Convention for signatures on June 5, 1992, and it became operative on December 29, 1993. Except for the United States, no other UN member state has ratified the convention. The Cartagena Protocol and the Nagoya Protocol are two further accords.

The Cartagena Protocol on Biosafety toward the Convention on Biological Diversity is an international treaty that governs the transfer of living-modified organisms (LMOs) created using contemporary biotechnology from one country to the next. It was added to CBD on January 29, 2000, and became effective on September 11, 2003.

The Nagoya Protocol on genetic resource access and the proper and equitable benefit distribution resulting from their application is another additional treaty to the CBD. It creates a clear legal precedent for the efficient execution of one of those CBD's three goals: the equitable and impartial allocation of benefits derived as a result of the exploitation of genetic resources. The Nagoya Protocol was signed on October 29, 2010, in Nagoya, Japan, and became effective on October 12, 2014.

10.7.2 Cartagena Protocol on Biosafety

The Cartagena Protocol on Biosafety as a Complement to the Convention on Biological Diversity (CBD) is a bio-safety international consensus that has been in effect since 2003. The primary objective of the Biosafety Framework is to safeguard biological diversity from possible threats posed by contemporary biotechnology-derived genetically engineered organisms. In accordance with the Biosafety Framework, the latest technological products must abide to the doctrine and enable developing nations to find an equilibrium among both economic and public health benefits. For instance, it will enable exporters to mark consignments comprising genetically engineered commodities such as cotton or maize, and it will enable countries to restrict the import of genetically engineered organisms if they genuinely think there seems to be limited scientific proof supporting the product's safety. The Cartagena Protocol on Biosafety, often known as the Biosafety Protocol, was approved in January 2000 after a temporary public task force on biosafety convened six times between July 1996 and February 1999.

At the first extraordinary session of the Conference of the Parties, the Taskforce provided an update on the Protocol's utterance to the Conference of the Parties to endorse a CBD bio-safety protocol. The Cartagena Protocol was officially adopted on January 29, 2000, after so many postponements. The Bio-Safety Protocol's purpose is to preserve and protect biological richness against possible threats likely to be faced by organisms that have been altered and deduced from contemporary biotechnology. The Protocol's purpose is to make a contribution to guaranteeing an adequate level of security in the safe transportation, handling, and utilization of "living modified organisms arising from modern biotechnology" which could have deleterious repercussions on environmental sustainability and conservation utilization of biodiversity while taking hazards in terms of human health into account and concentrating particularly addressing transboundary movements (Article 1 of the Protocol, SCBD, 2000; Pauchard, 2017).

Any biological thing that has the potential of transmitting or copying genetic information is regarded as a "living organism," which includes aseptic organisms, viruses, and viroids. A "living modified organism" is any natural organism with a distinguishable blend of genetic makeup collected because of the utilization of contemporary bioengineering. "Contemporary bioengineering" is described by the Protocol as "the implementation of in vitro oligonucleotides methodologies or cell fusion beyond the phylogenetic family" they do not have intrinsic physiological reproductive or recombination barriers conventional breeding and selection methods. Products derived from living modified organisms (LMOs) that include unique, distinguishable combinations of reproducible genetics obtained directly through modern biotechnology are referred to as LMO products. Agricultural crops that have undergone genetic modification to make them more productive or resistant to diseases or pests are examples of common LMOs. Maize, cotton, soybeans, tomatoes, and cassava are all genetically engineered crops. In addition to genetically engineered crops, agricultural commodities encompass living modified organisms meant for direct use such as food, feed, or processing (LMO-FFP). In general, no distinction was made between "living modified organisms" and "organisms that have been genetically modified," and the term "genetically engineered organism" was not used.

The framework encourages biosecurity by having established regulations and processes for the sake of safe transport, handling, and utilization of LMOs, with an emphasis on cross-border LMO mobility. It includes a number of operations, which include a pre-agreement procedure for purposefully releasing LMOs into the surroundings, and another for directly utilizing LMOs as food, feedstock, or processing. Signatories to the Convention must confirm the manner in which LMOs are handled, packed, and transported safely. Besides this, transboundary shipment of LMOs must be accompanied by suitable documentary evidence indicating the identity of the LMOs along with the point of contact for extra information.

These methods and techniques prerequisites are destined to just facilitate importing stakeholders with the data they require to arrive at informed choices regarding whether to allow LMO imports or not as well as how to safely manage them. The importing party builds its judgments on scientific evidence-based risk evaluation. The protocol outlines the fundamentals and processes for carrying out a risk evaluation. In the absence of relevant scientific knowledge and information, the importing party may exercise extreme caution in its import decisions. Parties may take into account socio-economic status factors when deciding whether to import LMO in conformance with their international commitments. Stakeholders must also clarify any threats discovered during the risk evaluation and safeguard themselves in the case of an unintentional release of LMOs. The Protocol defines a Bio-safety Clearing House in which Stakeholders can share information and encompasses a variety of significant terms and conditions factors such as capacity-building, a financing system, and compliance obligations prerequisites for community participation and knowledge.

10.7.3 Nagoya Protocol

The *Nagoya Protocol on Access to Genetic Resources and the Fair and Equitable Sharing of Benefits Arising from their Utilization to the Convention on Biological Diversity* is an international agreement which aims at sharing the benefits arising from the utilization of genetic resources in a fair and equitable way.. Their objective is to carry out one of the CBD's three objectives: equitable distribution of the advantages accruing from the use of genetic resources, and also to make contributions to bio-diversity preservation and environmentally friendly use. It requires its signatories to incorporate benefit-sharing, conformity, and access to genetic resources measures. The Nagoya Protocol appears to apply to CBD-covered biological resources and the benefits they provide. The framework also encompasses indigenous practices about CBD-covered genetic resources and the benefits of using them. Their goal is to implement one of the CBD's three goals: equitable and fair sharing of the benefits of genetic resource usage contribute to biodiversity conservation and long-term use.

10.8 Conclusion and possibilities

Biotechnology breakthroughs provide a plethora of opportunities to address global problems such as the transmission of infectious diseases, food insecurity, and ecological degradation. The same techniques, nevertheless, can be used by unscrupulous people or adversarial regions to generate potentially deadly pathogens that really can intentionally cause invasive diseases, obstruct agricultural distribution networks, or obstruct natural ecological balances. The world has already observed some worrisome historical precedents of deliberate misuse of genetic engineering to create biological weapons, instances of laboratory-acquired infections, and cases of accidental discharge of biological entities. To resolve such obstacles, most countries have implemented colloquial regulations or legislation to protect the confidentiality of the biotechnological investigation, mechanisms to prevent unauthorized access to biological matter, and export control mechanisms to monitor and control the transfer of delicate biological matter. Treaties, conventions, and recommendations have been developed at the worldwide level to safeguard equitable and open financial support of genetic engineering; however, these methods do not include the supervision and control requisite to greater collaboration in biotechnology research. This is due to the fact that all these global mechanisms either have not been updated on a regular basis or thus cannot keep up with the latest technological innovations, or they lack the experience and knowledge, and resource allocation required to track global biotechnological developments.

NBTs, or new breeding techniques, such as RNAi and CRISPR, have transformed plant breeding methodologies through accurate genetic manipulation, resulting in a transgene-free final product that is perfectly safe than those acquired by conventional genetic engineering technology. These methods have contributed to the creation of plant varieties that are resilient to both biotic and abiotic stresses. Although RNAi and CRISPR have made considerable valuable contributions to crop improvement, their potential toxicity and vulnerability assessments have deferred their adoption and implementation. Moreover, tightened regulatory regimes prevent massive agribusinesses, such as the EU, from maximizing the possibilities of these NBTs. There are major regulatory frameworks in place for these NBTs all around the globe. One viewpoint is process-based and is encouraged by the procedure used to develop the genetically -modified plant/plant. It is a very stringent regulatory platform that categorizes NBTs' products as transgenic-free or non-transgenic under genetically modified regulations. GMO regulations are rigorous, time-consuming, and expensive, demanding additional costs for pre-market labeling of products after four decades of threat evaluation and toxicity studies (in the EU). The end-product-based strategy is sometimes used for another fairly flexible approach where the toxicity and threat involved with the ultimate result are evaluated on an individual basis. If the final result of GE is transgene-free, it will not be classified; however, the risk is characterized by using a traditional breed-based risk assessment strategy (USA, Canada).

The EU and its allied countries should begin considering the interpretation of GMOs and improved legislation to recognize its most significant advancements in this field. The integration of NBT products in GMO regulatory requirements will be catastrophic for the EU and the economy of its connected nations. Biotech firms in the EU are relocating to the United States, causing financial losses. Moreover, when compared with the United States and Canada, the quantity of scientific research and advancement in the subject of biotechnology in the EU is steadily decreasing. On the contrary, the most important aspect is that numerous EU member countries are speaking out against existing legislation and that there are dissonances between many member nations. As a matter of fact of the debate, it is recommended that the EU emulate the examples of commercial exploitation of genetically engineered plants in the United States, Canada, and China and transformation derived from process-based regulatory oversight to end-product-based regulations in order for member nations to profit from the advantages of technology.

New breeding products are now being commodified in China. They have, however, specified in their legislative framework that imported genetically engineered crops to be marketed in China must first be marketed inside the nation where they originated. So, if a biotech company based in the United States desires to commercialize an activity in China for cultivation, it should first obtain acceptance in the United States prior to actually obtaining permission to grow in China. It imposes additional trade restrictions because used transgenic crops are becoming outdated in their home country once they enter China. China should help streamline its own regulation process by permitting concurrent registration in both the country of origin and China, or by allowing harvesting in China once the country of origin has conferred authorization without any further threat assessments or toxicology evaluations. This saves a lot of time, which accelerates the research and technology strategies.

The Cas9 protein's activity persists in successive generations, which needs to be acknowledged. Various transformation methodologies that do not result in plasmid assimilation for expression should be recognized, as well as plasmid-free transformation systems. Even though viral-based vector mechanisms have illustrated the ability to resolve these issues, they have also been found to have some constraints, including fewer haulage options, low processing efficiency, and poor inheritance of the preferred encoding in successive generations. These constraints must be overcome in order to develop Cas9-free desired plants in the future. Protoplast-based insertions of guide RNA and Cas9 Complexes may be employed in this regard. In all crop plants, optimized transformation systems on chloroplasts are still suffering from a lack. These issues must be dealt with in the future in order to take full advantage of the most recent GE tools.

Furthermore, there are no worldwide biosafety and bio-safety standards to which every research organization should indeed abide, and no framework in place to define responsibility and procedures for evaluating legal liability when experimental studies go wrong. In the absence of a verification and surveillance protocol, signatory states to the BTWC can only advise each other or lodge a complaint with the UN Security Council if, for example, government entities intentionally mishandle the biotechnology. The examples above demonstrate that these global mechanisms are ill-equipped to confront the dangers and challenges presented by bioengineering scientific discoveries. To ensure that biotechnology-related studies are carried out prudently, the global community must work together to create competent and qualified experiment safety, effectively communicate the long-debated validation and supervisory mechanism under the BWC, and include provisions introducing culpability and accountability measures for infringements. These steps would go a long way toward increasing the likelihood that the world can benefit from biotechnology advances while minimizing the dangers and downfalls.

Note

1 The European Commission harmonized the national regulations across Europe.

References

Abbas, M.S.T. 2018. Genetically engineered (modified) crops (Bacillus thuringiensis crops) and the world controversy on their safety. *Egypt J Biol Pest Control* 28, 52. 10.1186/s41938-018-0051-2

Ahmad S., Shahzad R., Jamil S., Tabassum J., Chaudhary, M.A.M., Atif, R.M., Iqbal, M.M., Monsur, M.B., Lv, Y., Sheng, Z., Ju, L., Wei, X., Hu, P., Shaoqing Tang - Regulatory Aspects, Risk Assessment, and Toxicity Associated with RNAi and CRISPR Methods 2021. "Kamel A. Abd-Elsalam, Ki-Taek Lim, In Nano biotechnology for Plant Protection, CRISPR and RNAi Systems, Elsevier, 687–721, ISBN 9780128219102.

Ahmad, S., Wei, X., Sheng, Z., Hu, P., Tang, S., 2020. CRISPR/Cas9 for development of disease resistance in plants: recent progress, limitations and future prospects. *Brief. Funct. Genomics* 19 (1), 6–39.

Aletor V.A. 1993. Allelochemicals in plant foods and feeding stuffs. Part I. nutritional, and physiopathological aspects in animal production. *Vet. Hum. Toxicol.* 35(1), 57–67.

Araki, M., Ishii, T. 2015. Towards social acceptance of plant breeding by genome editing. *Trends Plant Sci.* 20, 145–149.

Bain, C., Selfa, T., Dandachi, T., Velardi, S. 2017. 'Superweeds' or 'survivors'? Framing the problem of glyphosate resistant weeds and genetically engineered crops. *Journalof Rural Studies* 51, 211–221.

Bao, A., Burritt, D.J., Chen, H., Zhou, X., Cao, D., Tran, L.-S.P. 2019. The CRISPR/ Cas9 system and its applications in crop genome editing. *Crit. Rev. Biotechnol.* 39, 321–336.

Biswal, A.K., Mangrauthia, S.K., Reddy, M.R., Yugandhar, P. 2019. CRISPR mediated genome engineering to develop climate smart rice: challenges and opportunities. *Semin. Cell Dev. Biol.* 96, 100–106.

Bogdanove, A., Donovan, D., Elorriaga, E., Kuzma, J., Pauwels, K., Strauus, S., et al. 2018. Genome editing in agriculture: methods, applications and governance. *CAST Issue Paper 60*, 25–42.

Brooks, C., Nekrasov, V., Lippman, Z.B., Van Eck, J. 2014. Efficient gene editing in tomato in the first generation using the clustered regularly interspaced short palindromic repeats/CRISPR-associated9 system. *Plant Physiol.* 166, 1292–1297.

Bruetschy C. 2019. The EU regulatory framework on genetically modified organisms (GMOs). *Transgenic Research*, 28(Suppl 2), 169–174. 10.1007/s11248-019-00149-y

Burachik, M., Traynor, P.L. 2002. Analysis of a National Biosafety system: regulatory policies and procedures in Argentina. *ISNAR, The Hague.*

Burger, L., Evans, D. 2018. Bayer, BASF to pursue plant gene editing elsewhere after EU ruling. https://www.reuters.com/article/us-eu-court-gmo-companies/bayer-basf-to-pursue-plant-gene-editing-elsewhere-after-eu-ruling-idUSKBN1KH1NF.

Casacuberta, J.M., Devos, Y., Du Jardin, P., Ramon, M., Vaucheret, H., Nogué, F. 2015. Biotechnological uses of RNAi in plants: risk assessment considerations. *Trends Biotechnol.* 33, 145–147.

Char, S.N., Neelakandan, A.K., Nahampun, H., Frame, B., Main, M., Spalding, M.H., et al. 2017. An Agrobacterium-delivered CRISPR/Cas9 system for high-frequency targeted mutagenesis in maize. *Plant Biotechnol. J.* 15, 257–268.

Christiaens, O., Dzhambazova, T., Kostov, K., Arpaia, S., Joga, M.R., Urru, I., et al. 2018. Literature review of baseline information on RNAi to support the environmental risk assessment of RNAi-based GM plants. *EFSA Supporting Publ.* 15, 1424E.

Courtier-Orgogozo, V., Morizot, B., Boëte C. 2017. Agricultural pest control with CRISPR-based gene drive: time for public debate. *EMBO Rep.*, 18, 878–880.

Davison, J., Ammann, K. 2017. New GMO regulations for old: determining a new future for EU crop biotechnology. *GM Crops Food* 8, 13–34.

Devos, Y., Craig, W., Devlin, R.H., Ippolito, A., Leggatt, R.A., Romeis, J., et al. 2019. Using problem formulation for fit-for-purpose pre-market environmental risk assessments of regulated stressors. *EFSA J.* 17, e170708.

Devos, Y., Craig, W., Schiemann, J. 2012. Transgenic crops, risk assessment and regulatory framework in the European Union. *Encyl. Sustain. Sci. Technol.* 10765–10796.

Dubelman, S., Fischer, J., Zapata, F., Huizinga, K., Jiang, C., Uffman, J., et al. 2014. Environmental fate of double-stranded RNA in agricultural soils. *PLoS One* 9, e93155.

Duke, S.O., Powles, S.B. 2008. Glyphosate: a once-in-a-century herbicide. *Pest Manag. Sci.* 64, 319–325.

EFSA Panel on Genetically Modified Organisms Naegeli, H., Birch, A.N., Casacuberta, J., De Schrijver, A., Gralak, M.A., et al., 2018. Assessment of genetically modified maize MON 87411 for food and feed uses, import and processing, under Regulation (EC) No 1829/2003 (application EFSA-GMO-NL-2015-124). *EFSA J.* 16, e05310.

EFSA Panel on Genetically Modified Organisms Naegeli, H., Bresson, J.L., Dalmay, T., Dewhurst, I.C., Epstein, M.M., et al., 2019. Assessment of genetically modified maize MON 87427 MON 89034 MIR 162 NK 603 and subcombinations, for food and feed uses, under Regulation (EC) No 1829/2003 (application EFSA-GMO-NL-2016-131). *EFSA J.* 17, e05734.

Endo, M., Mikami, M., Toki, S. 2015. Multigene knockout utilizing off-target mutations of the CRISPR/Cas9 system in rice. *Plant Cell Physiol.* 56, 41–47.

Eriksson D. 2018. The Swedish policy approach to directed mutagenesis in a European context. *Physiologia plantarum* 164(4), 385–395. 10.1111/ppl.12740

Eriksson, D., Kershen, D., Nepomuceno, A., Pogson, B.J., Prieto, H., Purnhagen, K., et al., 2019. A comparison of the EU regulatory approach to directed mutagenesis with that of other jurisdictions, consequences for international trade and potential steps forward. *N. Phytol.* 222, 1673–1684.

Ermakova, I. 2006a. Genetically modified soy leads to the decrease of weight and high mortality of rat pups of the first generation. Preliminary studies. *Ecosinform* 1(2996), 4–9.

Ermakova, I.V. 2006b. Mine-field of genetics. *State Management of Resources* 2, 44–52.

Ermakova, I.V., Barskov, I.V. 2006. Influence of diet with the soy modified by the gene CP4 ePSPS on physiological state of rats and their offspring. *Agrarian Russia.*

Europejska, K. 2017. New techniques in agricultural biotechnology. High Level Group of Scientific Advisors. *Explanatory Note 02. Pobrane z.* http://ec.europa.eu/research/sam/pdf

Feng, C., Yuan, J., Wang, R., Liu, Y., Birchler, J.A., Han, F. 2016. Efficient targeted genome modification in maize using CRISPR/Cas9 system. *J. Genet. Genomics* 43, 37–43.

Fritsche, S., Poovaiah, C., Macrae, E., Thorlby, G. 2018. A New Zealand perspective on the application and regulation of gene editing. *Front. Plant Sci.* 9, 1323.

Gao, W., Xu, W.-T., Huang, K.-L., Guo, M.-Z., Luo, Y.-B. 2018. Risk analysis for genome editing-derived food safety in China. *Food Control* 84, 128–137.

Glass, Z., Lee, M., Li, Y., Xu, Q. 2018. Engineering the delivery system for CRISPR based genome editing. *Trends Biotechnol.* 36, 173–185.

Globus, R., Qimron, U. 2018. A technological and regulatory outlook on CRISPR crop editing. *J. Cell. Biochem.* 119, 1291–1298.

Gostek, K. 2015. Genetically modified organisms: how the United States' and the European Union's regulations affect the economy. *Mich. St. Int'l L. Rev.* 24, 761.

Horsch, R.B. 1993. Commercialization of genetically engineered crops. *Philos. Trans. R. Soc. Lond. B Biol. Sci.* 342, 287–291.

Huang, J., Hu, R., Rozelle, S., Pray, C. 2008. Genetically modified rice, yields, and pesticides: assessing farm-level productivity effects in China. *Econ. Dev. Cult. Change* 56, 241–263.

Ishii, T. 2017. Germ line genome editing in clinics: the approaches, objectives and global society. *Brief. Funct. Genomics* 16, 46–56.

Jin, Y., Drabik, D., Heerink, N., Wesseler, J. 2019. Getting an imported GM crop approved in China. *Trends Biotechnol.* 37, 566–569.

Jones, H.D. 2015. Regulatory uncertainty over genome editing. *Nat. Plants* 1, 1–3.

Jouanin, A., Boyd, L., Visser, R.G., Smulders, M.J. 2018. Development of wheat with hypoimmunogenic gluten obstructed by the gene editing policy in Europe. *Front. Plant Sci.* 9, 1523.

Kershen, D.L. 2015. Sustainability Council of New Zealand Trust v. The Environmental Protection Authority: gene editing technologies and the law. *GM Crops Food* 6, 216–222.

Kochhar, H., Adlakha-Hutcheon, G., Evans, B. 2005. Regulatory considerations for biotechnology-derived animals in Canada. *Rev. Sci. Tech. Off. Int. Épizoot.* 24, 117.

Li M., Liu Q., Teng Y., et al. 2019 Sep 11. The resistance mechanism of Escherichia coli induced by ampicillin in laboratory. *Infect. Drug Resist.* 12, 2853–2863. Published. doi:10.2147/IDR.S221212

Liener I.E. 1980. *Toxic constituents of plant food stuffs*. Second edition, New York and London, Academic Press, 502.

Liu, X., Xie, C., Si, H., Yang, J. 2017. CRISPR/Cas9-mediated genome editing in plants. *Methods* 121, 94–102.

Lück S., Kreszies T., Strickert M., Schweizer P., Kuhlmann M., Douchkov D. 2019. siRNA-Finder (si-Fi) software for RNAi-target design and off-target prediction. *Front. Plant Sci.*, 10, 1023.

Lundgren, J.G., Duan, J.J. 2013. RNAi-based insecticidal crops: potential effects on non-target species. *Bioscience* 63, 657–665.

Mallapaty, S. 2019. Australian gene-editing rules adopt 'middle ground'. *Nature.* Available from: 10.1038/d41586-019-01282-8.

Manyi-Loh, C., Mamphweli, S., Meyer, E., Okoh, A. 2018 Mar 30. Antibiotic use in agriculture and its consequential resistance in environmental sources: potential public health implications. *Molecules* 23(4), 795. Published. doi:10.3390/molecules23040795

Mchughen, A. 2006. *Plant genetic engineering and regulation in the United States.*

Papadopoulou, N., Devos, Y., Álvarez-Alfageme, F., Lanzoni, A., Waigmann, E. 2020. Risk assessment considerations for genetically modified RNAi plants: EFSA's activities and perspective. *Front. Plant Sci.* 11. Available from: 10.3389/fpls.2020.00445.

Pauchard, N. 2017. Access and benefit sharing under the convention on biological diversity and its protocol: what can some numbers tell us about the effectiveness of the regulatory regime? *Resources* 6(1), 11. 10.3390/resources6010011

Pauwels, K., Podevin, N., Breyer, D., Carroll, D., Herman, P. 2014. Engineering nucleases for gene targeting: safety and regulatory considerations. *N. Biotechnol.* 31, 18–27.

Peterson, B.A., Haak, D.C., Nishimura, M.T., Teixeira, P.J., James, S.R., Dangl, J.L., et al. 2016. Genome-wide assessment of efficiency and specificity in CRISPR/Cas9 mediated multiple site targeting in Arabidopsis. *PLoS One* 11, e0162169. Available from: 10.1371/journal.pone.0162169

Petrick, J.S., Frierdich, G.E., Carleton, S.M., Kessenich, C.R., Silvanovich, A., Zhang, Y., et al. 2016. Corn rootworm-active RNA DvSnf7: repeat dose oral toxicology assessment in support of human and mammalian safety. *Regul. Toxicol. Pharmacol.* 81, 57–68.

Pinzón, N., Li, B., Martinez, L., Sergeeva, A., Presumey, J., Apparailly, F., et al. 2017. microRNA target prediction programs predict many false positives. *Genome Res.* 27, 234–245.

Prince, M.J. 2000. The Canadian food inspection agency: modernizing science-based regulation. *Risky Business: Canada's Changing Science-Based Policy Regulatory Regime*. 209–233. Available from: 10.3138/9781442679399.

Ramon, M., Devos, Y., Lanzoni, A., Liu, Y., Gomes, A., Gennaro, A., et al. 2014. RNAibased GM plants: food for thought for risk assessors. *Plant Biotechnol. J.* 12, 1271–1273.

Ricroch, A., Harwood, W., Svobodová, Z., Sági, L., Hundleby, P., Badea, E.M., Rosca, I., Cruz, G., Salema Fevereiro, M.P., Marfà Riera, V., Jansson, S., Morandini, P., Bojinov, B., Cetiner, S., Custers, R., Schrader, U., Jacobsen, H.J., Martin-Laffon, J., Boisron, A., Kuntz, M. 2015. Challenges facing European agriculture and possible biotechnological solutions. *Crit. Rev. Biotechnol.* Jul 1, 1–9.

Sánchez, M., 2020. Chile as a key enabler country for global plant breeding, agricultural innovation, and biotechnology. *GM Crops Food* 11, 130–139.

Schiemann, J., Dietz-Pfeilstetter, A., Hartung, F., Kohl, C., Romeis, J., Sprink, T. 2019. Risk assessment and regulation of plants modified by modern biotechniques: current status and future challenges. *Annu. Rev. Plant Biol.* 70, 699–726.

Secretariat of the Convention on Biological Diversity 2000. Cartagena protocol on biosafety to the convention on biological diversity: text and annexes. https://www.cbd.int/doc/legal/cartagena-protocol-en.pdf.

Séralini, G.E., Clair, E., Mesnage, R., Gress, S., Defarge, N., Malatesta, M., Hennequin, D., de Vendômois, J.S. 2012. Long term toxicity of a Roundup herbicide and a Roundup-tolerant genetically modified maize. *Food Chem. Toxicol.: Int. J. Published Br. Ind. Biol. Res. Assoc.*, 50(11), 4221–4231. 10.1016/j.fct.2012.08.005 (Retraction published Food Chem Toxicol. 2014 Jan;63:244)

Séralini, G.E., Clair, E., Mesnage, R., Gress, S., Defarge, N., Malatesta, M., Hennequin, D., de Vendômois, J.S. 2014. Republished study: long-term toxicity of a Roundup herbicide and a Roundup-tolerant genetically modified maize. *Environ. Sciences Europe*, 26(1), 14. 10.1186/s12302-014-0014-5

Séralini, G.E., Mesnage, R., Clair, E. et al. 2011. Genetically modified crops safety assessments: present limits and possible improvements. *Environ. Sci. Eur.* 23, 10. 10.1186/2190-4715-23-10

Shan, Q., Wang, Y., Li, J., Gao, C. 2014. Genome editing in rice and wheat using the CRISPR/Cas system. *Nat. Protoc.* 9, 2395.

Simee W. 2011. Isolation and determination of anti-nutritional compounds from root to shells of peanut (Arachis Hypogaea). *J Disper. Sci. Technol.* 28, 341–347.

Smyth, S., Mchughen, A. 2008. Regulating innovative crop technologies in Canada: the case of regulating genetically modified crops. *Plant Biotechnol. J.* 6, 213–225.

Spicer, A., Molnar, A. 2018. Gene editing of microalgae: scientific progress and regulatory challenges in Europe. *Biology* 7, 21.

Sprink, T., Eriksson, D., Schiemann, J., Hartung, F. 2016. Regulatory hurdles for genome editing: process-vs. product-based approaches in different regulatory contexts. *Plant Cell Rep.* 35, 1493–1506.

Sultan, I., Rahman, S., Jan, A.T., Siddiqui, M.T., Mondal, A.H., Haq, Q.M.R. 2018. Antibiotics, resistome and resistance mechanisms: a bacterial perspective. *Front. Microbiol.* 9, 2066. doi: 10.3389/fmicb.2018.02066

Tagliabue, G. 2017. Product, not process! Explaining a basic concept in agricultural biotechnologies and food safety. *Life Sci. Soc. Policy* 13, 3.

Taning, C.N., Arpaia, S., Christiaens, O., Dietz-Pfeilstetter, A., Jones, H., Mezzetti, B., et al. 2019. RNA-based biocontrol compounds: current status and perspectives to reach the market. *Pest Manage. Sci.* 76(3), 841–845.

Velimirov, A., Binter, C., Zentek, J. 2008. Biological effects of transgenic maize NK603xMON810 fed in long-term reproduction studies in mice. *Food Standards Australia New Zealand*, April 2010. Available from:http://www.biosicherheit.de/pdf/aktuell/zentek_studie_2008.pdf

Vogt, D.U., Parish, M., Division, D.S.P. 2001. Food biotechnology in the United States: Science, regulation and issues: Congressional Research Service, Library of Congress. https://www.everycrsreport.com/reports/RL30198.html.

Wang, P., Zhang, J., Sun, L., Ma, Y., Xu, J., Liang, S., et al. 2018. High efficient multisites genome editing in allotetraploid cotton (Gossypium hirsutum) using CRISPR/Cas9 system. *Plant Biotechnol. J.* 16, 137–150.

Whelan, A.I., Lema, M.A. 2015. Regulatory framework for gene editing and other new breeding techniques (NBTs) in Argentina. *GM Crops Food* 6, 253–265. doi: 10.1080/21645698.2015.1114698.

Wolt, J.D. 2017. Safety, security, and policy considerations for plant genome editing. *Progress in Molecular Biology and Translational Science.* Elsevier, pp. 215–241.

Wolt, J.D., Keese, P., Raybould, A., Fitzpatrick, J.W., Burachik, M., Gray, A., et al. 2010. Problem formulation in the environmental risk assessment for genetically modified plants. *Transgenic Res.* 19, 425–436.

Wolt, J.D., Wang, K., Yang, B. 2016. The regulatory status of genome-edited crops. *Plant Biotechnol. J.* 14, 510–518.

Woo, J.W., Kim, J., Kwon, S.I., Corvala´n, C., Cho, S.W., Kim, H., et al. 2015. DNA free genome editing in plants with preassembled CRISPR-Cas9 ribonucleoproteins. *Nat. Biotechnol.* 33, 1162–1164.

Yousuf, N., Dar, S.A., Gulzar, S., Nabi, S.U., Mukhtar, S., Lone, R.A. 2017. Terminator technology: perception and concerns for seed industry. *Int. J. Pure App. Biosci.* 5(1), 893–900.

Zannoni, L. 2019. Evolving regulatory landscape for genome-edited plants. *CRISPR J.* 2, 3–8.

Zhang, D., Hussain, A., Manghwar, H., Xie, K., Xie, S., Zhao, S., et al. 2020. Genome editing with the CRISPR-Cas system: an art, ethics and global regulatory perspective. *Plant Biotechnol. J.* 18, 1651–1669.

Zhang, Y., Liang, Z., Zong, Y., Wang, Y., Liu, J., Chen, K., et al. 2016. Efficient and transgene-free genome editing in wheat through transient expression of CRISPR/ Cas9 DNA or RNA. *Nat. Commun.* 7, 1–8.

Abbreviations

GE	Genome Editing
NBT	New Breeding Technologies
EFSA	European Food Safety Authority
GM	Genetically modified
RNAi	RNA interference
miRNA	MicroRNA
dsRNA	Double-stranded RNA
TALENs	Transcription activator-like effector nucleases
ZFNs	Zinc finger nucleases
CRISPR	Clustered Regularly Interspaced Short Palindromic Repeats
EPA	Environmental Protection Agency
EFSA	The European Food Safety Authority

11 Role of bioinformatics databases in functional genomics and metabolic engineering researches

Mohammed Ali, M.A. Al-kordy, and Ahmad M. Alqudah

11.1 Introduction

Many years ago, scientists had limited knowledge of the primary and secondary metabolic pathways in many plants, with few known genes encoding enzymes in the pathways. Until now few genes were cloned from a more traditional approach to identify and purify the enzyme, followed by cloning these encoding genes. In 1999, Somerville and Somerville predestined that about 1,000 plant genes have a function that was already known at this time (Somerville and Somerville, 1999). Moreover, a review by Henry (2022) reported that 1,035 plant genome sequences had been reported (NCBI: https://www.ncbi.nlm.nih. gov/genome/browse#!/overview/plants) (accessed 27 November 2022), which shows that a lot of plant genomics have been isolated and sequenced, and read and analyzed with high quality until now. The NCBI database contains ~986 genomes from flowering plant, ~6,449 at the chromosome level, ~236 organelles, and ~2243 assemblies (https://www.ncbi.nlm.nih.gov/genome/browse# !/overview/flowering%20plants) (accessed 27 November 2022). Furthermore, the role of functional genomics demands to decode the function of the gene is unknown. Moreover, prediction and estimation of the functions of unknown genes found in large-scale sequencing projects (Genomic DNA sequencing) by alignments between the amino acid and nucleotide sequence and other documented sequences at various genomics databases (Chen et al. 2018) (see Figure 11.1). However, so far, we have found many sequences that do not have a function. As such, several advanced tools can be used to predict gene roles. For example, aligning proteins and peptides, comparing 3D-folded proteins, and determining the location of active domains within proteins (Ali et al. 2021). Finally, by using bioinformatics tools and computation biology programming for data analysis we can predict the function of target genes. In the last decade, numerous studies have used Next-Generation-Sequencing (NGS) technology for DNA (DNA-Seq) and RNA sequencing (RNA-Seq) as a robust method for identifying and discovering many genes, related to various primary and secondary pathways in models and non-model plants with unknown genomic sequences (Ward et al. 2012) (see Figure 11.1). For example, terpenes and terpenoids biosynthesis pathway in *Salvia officinalis* L,

DOI: 10.4324/9781003382102-11

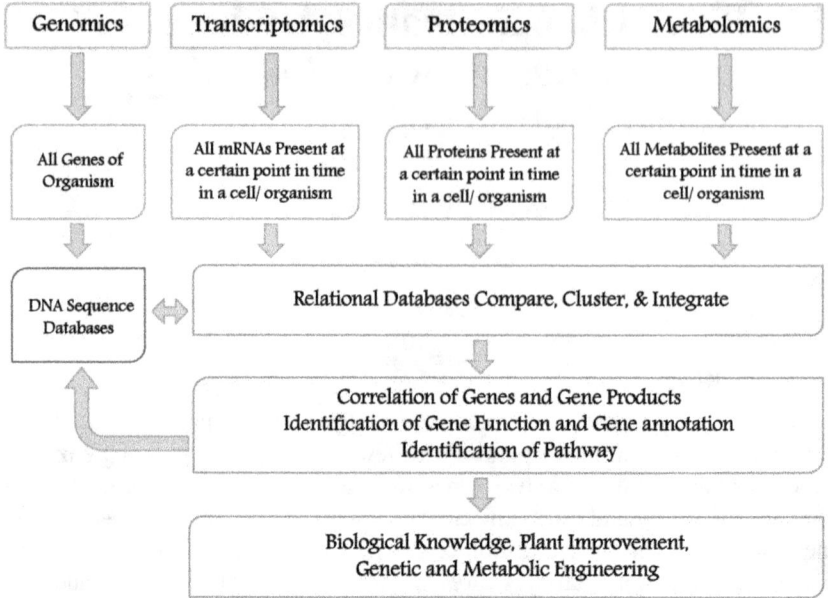

Figure 11.1 Functional genomics concerns the determination of the function of each gene; it thus should teach us the relationship between the metabolite and the genes.

Salvia guaranitica L., and *Fragaria nilgerrensi* (Mehmood et al. 2021; Ali et al. 2018, 2017), biosynthesis of flavonoid and carotenoid in Safflower (*Carthamus tinctorius*) and *Momordica cochinchinensis* (Hyun et al. 2012; Li et al. 2012), Cellulose and Lignin biosynthesis in Chinese fir (*Cunninghamia lanceolata*) (Huang et al. 2012).

In recent years, bioinformatics databases have greatly developed and proliferated rapidly, and become a key for the biologist's everyday toolbox (Chen et al. 2018; Choudhuri, 2014). With the advent of fast Internet connections and the World Wide Web, the data and information presented in the databases and analysis tools can be accessed freely, easily, and quickly (Dash et al. 2016; Poliakov et al. 2014). Bioinformatics is a rapidly evolving discipline depending on the use of computers in biology research for big data analysis (Spannagl et al. 2016; Kersey et al. 2014; Proost et al. 2009; Duvick et al. 2008; Lyons, 2008).

Moreover, there are several reasons to search on bioinformatics databases, for instance: (i) when obtaining a new DNA, RNA, and protein sequence, one needs to know whether it has already been submitted in the databanks fully or partially, or whether they contain any homologous sequences with other sequences which have already been deposited in the databanks, (ii) search for homologous protein domains – protein domains similar in their sequence and active cite, and therefore also in their presumed folding or function or 2D- and

3D structure, (iii) some of the bioinformatics databases contain fully or partially annotation which has already been added to a specific sequence. This annotation can provide informative information for the searched sequence or its homologous sequences, (iv) prediction of the putative expression in cell and tissue for candidate genes involved in major primary and secondary metabolite biosynthetic pathways, (v) annotation of the functional classification for any sequences by assembled and mapped the transcripts into the reference KEGG pathways database, (vi) redraw the metabolic networks maps based on the metabolic encyclopedia of enzymes and various metabolic pathways, and (vii) prediction the cytotoxicity for phytochemical compounds by validation of the interactions between these phytochemical compounds and human proteins and metabolites. This chapter focuses on the use of bioinformatics databases in research related to functional genomics and metabolic engineering. In the end, this chapter will provide vast informative data besides facilitating functional genomics and genome editing studies in various plant species that are used as a source of global food security.

11.1.1 The bio-analytic resource database

The Bio-Analytic Resource (BAR; http://bar.utoronto.ca) is a portal for accessing large data sets from more than 30 different plant species, Mouse and Human. These data were collected from a friendly web based on transcriptomic, protein-protein interaction, functional genomics, promoter data, and other data for hypothesis generation and confirmation (Toufighi et al. 2005). Data sets include: ~150 million gene expression measurements, 70,944 predicted protein-protein interactions, ~2.8 million protein-DNA interactions, 29,180 predicted protein tertiary structures, and 11, 700 proteins with documented subcellular localizations. The BAR's "electronic Fluorescent Pictograph" (eFP) Browser (Winter et al. 2007) for exploring gene expression data initially from Arabidopsis (*Arabidopsis thaliana*), then the eFP has been adapted for more than 30 other plant species such as (Maize, Poplar, Tomato, Camellia, Soybean, Potato, Barley, Medicago, Eucalyptus, Rice, Willow, Sunflower, Cannabis, Wheat, Sugarcane, Arachis, Grape, Kalanchoe, Actinidia, Brassica, Strawberry, Triticale, Brachypodium, Little Millet, Physcomitrella, Selaginella, Phelipanche, Striga, Triphysaria, *Eutrema salsugineum,* and *Cannabis sativa* Plants (see Table 11.1). Here, this chapter will highlight the BAR tools for querying proteomics and transcriptomic data sets, which permit the exploration of linked transcriptomic, metabolomic, and enzymatic activity data in different plant species at various developmental stages of tissues and organs. A recent BAR tool, the "expressologs", link between sequence and transcriptional data in a translational manner to identify the most likely "expressologs" for our target query gene (plant homologs showing the most similar pattern of expression in equivalent tissues) (Patel et al. 2012). This chapter also focuses on other BAR tools for exploring protein-protein interactions, promoter analyses, and improving function prediction in Arabidopsis and tomato for extracting even

Table 11.1 Data visualization and analytic tools available on the BAR

Tool	URL	Reference
ePlants		
ePlant Arabidopsis	https://bar.utoronto.ca/eplant/	Waese et al., 2017
ePlant Maize	https://bar.utoronto.ca/eplant_maize/	Waese et al., 2017
ePlant Poplar	https://bar.utoronto.ca/eplant_poplar/	Waese et al., 2017
ePlant Tomato	https://bar.utoronto.ca/eplant_tomato/	Waese et al., 2017
ePlant Camelina	https://bar.utoronto.ca/eplant_camelina/	Waese et al., 2017
ePlant Soybean	https://bar.utoronto.ca/eplant_soybean/	Waese et al., 2017
ePlant Potato	https://bar.utoronto.ca/eplant_potato/	Waese et al., 2017
ePlant Barley	https://bar.utoronto.ca/eplant_barley/	Waese et al., 2017
ePlant Medicago	https://bar.utoronto.ca/eplant_medicago/	Waese et al., 2017
ePlant Eucalyptus	https://bar.utoronto.ca/eplant_eucalyptus/	Waese et al., 2017
ePlant Rice	https://bar.utoronto.ca/eplant_rice/	Waese et al., 2017
ePlant Willow	https://bar.utoronto.ca/eplant_willow/	Waese et al., 2017
ePlant Sunflower	https://bar.utoronto.ca/eplant_sunflower/	Waese et al., 2017
ePlant Cannabis	https://bar.utoronto.ca/eplant_cannabis/	Waese et al., 2017
ePlant Wheat	https://bar.utoronto.ca/eplant_wheat/	Waese et al., 2017
ePlant Sugarcane	https://bar.utoronto.ca/eplant_sugarcane/	Waese et al., 2017
Arabidopsis eFP Browsers	https://bar.utoronto.ca/efp/cgi-bin/efpWeb.cgi	Winter et al. 2007
Arabidopsis eFP Browser	https://bar.utoronto.ca/efp_arabidopsis_lipid/cgi-bin/efpWeb.cgi	Kehelpannala et al. 2021
Arabidopsis Lipid Map eFP Browser		
Cell eFP Browser	https://bar.utoronto.ca/cell_efp/cgi-bin/cell_efp.cgi	Winter et al. 2007
Arabidopsis Seed Coat eFP Browser	https://bar.utoronto.ca/efp_seedcoat/cgi-bin/efpWeb.cgi	Dean et al. 2011
Arabidopsis Translatome eFP Browser		Mustroph et al. 2009
Arabidopsis Spatio-Temporal Root Stress eFP Browser	https://dinmenylab.info/browser/query	Geng et al. 2013

Expression Anglers and other Expression Browsers

Name	URL	Reference
Expression Angler 2016	https://bar.utoronto.ca/ExpressionAngler/	Austin et al. 2016
Legacy Expression Angler	https://bar.utoronto.ca/ntools/cgi-bin/ntools_expression_angler.cgi	Toufighi et al. 2005
Poplar Expression Angler	https://bar.utoronto.ca/eapop/cgi-bin/ntools_expression_angler.cgi?pub=	
Sample Angler	https://bar.utoronto.ca/ntools/cgi-bin/ntools_sample_angler.cgi	
e-Northerns w. Expression Browser	https://bar.utoronto.ca/affydb/cgi-bin/affy_db_exprss_browser_in.cgi	Toufighi et al. 2005
Poplar Expression Browser	https://bar.utoronto.ca/ebpop/cgi-bin/pop_db_exprss_browser_in.cgi?pub=	

Monocot eFP Browsers

Name	URL	Reference
Maize eFP Browser	http://bar.utoronto.ca/maizeefp/	Wang et al. 2014 & Li et al. 2010
Rice eFP Browser	http://bar.utoronto.ca/efprice/cgi-bin/efpWeb.cgi	
Barley eFP Browser	http://bar.utoronto.ca/efpbarley/cgi-bin/efpWeb.cgi	

Monocot eFP Browsers

Name	URL	Reference
Brachypodium eFP Browser	https://bar.utoronto.ca/efp_brachypodium/cgi-bin/efpWeb.cgi	Sibout et al. 2017
Wheat eFP Browser	https://bar.utoronto.ca/efp_wheat/cgi-bin/efpWeb.cgi	Ramírez-González et al. 2018
Little Millet eFP Browser	https://bar.utoronto.ca/efp_little_millet/cgi-bin/efpWeb.cgi	

Dicot eFP Browsers

Name	URL	Reference
Poplar eFP Browser	http://bar.utoronto.ca/efppop/cgi-bin/efpWeb.cgi	Champigny et al. 2013
Medicago eFP Browser	http://bar.utoronto.ca/efpmedicago/cgi-bin/efpWeb.cgi	Kagale et al. 2016
Soybean eFP Browser	http://bar.utoronto.ca/efpsoybean/cgi-bin/efpWeb.cgi	Clevenger et al. 2016
Potato eFP Browser	http://bar.utoronto.ca/efp_potato/cgi-bin/efpWeb.cgi	Fasoli et al. 2012
Tomato eFP Browser	http://bar.utoronto.ca/efp_tomato/cgi-bin/efpWeb.cgi	Laverty et al. 2019 & Van Bakel et al. 2011
E. salsugineum eFP Browser	http://bar.utoronto.ca/efp_eutrema/cgi-bin/efpWeb.cgi	
C. sativa eFP Browser	https://bar.utoronto.ca/efp_camelina/cgi-bin/efpWeb.cgi	
Arachis eFP Browser	https://bar.utoronto.ca/efp_arachis/cgi-bin/efpWeb.cgi	
Grape eFP Browser	https://bar.utoronto.ca/efp_grape/cgi-bin/efpWeb.cgi	
Cannabis eFP Browser	https://bar.utoronto.ca/efp_cannabis/cgi-bin/efpWeb.cgi	

(Continued)

Table 11.1 (Continued)

Tool	URL	Reference
Kalanchoe eFP Browser	https://bar.utoronto.ca/efp_kalanchoe/cgi-bin/efpWeb.cgi	Zhang et al. 2020
Actinidia eFP Browser	https://bar.utoronto.ca/efp_actinidia/cgi-bin/efpWeb.cgi	Brian et al. 2021
Brassica eFP Browser	https://bar.utoronto.ca/efp_brassica_rapa/cgi-bin/efpWeb.cgi	
Strawberry eFP Browser	https://bar.utoronto.ca/efp_strawberry/cgi-bin/efpWeb.cgi	
Other Gene Expression and Protein Tools		
ThaleMine	https://bar.utoronto.ca/thalemine/begin.do	Krishnakumar et al. 2015
eFP-Seq Browser	https://bar.utoronto.ca/eFP-Seq_Browser/	Sullivan et al. 2019
Expressolog Tree Viewer	https://bar.utoronto.ca/expressolog_treeviewer/cgi-bin/expressolog_treeviewer.cgi	Patel et al. 2012
Promomer	https://bar.utoronto.ca/ntools/cgi-bin/BAR_Promomer.cgi	Patel et al. 2012
Cistome	https://bar.utoronto.ca/cistome/cgi-bin/BAR_Cistome.cgi	Austin et al. 2016
PMET	http://nero.wsbc.warwick.ac.uk/tools/user_cases.php	
Arabidopsis Interactions Viewer	https://bar.utoronto.ca/interactions/cgi-bin/arabidopsis_interactions_viewer.cgi	Geisler-Lee et al. 2007
Arabidopsis Interactions Viewer 2.0	https://bar.utoronto.ca/interactions2/	
Rice Interactions Viewer	https://bar.utoronto.ca/interactions/cgi-bin/rice_interactions_viewer.cgi	Ho et al. 2012
Gene Slider	https://bar.utoronto.ca/geneslider/	Waese et al. 2016
Physcomitrella eFP Browser	https://bar.utoronto.ca/efp_physcomitrella/cgi-bin/efpWeb.cgi	
Selaginella eFP Browser	https://bar.utoronto.ca/efp_selaginella/cgi-bin/efpWeb.cgi	
Mouse eFP Browser	https://bar.utoronto.ca/efp_mouse_efp/cgi-bin/efpWeb.cgi	Zhang et al. 2004
Human eFP Browser	https://bar.utoronto.ca/efp_human/cgi-bin/efpWeb.cgi	
Other Gene Expression and Protein Tools		
Phelipanche eFP Browser	https://bar.utoronto.ca/efp_phelipanche/cgi-bin/efpWeb.cgi	
Striga eFP Browser	https://bar.utoronto.ca/efp_striga/cgi-bin/efpWeb.cgi	
Triphysaria eFP Browser	https://bar.utoronto.ca/efp_triphysaria/cgi-bin/efpWeb.cgi	

Molecular Markers and Mapping Tools

Tool	URL	Reference
Next Generation Mapping	https://bar.utoronto.ca/ngm/	Austin et al. 2011
Marker Tracker	http://bar.utoronto.ca/markertracker/	
Blast Digester	http://bar.utoronto.ca/ntools/cgibin/ntools_blast_digester.cgi	Ilic et al. 2004
CapsID	http://bar.utoronto.ca/unavailable.htm	Taylor and Provart, 2006

Other Genomic Tools and Widgets

Tool	URL	Reference
Arabidopsis Citation Network Viewer	http://bar.utoronto.ca/50YearsOfArabidopsis/	Provart et al. 2016
ClustalW with MView Output	https://bar.utoronto.ca/ntools/cgibin/ntools_multiplealign_w_mview.cgi	Provart and Zhu, 2003
DataMetaFormatter	https://bar.utoronto.ca/ntools/cgibin/ntools_treeview_word.cgi	
Heatmapper	https://bar.utoronto.ca/ntools/cgibin/ntools_heatmapper.cgi	
Heatmapper Plus	https://bar.utoronto.ca/ntools/cgibin/ntools_heatmapper_plus.cgi	
Duplicate Remover	https://bar.utoronto.ca/ntools/cgibin/ntools_duplicate_remover.cgi	
Venn Selector	https://bar.utoronto.ca/ntools/cgibin/ntools_venn_selector.cgi	
Venn SuperSelector	https://bar.utoronto.ca/ntools/cgibin/ntools_venn_superselector_geneview_values.cgi	
Random ID list generator	https://bar.utoronto.ca/ntools/cgibin/ntools_random_list_generator.cgi	
AGURR	https://bar.utoronto.ca/agurr/cgi-bin/agurr_main_driver.cgi	
Classification Super Viewer	https://bar.utoronto.ca/ntools/cgibin/ntools_classification_superviewer.cgi	Provart et al. 2003
Medicago Classification Super Viewer	https://bar.utoronto.ca/ntools/cgibin/ntools_classification_superviewer_medicago.cgi	Herrbach et al. 2017
_at to AGI converter	https://bar.utoronto.ca/ntools/cgi-bin/ntools_agi_converter.cgi	
MASTA	https://bar.utoronto.ca/masta/masta.html	
GeneMANIA	https://genemania.org/	
Topo-phylogeny	https://bar.utoronto.ca/Topo-phylogeny/	Waese et al. 2017

more knowledge from the data sets that are available in this database. For example, gene 1, 8-cineole synthase, AT3G25830/TPS-CIN (Ali et al. 2022a), will be used as our "gene of interest" for most example queries presented in this chapter. The BAR database tools are organized into three main sections: Other Genomic Tools and Widgets, Molecular Markers and Mapping Tools, Gene Expression and Protein Tools. The tools are displayed in small tiles and each tile contains three icons, an "Info" button that contains information about the tool, a "Pub" button that is associated with related articles and journal publications, and a "Go" button for applying to that tool. This database used the eFP ("electronic Fluorescent Pictograph") browser for provides easy access to more than 150 million expression measurements from Arabidopsis thaliana samples and other plant species that are presented in Table 11.1. Small pictographs represent the samples (e.g. tissues, organs, and cells) and conditions (e.g. abiotic, biotic, chemical, and hormones) from which the expression data were generated, while expression level values for a given gene within these samples are represented by a color scale.

11.1.1.1 Methods

1 Navigate to https://bar.utoronto.ca and choose the "Arabidopsise Plants Browser" from the BAR's home page.
2 Enter your query gene name or ID. In our case, we enter "AT3G25830" or type "TPS-CIN" for the 1, 8-cineole synthase gene into the Enter a gene name box. Click on "Go".
3 Figure 11.2 shows the output when querying the e Plants Browse using TPS-CIN and the default settings. The expression data at various samples are depicted in a pictographic manner. Higher levels of TPS-CIN steady-state mRNA abundance are detected in green cotyledons followed by curled, cotyledons, and 24 imbibed seeds.
4 With the options available in the data visualization tool with these icons (see Figure 11.2), it is possible to view and predict the expression level of our query gene "TPS-CIN" (or any other gene of interest) in many visualization tools (see Figure 11.2a–k). For example, the tissue & experiment eFP viewer icon tool displays the results of our target gene expression profiling at different tissues, organ and stress condition (e.g. Tissue Specific Root, Abiotic Stress, Abiotic Stress II, Chemical, Tissue Specific Xylem And Cork, Guard Cell Meristemoids, Tissue Specific Embryo Development, Tissue Specific Microgametogenesis, Tissue Specific Pollen Germination, Tissue Specific Stigma And Ovaries, Tissue Specific Guard And Mesophyll Cells, Biotic Stress *Golovinomyces orontii*, Tissue Specific Trichomes, Biotic Stress *Pseudomonas syringae*, Biotic Stress *Hyaloperonospora arabidopsidis*, Biotic Stress *Myzus persicaere*, Biotic Stress Elicitors, Biotic Stress Erysiphe orontii, Biotic Stress *Phytophthora infestans*, Tissue Specific Shoot Apical Meristem, Guard Cell Suspension Cell ABA Response With ROS Scavenger, Biotic Stress *Botrytis cinerea*, Tissue Specific Stem Epidermis,

Select a data visualization tool with these icons.

Buttons to view the raw data and citation and experiment information for this view

Zoom in and zoom out

(b) Gene Information viewer

(c) Publication viewer

(d) Navigator viewer

(e) This Heat Map displays the expression levels across 350+ samples of all genes that are loaded along with the corresponding subcellular localizations of the gene products. Hover over a heat map cell to see associated metadata. Click on a sample to open the relevant eFP viewer.

(f) The Tissue & Experiment eFP viewer displays the results of numerous gene expression profiling experiments. Select a view by clicking on a thumbnail

(g) Pictograph shows subcellular compartments. Locations that are documented or predicted are colored depending on confidence of localization in a given compartment (red = highest confidence)

(h) The Chromosome viewer shows where genes are situated on chromosomes.

Hover over the samples to see their expression level, sample size and standard deviation

Enter an AGI ID or genealias in this box. In our example, the gene 1,8 -cineole synthase, AT3G25830/TPS-CIN

Load coexpressed genes with Expression Angler.

Button allows viewing the AtGenExpress eFP file and Klepikova eFP (RNA-Seq data) file.

Expression Level scale red=higher expression level yellow= lower expression level

The gene panel lists all the currently loaded genes. The green one is active.

The expression level of the currently selected gene is greater than 94% of all other genes.

(k) This is an implementation of JBrowse, a genome browser that lets you explore sequence data from a broad overview all the way down to individual nucleotides. Zoom in and out of your selected gene to explore gene binding sites, polymorphisms and other gene features provided by Araport

(j) This view displays the 3D molecular structure of the protein associated with the selected gene.

(i) The Interactions viewer displays protein interactors for the currently selected gene product from our database of ~80k predicted and ~100k confirmed protein interactions, and 2.8M protein-DNA interactions. Experimentally-determined interactions are shown with green edges. Mouse over edges for more info.

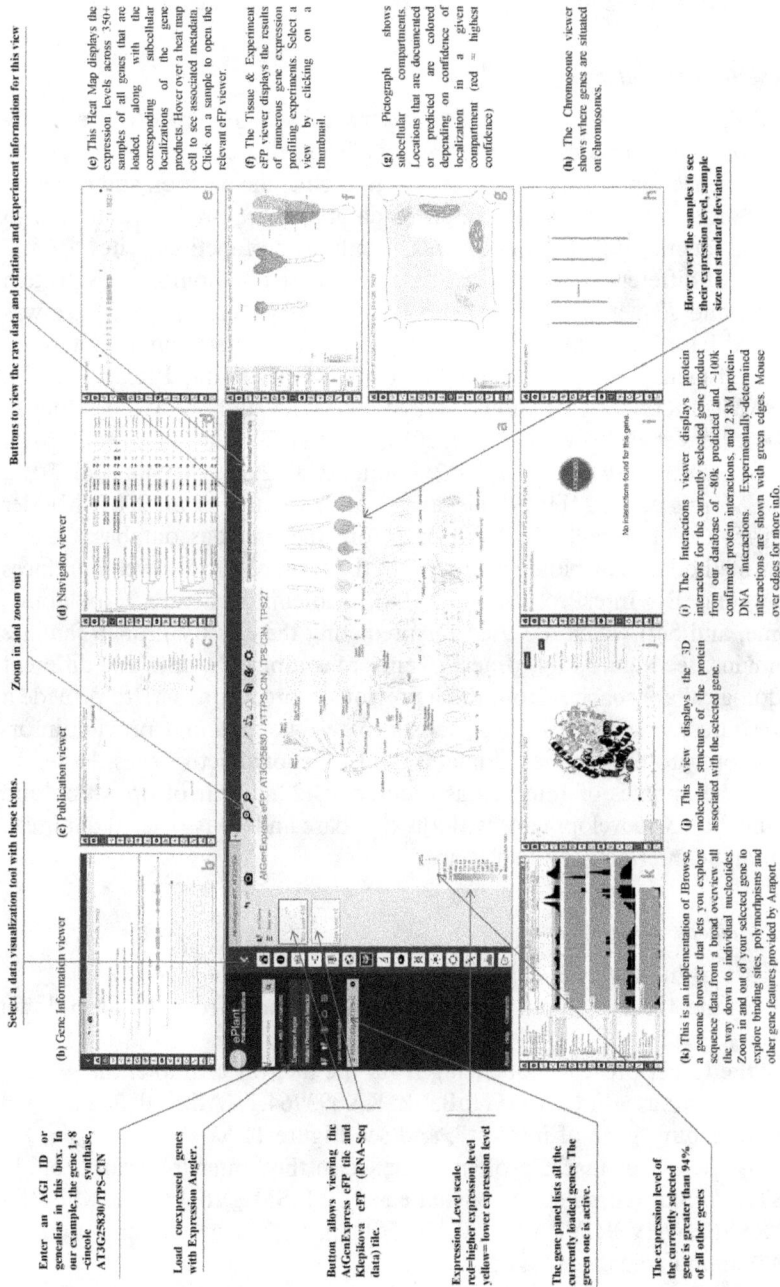

Figure 11.2 Default ePlants Browser views the expression levels of TPS-CIN (AT3G25830) in *Arabidopsis thaliana*. Stronger expression is represented by a darker color. Options for exploring the expression data are highlighted by the callout boxes.

Guard Cell Drought, Guard Cell Mutant And Wild Type Guard Cell ABA Response, *Heterodera schachtii*, DNA Damage, Germination, Root Immunity Elicitation, and Shoot Apex and Single Cell). These previous icon tools can be used to identify where the gene of interest is most strongly expressed (Usadel et al. 2009; Schmid et al. 2005; Nakabayashi et al. 2005).

11.1.2 The InterPro database

The InterPro database (InterPro, http://www.ebi.ac.uk/interpro/) is a freely available web-based encompassing a variety of tools that can be used to classify protein sequences into protein families and to predict the position and presence of important domains and sites. InterPro databases are predictive models to predict and identify the protein domains, families, and functional sites, from a wide range of different protein family databases based on homologous protein sequences, and sequence similarity and function. The InterPro database was constructed based on a large data set from various protein family databases (e.g. CATH-Gene3D, CDD, HAMAP, PANTHER, Pfam, PIRSF, SFLD, TIGRFAMs, PRINTS, PROSITE, SUPERFAMILY and SMART) (Mitchell et al. 2015; Lees et al. 2014; Finn et al. 2014; Mi et al. 2013; Pedruzzi et al. 2013; Sigrist et al. 2013; Attwood et al. 2012; Hunter et al. 2012; Letunic et al. 2012; de Lima Morais et al. 2011; Nikolskaya et al. 2007; Bru et al. 2005; Mulder et al. 2005) (see Figure 11.3). Each one from the previous databases has a different method and/or biological focus, in the form of various web interfaces browsers (e.g By InterPro, Member DB, Protein, Structure, Taxonomy, Proteome, and Set) which are used for predicting the classify protein families and domains (see Figure 11.4). InterPro aims to combine the data and different methodological approaches from these previous source databases to provide a single resource, which helps the scientists to easily access and provide information about protein families, domains, and functional active sites. Here, we report on the updates of InterPro as it enters its 22nd year of operation and give highlight new developments with the database and its associated data sets, programmatic interfaces, and software.

11.1.2.1 Methods

1 Navigate to https://www.ebi.ac.uk/interpro/search/sequence/.
2 Click on the "Search" icon, and select "By sequence" from the drop-down menu
3 Copy the target protein sequencing from the file, for example, the gene 1, 8-cineole synthase, *SgCINS* (GenBank: KX893964.1) (Ali et al. 2022a), will be used as our "gene of interest", and see Figure 11.5a.
4 Enter or paste our target protein sequence in the "InterPro Scan tool" in FASTA format (complete or partial e.g. MCTISMHVSILSKPLNSLHR-SERRSSNSWPVSRIVPAARLRASCSSQL), with a maximum length of 40,000 amino acids, see Figure 11.5b.

The CATH-Gene3D database describes protein families and domain architectures in complete genomes. Protein families are formed using a Markov clustering algorithm, followed by multi-linkage clustering according to sequence identity. Mapping of predicted structure and sequence domains is undertaken using hidden Markov models libraries representing CATH and Pfam domains. CATH-Gene3D is based at University College, London, UK.

Pfam is a large collection of multiple sequence alignments and hidden Markov models covering many common protein domains. Pfam is based at EMBL-EBI, Hinxton, UK.

PIRSF protein classification system is a network with multiple levels of sequence diversity from superfamilies to subfamilies that reflects the evolutionary relationship of full-length proteins and domains. PIRSF is based at the Protein Information Resource, Georgetown University Medical Centre, Washington DC, US.

SFLD (Structure-Function Linkage Database) is a hierarchical classification of enzymes that relates specific sequence-structure features to specific chemical capabilities.

CDD is a protein annotation resource that consists of a collection of annotated multiple sequence alignment models for ancient domains and full-length proteins. These are available as position-specific score matrices (PSSMs) for fast identification of conserved domains in protein sequences via RPS-BLAST. CDD content includes NCBI-curated domain models, which use 3D-structure information to explicitly define domain boundaries and provide insights into sequence/structure/function relationships, as well as domain models imported from a number of external source databases.

HAMAP stands for High-quality Automated and Manual Annotation of Proteins. HAMAP profiles are manually created by expert curators. They identify proteins that are part of well-conserved protein families or subfamilies. HAMAP is based at the SIB Swiss Institute of Bioinformatics, Geneva, Switzerland.

PANTHER is a large collection of protein families that have been subdivided into functionally related subfamilies, using human expertise. These subfamilies model the divergence of specific functions within protein families, allowing more accurate association with function, as well as inference of amino acids important for functional specificity. Hidden Markov models (HMMs) are built for each family and subfamily for classifying additional protein sequences. PANTHER is based at the University of Southern California, CA, US

PRINTS is a compendium of protein fingerprints. A fingerprint is a group of conserved motifs used to characterise a protein family or domain. PRINTS is based at the University of Manchester, UK.

PROSITE is a database of protein families and domains. It consists of biologically significant sites, patterns and profiles that help to reliably identify to which known protein family a new sequence belongs. PROSITE is based at the Swiss Institute of Bioinformatics (SIB), Geneva, Switzerland.

SUPERFAMILY is a library of profile hidden Markov models that represent all proteins of known structure. The library is based on the SCOP classification of proteins: each model corresponds to a SCOP domain and aims to represent the entire SCOP superfamily that the domain belongs to. SUPERFAMILY is based at the University of Bristol, UK.

SMART (a Simple Modular Architecture Research Tool) allows the identification and annotation of genetically mobile domains and the analysis of domain architectures. SMART is based at EMBL, Heidelberg, Germany.

TIGRFAMs is a collection of protein families, featuring curated multiple sequence alignments, hidden Markov models (HMMs) and annotation, which provides a tool for identifying functionally related proteins based on sequence homology. TIGRFAMs was formerly based at the J. Craig Venter Institute (Rockville, MD, US) and is now maintained at NCBI.

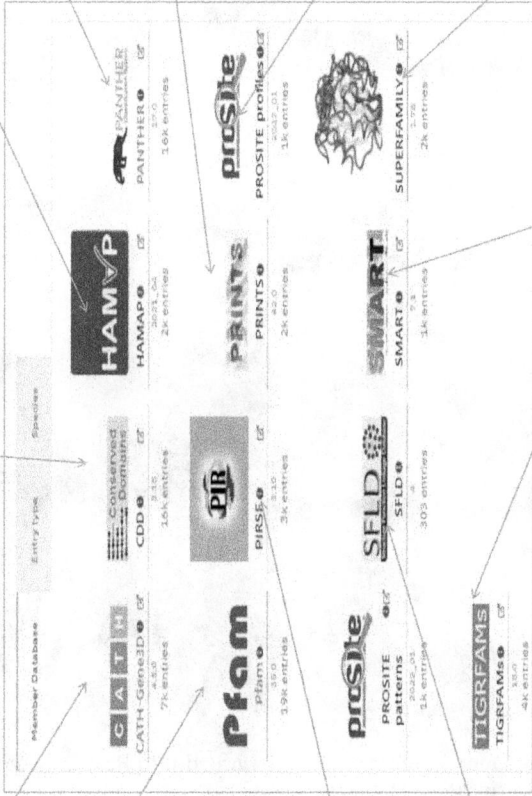

Figure 11.3 *Number* of member database signatures integrated into Protein Sequence Analysis and Classification Database (InterPro).

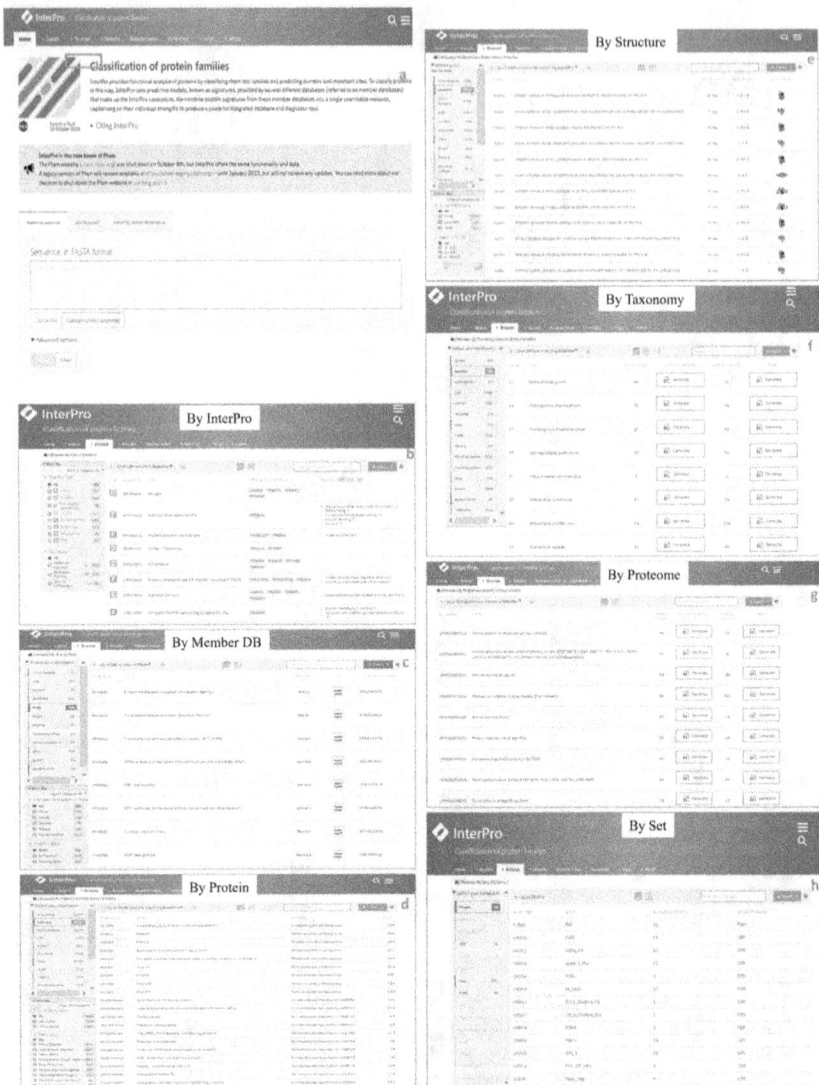

Figure 11.4 Different search browsers integrated into Protein Sequence Analysis and Classification Database (InterPro).

5 Click on the "Search" icon to perform the search, or click on the "Clear" icon to clean the "Enter your sequence" box, see Figure 11.5b.

6 After a few mines, the "Your InterProScan Search Results" web page will present with the green right symbol "√" in the status Column, which means the search was successful and the job finished, see Figure 11.5c.

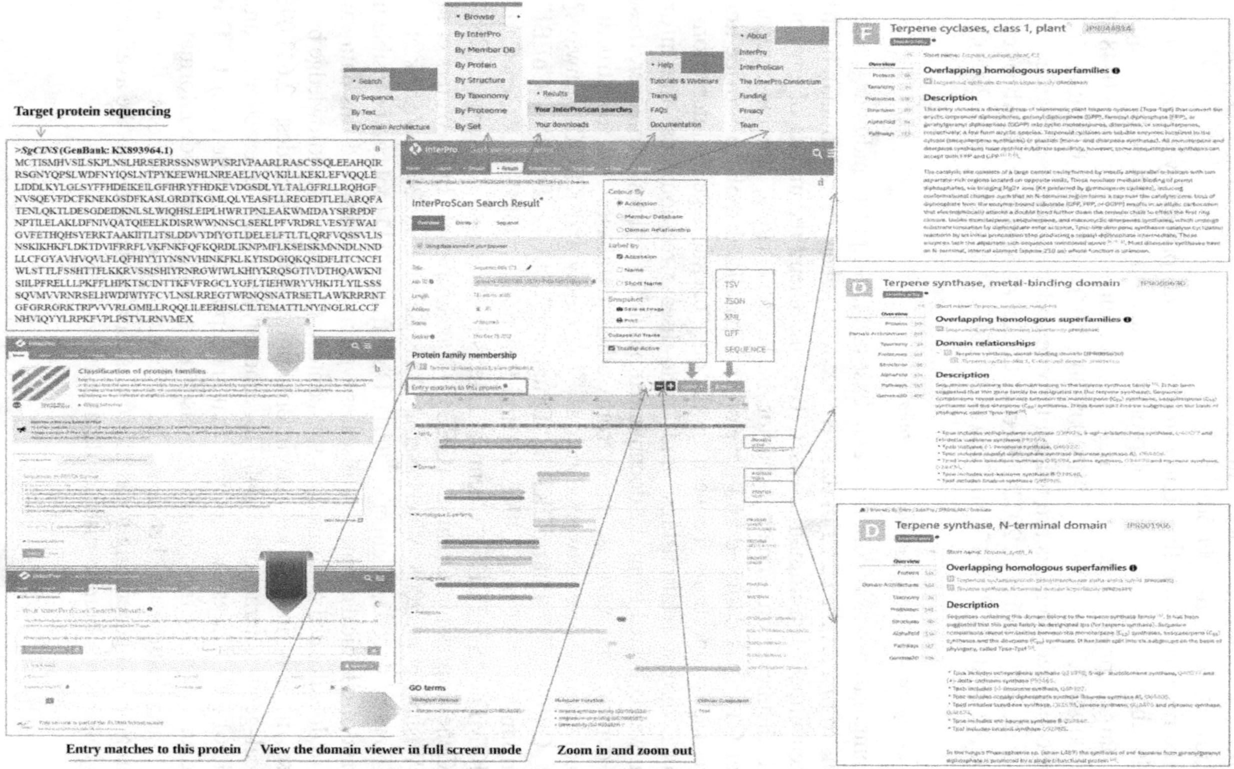

Figure 11.5 InterPro matches for UniProtKB entry 1, 8-cineole synthase, SgCINS (GenBank: KX893964.1) showing predicted protein family membership, domains, and sites.

7 By chick in the green right symbol "√" or in the result title, another web page will open, which provides more information about the length of amino acids "in our case it was 731 aa" and see Figure 11.5d.

8 In the part of "Entry matches to this protein" it has many options such as; viewing the domain viewer in full-screen mode, zoom in and zoom out, and more options to customize the protein viewer (e.g Colour By, Label by, and Snapshot), Export data with different extensions (e.g. TSV, JSON, XML, and GFF), see Figure 11.5d.

9 On the right side from the "Entry matches to this protein" you will find the results about the classification of protein families and domains, for example, our target protein sequence has six domains such as; "IPR044814", "IPR001906", "IPR005630", "IPR008930", "IPT08949", and "IPR036965", these conserved domains are related terpene and terpenoid, see Figure 11.5d (Ali et al. 2022a; Ali et al. 2022b; Ali et al. 2022c; Zhao et al. 2021; Abbas et al. 2019; Ali et al. 2018, 2017; Su-Fang et al. 2014; Degenhardt et al. 2009).

11.1.3 Comparative Toxicogenomics Database

The Comparative Toxicogenomics Database (CTD; http://ctd.mdibl.org) offers a means to predict the effect of different recognize chemicals on human health and their impacts on human diseases. To comprehend the impact of environmental chemicals on human behavior health and safety, the bio-curators team at CTD has designed and developed unparalleled tools to provide and predict the connections between both of: "chemical–gene interactions", "chemical–disease relationships" and "gene–disease relationships" this information was collected from the documented published articles. Moreover, these tools depend on various data from other databases for constructing chemical–gene–disease networks such as "Taxonomy and KEGG pathways", "Gene Ontology", "PharmGKB", "DrugBank", "ArrayTrack", "ChEBI", "ChemBank", "PubChem", "KEGG and Reactome", "OMIM", "GeneCards and HGMD", "STITCH", "Chemical Effects in Biological Systems (CEBS)", and "Gene Expression Omnibus (GEO)" (Kanehisa et al. 2008; Waters et al. 2008; Wishart et al. 2008; Wheeler et al. 2008; Seiler et al. 2008; Vastrik et al. 2007; Tong et al. 2003; Klein et al. 2001). Data from CTD tool can be visible in the form of chemicals, genes, or diseases with more details, and this data can easily to downloaded and saved. In this context, CTD acts as a knowledge base because it contains a lot of information, which includes: 2,643,098 chemical–gene interactions, 348,716 phenotype–based interactions, and 31,394,161 gene–disease associations that are captured (see Table 11.2). CTD is unique and different from other databases in its ability to provide information about the relationship between the chemical–gene interactions through reporting or/and generating novel inferences data, which may not be available through the other human diseases databases.

Table 11.2 CTD curated data status

Curated data	Number[a]
Chemical–gene interactions (curated)	2,643,098
Unique chemicals	14,396
Unique genes	54,190
Unique organisms	622
Phenotype-based interactions (curated)	348,716
Unique chemicals	9,392
Unique genes	3,674
Unique GO terms	6,276
Unique anatomical terms	961
Unique organisms	342
Gene–disease associations	31,394,161
Curated	41,728
Unique genes	8,886
Unique diseases	5,862
Inferred	31,352,433
Unique genes	53,899
Unique diseases	3,243
Chemical–disease associations	3,250,731
Curated	228,315
Unique chemicals	10,263
Unique diseases	3,287
Inferred	3,022,416
Unique chemicals	14,133
Unique diseases	5,852
Chemical–GO associations (enriched)	6,210,137
Chemical–pathway associations (enriched)	1,472,422
Disease–pathway associations (inferred)	303,441
Gene-gene interactions	1,229,718
Gene–GO annotations	1,429,404
Gene–pathway annotations	135,783
GO–disease associations (inferred)	2,877,702
Chemicals with curated data	17,174
Diseases with curated data	7,281
Via OMIM curation	4,135
Genes with curated data	54,279
Via OMIM curation	3,629
Curated references	141,522
Curated Exposure Statements	204,466
Curated Exposure References	3,300

Note
[a] As of November 2022.

11.1.3.1 Methods

1 Go to https://ctdbase.org/;jsessionid=9815EAE375F17226BA1AF6FCD-01EC44F and click on "Keyword Search" and from the drop-down pick-list at the Keyword Search Box you can select any term-of-interest (see Table 11.3), for example, in our case we select "Chemical" and we write the

Table 11.3 The vocabularies used by bio-curators for search/browsing in the CTD
database to facilitate querying

Vocabularies	Description
Chemicals	The CTD chemical vocabulary was derived from a modified subset of the chemicals and supplementary concepts in the 'Drugs and Chemicals' category of Medical Subject Headings (MeSH) from the U.S. National Library of Medicine (Sewell, 1964). You can view diverse relationships in the chemical hierarchy, and access detailed information, including structure, toxicology data, and associated genes, GO terms, diseases, pathways, and references
Diseases	The CTD disease vocabulary comprises terms from the disease subset of the U.S. National Library of Medicine (NLM), Medical Subject Headings (MeSH®), combined with genetic disorders from the Online Mendelian Inheritance in Man® (OMIM®) database. CTD biocurators mapped OMIM diseases to terms within the hierarchical MeSH disease vocabulary to expand our disease representation. The disease term is used to browse relationships in the human disease hierarchy and access detailed information, including associated chemicals, genes, pathways, and references.
Genes	CTD uses official gene vocabulary (e.g; symbol, synonym, accession ID, organism taxon, chemical, interaction type, disease, or Gene Ontology annotation) from the National Center for Biotechnology Information's (NCBI) Entrez-Gene database (Wheeler et al. 2008). You can view diverse information about genes/proteins from diverse vertebrates and invertebrates, including curated interacting chemicals, curated and inferred disease relationships, and associated pathways and functional annotations. You can browse genes, or access them using the Keyword search or by formulating advanced queries.
Chemical–Gene/Protein Interactions	To improve understanding about the mechanisms of chemical actions, we manually curate chemical–gene and –protein interactions in vertebrates and invertebrates from the published literature. These interactions are both direct (e.g. "chemical binds to protein") and indirect (e.g. "chemical results in increased phosphorylation of a protein" via intermediate events). You can search for cross-species chemical–gene and protein interactions curated from the published literature. Interactions may be retrieved by chemical, interaction type, gene, organism, or Gene Ontology annotation.
Chemical–Phenotype Interactions	A CTD chemical-phenotype interaction statement includes 8 types of data (C-Q-E-A-T-M-S-P) annotated using 8 controlled vocabularies, including, at a minimum: C, a chemical from the CTD Chemical Vocabulary; Q, a CTD action qualifier that reflects the direction of the

(Continued)

Table 11.3 (Continued)

Vocabularies	Description
	interaction ("increases", "decreases", or "affects", when not specified by the authors); E, the entity phenotype from GO; A, an anatomical term from the MeSH "Anatomy [A]" branch; T, an organism from NCBI Taxonomy; M, a CTD method code (in vivo, in vitro); S, the CTD information source code (abstract, full text); and P, the article identifier (PMID) from NCBI PubMed. "Not reported" is allowed for both taxon and anatomy fields if the authors do not provide this information. By this term, you can search for cross-species chemical–phenotype interactions curated from the published literature. Interactions may be retrieved by chemical, interaction type, Gene Ontology (GO) termed as a phenotype vocabulary term, and/or organism.
Gene Ontology (GO)	Gene Ontology (GO) annotations are integrated with gene data in CTD. In addition, you can browse GO and use it to access detailed information about genes/proteins that interact with a chemical displayed for each chemical, and annotated genes and associated diseases.
Pathways	KEGG and REACTOME pathway data describe known molecular interaction and reaction networks. These data are integrated with chemicals, genes, and diseases in CTD to provide insights into molecular networks that may be affected by chemicals, and possible mechanisms underlying environmental diseases. You can browse pathways and access detailed information, including annotated genes and associated diseases.
Organisms	The CTD organism vocabulary consists of the Eumetazoa portion (vertebrates and invertebrates) of the NCBI Taxonomy database. You can browse organisms, or use them to formulate gene, access associated genes, interactions, or reference queries.
Anatomy	Curating of chemical-induced phenotypes in CTD use controlled vocabularies and a structured format of seven components (chemical, action qualifier, phenotype entity, taxon, anatomy, PubMed reference, and inference network). The anatomy term is used to browse relationships in the anatomy hierarchy and access associated data, including chemical-phenotype interactions for each anatomical term.
References	CTD contains reference articles related to toxicologically significant vertebrate and invertebrate genes, diseases, and associated chemicals. These reference articles are derived from the literature in PubMed and are associated with a unique PubMed identifier. You can browse references and use them to search for reference articles by gene, organism taxon, chemical, chemical–gene interaction type, disease, citation information, or accession ID.

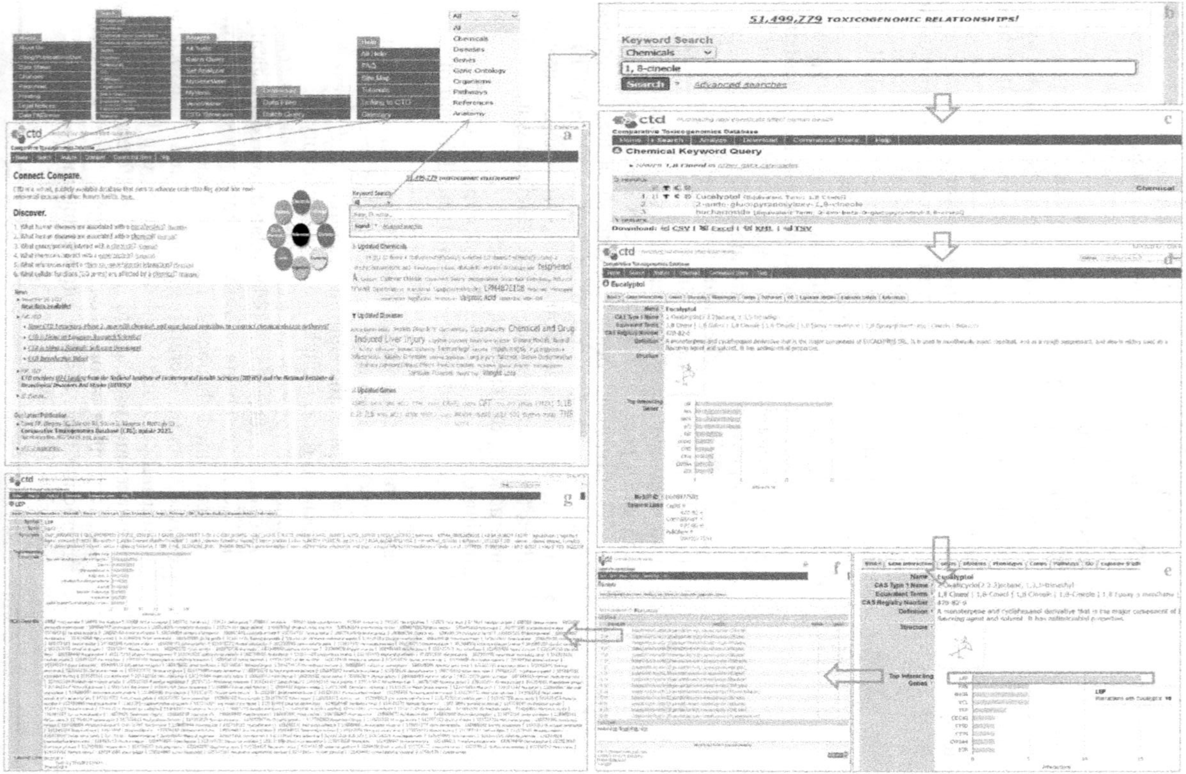

Figure 11.6 The flowchart outline using the Comparative Toxicogenomics Database (CTD) search query tool (https://ctdbase.org/jsessionid=9815EAE375F17226BA1AF6FCD01EC44F) to identify the impact of 1, 8-cineole as a chemical compound on human gene sets which related to human health and diseases.

compound word"1, 8-cineole" that will be used as our "Chemical of interest" see Figure 11.6a and click on "Search".

2 From the Chemical Keyword Query window we select the result number three "Eucalyptol" and click on it see Figure 11.6b and c.

3 Variable selections: clicking on the "Basics" tab shows much information about our chemical compound such as "Name", "CAS Type 1 Name", "Equivalent Terms", "CAS Registry Number", "Definition", "Structure", "Top Interacting Genes", "MeSH® ID", "External Links" and "Ancestors", see Figure 11.6d. Moreover, you can get more information by transitioning between the different windows to another in the horizontal bar.

4 As you see in Figure 11.6e we found our chemical compound "1, 8-cineole" interact with many genes, and the CDT database results listed the most important top ten genes (e.g, *LEP, PXN, RHOA, VCL, TNF, CDC42, CTSD, CTTN, CYP3A4,* and *EZR*), and we can get more information about any gene that have interaction with our target compound by click on the gene name.

5 Gene-interactions content information can be accessed by a click on the gene name *"LEP"*, see Figure 11.6f. A gene-interactions icon identifies valuable information that have chemical-gene-interactions associated data such as the type of interaction which means the effect of this chemical compound on the genes for example "Eucalyptol can inhibit the reaction of *LEP* gene mutant which decreased the expression of CTTN protein", on the other hand, "Eucalyptol can inhibit the reaction of LEP gene mutant and increased the abundance of Blood Glucose level", see Figure 11.6f. And all these data have been supported by many references.

6 Figure 11.6g shows much information about our selected gene such as gene symbol, gene name, synonyms, top interacting chemicals, NCBI Gene IDs for each organism, and external links. And through the previous steps, we can predict the molecular interaction, and reaction networks and provide information about chemical-gene interactions and their roles in various environmental diseases.

11.1.4 Phytozome

Phytozome is a free database for presenting large genetic data sets from ~274 assembled and annotated genomes. This database was designed based on the plant genomics, proteomics, and metabolomics data, collected and sorted from ~139 of different plant species. This Plant Comparative Genomics portal (Phytozome) was designed and created by the Department of Energy's Joint Genome Institute (JGI) at the University of California, Berkeley, California, USA (Goodstein et al. 2012). Phytozome offers various annotations for each gene based on the promoter, protein-protein interaction, and transcriptomic data. The current version of Phytozome (v13.0) has ~103 plant genomics sequences, and the evolutionary history of each gene at the gene structure, enzyme, expression, nucleotide, and amino acids sequence level (https://phytozome-next.jgi.doe.gov/). The phytozome database has a large data set from different plants genomes associated with various databases such as LIS for legumes, Gramene for grasses, TAIR for Arabidopsis, SGN for Solanaceae, and GDR for Rosaceae, GreenPhylDB, Plaza and PlantGDB (Bombarely et al. 2011; Proost et al. 2009; Conte et al. 2008; Swarbreck et al. 2008; Jung et al. 2008; Liang et al. 2008; Gonzales et al. 2005). Furthermore, all gene sets in Phytozome database are annotated using different databases (e.g. KOG, KEGG, PANTHER, ENZYME, PFAM, Pathway, GO, TMHMM, GENE3D, TIGRFAMs, PIRSF and InterPro).

11.1.4.1 Methods

1 From https://phytozome-next.jgi.doe.gov, click on the plant genomic list to select which plant genomic you need to search on it and see Figure 11.7a.
2 If we have the ID name of your target gene for example "*Solyc03g114340*" from *Solanum lycopersicum*, we can use "Find Genes by Keyword" and click on search "magnifying glass mark" to get detail about our target genes at a new window and see Figure 11.7b. At the new window the "Views" we have more details information such as (View gene report; "**G**letter mark"), (Browse genome; "**B** letter mark"), organism "*S. lycopersicum*" and Description "Solyc03g114340.3-(1 of 1) (EC:1.1.1.267" "1-deoxy-D-xylulose-5-phosphate-reeducates", see Figure 11.7c.
3 On the other hand, if we did not have the gene ID and we need to search for a homologous sequence from other organisms in our target plant genomics, we can use "Search by BLAST" and see Figure 11.7a, then we can select and copy our query sequence in FASTA (nucleotide or amino acid) format, for-example (GenBank: FJ476255.1; *S. miltiorrhiza* 1-deoxy-d-xylulose 5-phosphate-reductoisomerase) in form of mRNA as complete cds. https://www.ncbi.nlm.nih.gov/nuccore/FJ476255.1?report=fastaseeFigure11.7d, and past our query sequence in "BLAST Search" box and click on "GO" and see Figure 11.7e.
4 At the BLAST results page, we have many data and information related to our query sequence search results at "HSP Table View" such as "View genome hit in JBrowser, Species, E-value, % Identity, Align len, Strands, Query ID, Query from, Query to, Target from, Tatget to, Bitscore, # Identical, Positives, Gaps, Query len. and Target len". Furthermore, all these previous data can be downloaded by clicking on "Export" icon and see Figure 11.7f.
5 By clicking on the view genome hit in JBrowser "B letter mark" from the "HSP Table View" we can go to a new window to see our target gene at the view genome hit in JBrowser and see Figure 11.7g. And by clicking on "Transcript" from the right window we will get more information and details related to our query sequence and see Figure 11.7h.

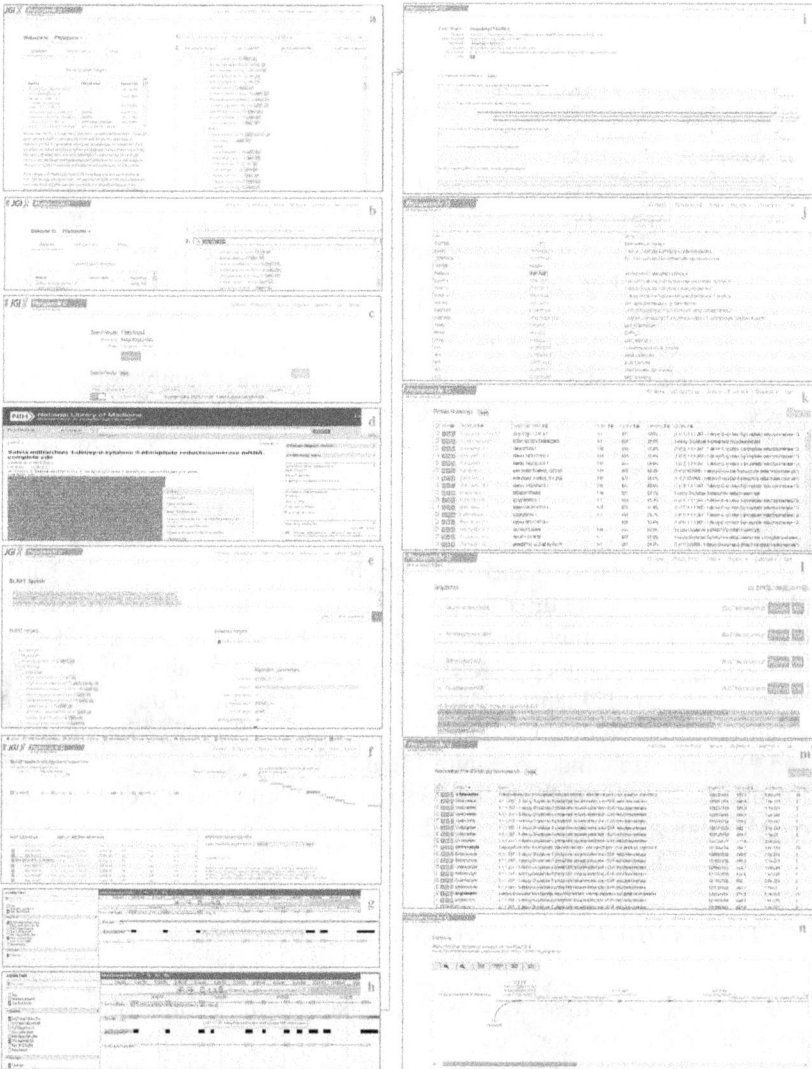

Figure 11.7 The flowchart of the Phytozome database approach to annotate the function of "Solyc03g114340" from Solanum lycopersicum based on the promoter, protein-protein interaction, and transcriptomic data stored in this database.

6 After clicking on "Transcript" a new window enroll "Gene Report" will present informative information for example, Functional Annotations which show the type, position, and the numbers of domains and motifs (e.g, IPR013512, IPR003821, IPR013644, IPR016040, and IPR026877) and see Figure 11.7i.

7 By scrolling the ruler down on the right side of the page, small windows containing information will appear as you see in Figure 11.7j. For example, the window in Figure 11.7j has information related to our target gene on various databases such as "ENZYME, PIRSF, TMHMM, TIGRFAMs, Pathway, InterPro,... ... ...".

8 To show "Protein Homologs" term details, scroll down, Click "Views" on one of the three types of letters [(View gene report; "**G** letter mark"), (Browse genome; "**B** letter mark") and (View pair wise BLAST alignment; "**A** letter mark")] to get detailed information about Protein Homologs, the data have been arranged into "Organism", "Transcript Name", "Ortho", "Score", "Similarity" and "Define" and see Figure 11.7k.

9 Then by scrolling the ruler down, we can present more details about the term "Sequence". The target sequence is shown with a number of engaged bases pare such as "Genomic sequence [5546]", "Transcript sequence [1905]", "CDS sequence [1428]" and "Peptide sequence [476]" (see Figure 11.7). The Exons "CDS" zone of our target gene is marketed in blue color, and the "5'UTR" and "3'UTR" are marketed in green and peach color, respectively.

10 To show "Associated PlantFAMs via hmmsearch" term details, Click "Views" on one of the three types of letters [(View family report; "**F** letter mark"), (View family membership; "**M** letter mark"), and (View hmmsearch text alignment; "**A** letter mark")] to get a detailed information about Associated PlantFAMs via hmmsearch, the data have been arranged into "Node", "Name", "Family ID", "Score", "Significance" and "Family size" and see Figure 11.7m.

11 Finally, we have a small window to show the "Pathway". And by clicking on "Attributes" we can select which type of data we need to present in our results (see Figure 11.7n).

11.1.5 *KEGG PATHWAY: Kyoto encyclopedia of genes and genomes pathway*

KEGG PATHWAY (https://www.genome.jp/kegg/pathway.html#genetic) is a freely available database for the analysis of gene and protein functions based on the genomics proteomics and metabolomics information stored in this database. This analysis information is presented in graphical representations such as cell cycle, membrane transport, signal transduction, and cellular processes. (Tanabe and Kanehisa, 2012; Schmid and Blaxter, 2008; Aoki-Kinoshita and Kanehisa, 2007; Moriya et al. 2007; Masoudi-Nejad et al. 2007; Kanehisa et al. 2006; Hashimoto et al. 2006; Wu et al. 2006; Mao et al. 2005; Kanehisa, 2002; Kanehisa and Goto, 2000). KEGG PATHWAY is categorized into seven major divisions, known as "Metabolism", "Genetic Information Processing", "Environmental Information Processing", "Cellular Processes", "Organismal Systems", "Human Diseases", and "Drug Development". Each of the previous divisions provides many details about genomes information and their relationships with biological systems and their interactions with the environment, in the form of collection pathway maps which were drawn by Java graphics tools for the presentation of genome maps. These maps will help us for understand the manipulating expression maps, comparing two genome maps, path computation, sequence, and graph comparison, the molecular interaction, reaction, and relation networks between the genes and other molecules. This subchapter introduces a brief description about KEGG PATHWAY and its various tools which can be used for understanding the functions and utilities of genomics in cells and organisms.

11.1.5.1 Methods

1 Go to the KEGG webpage (https://www.genome.jp/kegg/) and see Figure 11.8a. And from the "Databases" drop-down list, select the pathway to go forward to KEGG PATHWAY Database (https://www. genome.jp/kegg/pathway.html) and see Figure 11.8b.
2 In the "KEGG PATHWAY Database" page you can search for the name of any Gene or protein or enzyme, by pasting the name into the "Enter keywords" box, then press the "Go" button to proceed with the search.
3 As shown in Figure 11.8c, users can browse a pathway for any Gene or protein or enzyme; by following a hierarchical classification scheme that depends on ontology and enzyme function. For example, for searching on DXR gene "1-deoxy-D-xylulose-5-phosphate reductoisomerase", we write the short name of this gene on the "Enter keywords" box, then press the "Go" button to precede the search. After that, you can get several pathways that represent the presence of this gene such as; "Terpenoid backbone biosynthesis (map00900)", "Metabolic pathways (map01100)", and "Biosynthesis of secondary metabolites (map01110)".
4 By clicking on the "Terpenoid backbone biosynthesis (map00900)" map, a new window with the title "Terpenoid backbone biosynthesis - Reference pathway", which contains our target gene in highlighted crimson box with the Enzyme Commission number (EC number; 1.1.1.267) and see Figure 11.8c. Through this pathway map, we can determine the sequence and position of our target gene in the pathway of terpene synthesis, and their relationship with other genes in the same pathway.
5 Clicking on the highlighted crimson box "1.1.1.267" will open three tiles containing more information related to "ORTHOLOGY: K00099", "ENZYME: 1.1.1.267" and "REACTION: R05688" and see Figure 11.8d–f.
6 The first tile "ORTHOLOGY: K00099" contains information that describes our target gene such as Entry KO"K00099", Symbol "dxr; EC:1.1.1.267", Name "1-deoxy-D-xylulose-5-phosphate-reductoisomerase", Pathway "map00900, map01100 and map01110", Module "M00096 C5 isoprenoid biosynthesis, non-mevalonate pathway", Brite "KEGG Orthology (KO) [BR:ko00001] and Enzymes [BR:ko01000]", Other databases "RN:R05688,

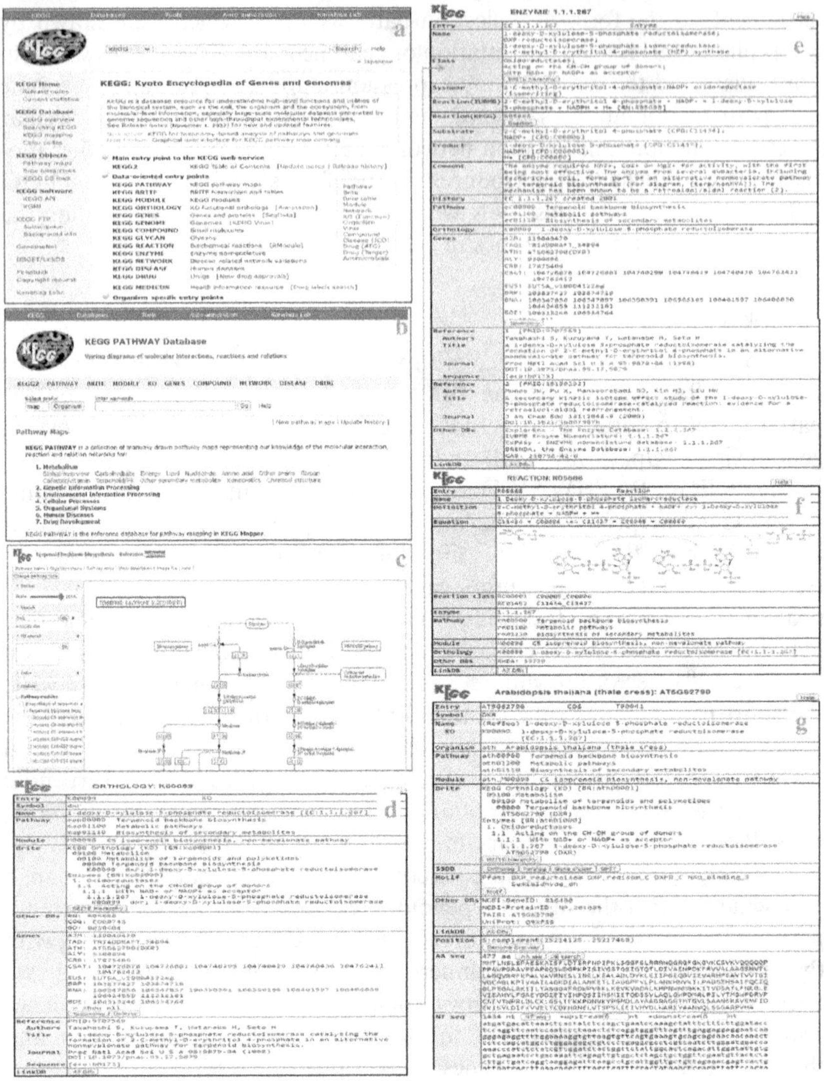

Figure 11.8 Leveraging Kegg "kyoto encyclopedia of genes and genomes pathway" content to analyze the gene and protein functions based on the genomics proteomics and metabolomics information that are stored in this database.

COG: COG0743, and GO: 0030604", Genes "e.g, ATH: AT5G62790 (DXR)", Reference "PMID:9707569" and LinkDB and see Figure 11.8d.

7 Then move to the second tile "ENZYME: 1.1.1.267" which contains other information such as Entry enzyme "EC 1.1.1.267", Name "1-deoxy-D-xylulose-5-phosphate-reductoisomerase", Class "Oxidoreductases", Sysname

"2-C-methyl-D-erythritol-4-phosphate: NADP+ oxidoreductase (isomer-izing)", Reaction(IUBMB) "2-C-methyl-D-erythritol 4-phosphate + NADP + = 1-deoxy-D-xylulose 5-phosphate + NADPH + H+ [RN:R05688]", Reaction(KEGG) "R05688", Substrate, Product, Comment, History, Pathway, Orthology, Genes, Reference, and Other DBs and see Figure11.8e.

8 To peruse the gene reaction in the graphical onlooker we go to the third tile "REACTION: R05688" which contains information to explain the gene reaction equation such as; Entry Reaction, Name, Definition, Equation, Reaction class, Enzyme, Pathway, Module, Orthology, Other DBs, and LinkDB and see Figure 11.8f.

9 When users would like to collect more information about the source of the genes, gene amino acid (AA seq), or nucleotide (NT seq) sequences, Motif, and position of our target gene on which chromosome. For example, by clicking on "ATH: AT5G62790 (DXR)" from the gene list at "Genes" (see Figure 11.8d), we can see a graphical window that displays the previous information with more details (see Figure 11.8g).

11.1.6 The arabidopsis information resource database

The TAIR (https://www.arabidopsis.org/index.jsp) is a freely available and integrative database presenting genome-scale information related to *Arabidopsis thaliana* plant. Moreover, the presented sequencing data and information at the TAIR database has grown dramatically, due to sharing of data from various other genomics projects databases such as Plant Ontology consortium, Cereon Genomics, Meinke Lab Website, DNA sequences, AtDB, GenBank, ABRC, NASC, TIGR, AFGC, Websites, Literature, ATGenExpress, NASCArrays, Kazusa, Stanford Genome Center, SALK, Arabidopsis Tilling Project, PubMed, Agricola, MetaCyc, Biosis and Growth & Developmental Stages Ontologies POC (https://www.arabidopsis.org/about/datasources.jsp). In the last decade, researchers and developers have been focused on enriching the content of data at the TAIR database by improving many tools such as (e.g. visualization, storage, annotation, and retrieval tools). These tools will facilitate the search process and easily display many data related to sequences such as gene expression, phenotypes, number of alleles, gene ontology, gene and protein families, pathway, DNA and seed stocks, genetic and physical markers, gene product function data and sequencing data from microarray experiments.

11.1.6.1 Methods

1 Navigate to https://www.arabidopsis.org/index.jspwebsite and see Figure 11.9a.
2 Select "BLAST" from the "Tools" drop-down menu, then copy the target protein sequencing from the file, for example, the gene, *SoFLDH* (EC: 1.1.1.354; NAD$^+$-dependent-farnesol-dehydrogenase) (Ali et al. 2022b), will be used as our "gene of interest" and see Figure 11.9b.

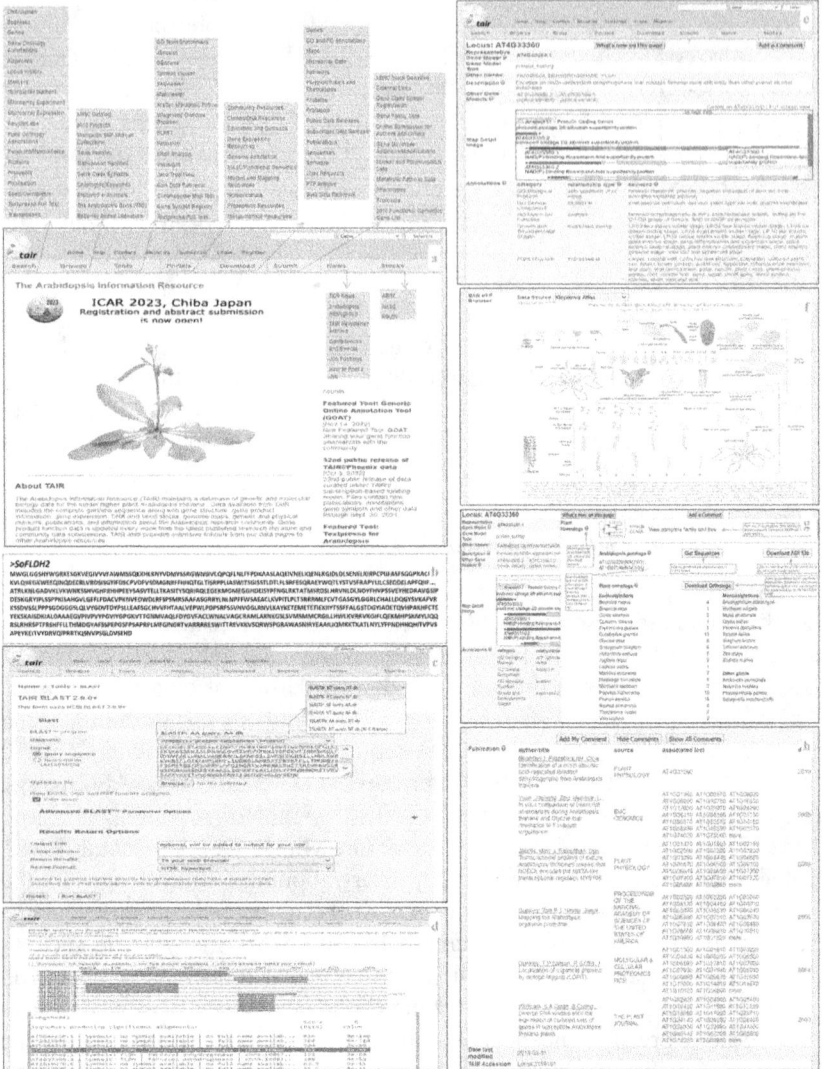

Figure 11.9 The flowchart outline using the Arabidopsis Information Resource (TAIR) to find information about arabidopsis genes and showing the major data related to these genes.

3 Enter or paste our target protein sequence in the "Input query sequence" box, then from the " BLAST™ program" drop-down menu you we can select "BALSTN: AA query, AA db" for searching on homologous protein sequence, then click on "Run BLAST" button to proceed with the search, and see Figure 11.9c.

4 A new page with the title "blastp query on Araport11 protein sequences (protein) sequences" will open, which has a summary of BLAST results, then we can choose any Arabidopsis gene ID (e,g; AT4G33360.2), see Figure 11.9d.

5 By clicking on the Arabidopsis gene ID (e,g; AT4G33360.2), another web page will open, which provides more information related to this gene such as Representative Gene Model "AT4G33360.1", Gene Model Type "protein_-coding", Other names "FARNESOL DEHYDROGENASE, FLDH", Description "Encodes an NAD+-dependent-dehydrogenase that can oxidize other prenyl alcohol less than farnesol substrates", Other Gene Models "AT4G33360.2 and AT4G33360.3", Map Detail Image, Annotations, BAR eFP Browser, the position of a gene on the chromosome, gene in other plant homologs, Search Gene Families, Map Locations, Map Links, polymorphism site and type, External Link and Publication and see Figure 11.9e–h.

11.1.7 *ExplorEnz-enzyme database*

ExplorEnz-Enzyme database (ExplorEnz;https://www.enzyme-database.net/) is a freely available database for presenting legalized information about Nomenclature and Classification of Enzymes based on the Nomenclature Committee of the International Union of Biochemistry and Molecular Biology (NC-IUBMB), the International Union of Pure and Applied Chemistry (IUPAC) and Joint Commission on Biochemical Nomenclature (JCBN). ExplorEnz is designed and created by Andrew McDonald for the IUBMB (School of Biochemistry & Immunology, Trinity College, Dublin, Ireland. Enzymes List at ExplorEnz has been functional classification based on the Reactions they Catalyze. At ExplorEnz, the Enzymes List has been functionally classified based on the Reactions they Catalyze into seven sections, which include EC 1 (Oxidoreductases), EC 2(Transferases), EC 3 (Hydrolases), EC 4 (Lyases), EC 5 (Isomerases), EC 6 (Ligases), and EC 7 (Translocases). Moreover, each type of enzyme has a unique Enzyme Commission number (EC number) which assignment and preparation by ExplorEnz group at Trinity College Dublin. The EC number for each enzyme or a group of enzymes that perform the same reaction consists of four-part, and each part represents the type of reaction. In addition to the "EC number", an accepted name 'not misleading or ambiguous' is assigned. This name is used usually as a common name in different databases and publications (McDonald et al. 2009).

11.1.7.1 *Methods*

1 Go to the ExplorEnz-Enzyme database (https://www.enzyme-database.net/) and see Figure 11.10a. And from the horizontal menu, there are many options you can select for enzyme search based on the EC number or common name, see Figure 11.10b–d.

302 *Mohammed Ali et al.*

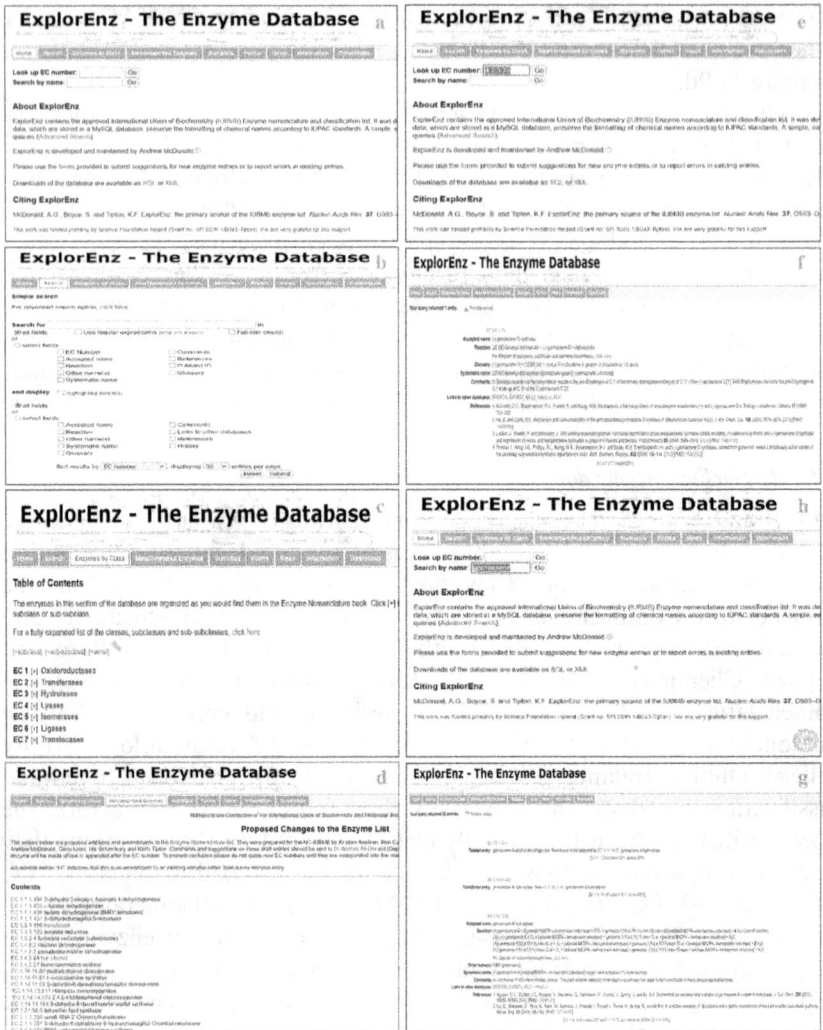

Figure 11.10 Steps involved in the search of enzyme name using the ExplorEnz-Enzyme database.

2 For a general search, you have two ways which depend on the information available about the enzyme, the first way is by using the EC number, while the second way is searching by enzyme name, for example, the enzyme (-)-germacrene-D-synthase [EC:4.2.3.75] (Ali et al. 2018), will be used as our "enzyme of interest" and see Figure 11.10e.

3 Choose "Home" from the horizontal menu, and enter or paste EC number of our target enzyme in the "Look up EC number" then click on "Go" button to proceed with the search and see Figure 11.10e.

4 A new page with the title "Your query returned 1 entry" will open, which has information about our target enzyme such as, Accepted name, Reaction, Glossary, ..., and References and see Figure 11.10f.
5 For searching an enzyme by its name we can use "Search by name:" from the "Home" page, and we can enter or paste our enzyme name (e.g, germacrene) in the "Search by name:" then click on "Go" button to proceed with the search and see Figure 11.10h.
6 As a research result, a new page with the title "Your query returned 32 entries" will open, which has information about our target enzyme such as accepted name, Reaction, Glossary,... ...., and References and see Figure 11.10g.
7 On the other hand, you will see other search tools in the horizontal menu such as "Enzymes by Class" which can be used for searching any enzyme in the expanded list of the classes, subclasses, and sub-subclasses. And "New/ Amended Enzymes" are used for searching for enzymes from a new list of enzyme entry.

References

Abbas, F., Yanguo, K., Rangcai, Y., & Yanping, F. (2019). Functional characterization and expression analysis of two terpene synthases involved in floral scent formation in Lilium 'Siberia'. *Planta* 249, 71–93. doi: 10.1007/s00425-018-3006-7

Ali, M., Alshehri, D., Alkhaibari, A. M., Elhalem, N. A., & Darwish, D. B. E. (2022a). Cloning and Characterization of 1,8-Cineole Synthase (*SgCINS*) Gene From the Leaves of *Salvia guaranitica* Plant. *Frontiers in Plant Science* 13, 869432. 10.3389/fpls.2022.869432

Ali, M., Nishawy, E., Ramadan, W.A., Ewas, M., Rizk, M. S., Sief-Eldein, A. G. M., et al. (2022b) Molecular characterization of a Novel NAD+-dependent farnesol dehydrogenase SoFLDH gene involved in sesquiterpenoid synthases from *Salvia officinalis*. *PLoS ONE* 17(6):e0269045. 10.1371/journal.pone.0269045.

Ali, M., Miao, L., Soudy, F. A., Darwish, D. B. E., Alrdahe, S. S., Alshehri, D., Benedito, V. A., Tadege, M., Wang, X., & Zhao, J. (2022c). Overexpression of terpenoid biosynthesis genes modifies root growth and nodulation in soybean (*Glycine max*). Cells, 11, 2622. 10.3390/cells11172622

Ali, M., Hussain, R. M., Rehman, N. U., She, G., Li, P., Wan, X., Guo, L., & Zhao, J. (2018). De novo transcriptome sequencing and metabolite profiling analyses reveal the complex metabolic genes involved in the terpenoid biosynthesis in blue anise sage (*Salvia guaranitica* L.). *DNA Research*, 25, 597–617. doi: 10.1093/dnares/dsy028.

Ali, M., Li, P., She, G., Chen, D., Wan, X., & Zhao, J. (2017). Transcriptome and metabolite analyses reveal the complex metabolic genes involved in volatile terpenoid biosynthesis in garden sage (*Salvia officinalis*). *Scientific Reports*. 7:16074. doi: 10.1038/s41598-017-15478-3

Ali, M., Miao, L., Hou, Q., Darwish, D. B., Alrdahe, S. S., Ali, A., Benedito, V. A., Tadege, M., Wang, X., & Zhao, J. (2021). Overexpression of terpenoid biosynthesis genes from garden sage (*Salvia officinalis*) modulates rhizobia interaction and nodulation in soybean. *Frontiers in Plant Science*, 12, 783269. doi: 10.3389/fpls.2021.783269.

Aoki-Kinoshita, K. F., & Kanehisa, M. (2007). Gene annotation and pathway mapping in KEGG. *Methods in Molecular Biology (Clifton, N.J.)*, 396, 71–91. 10.1 007/978-1-59745-515-2_6

Attwood, T. K., Coletta, A., Muirhead, G., Pavlopoulou, A., Philippou, P. B., Popov, I., Romá-Mateo, C., Theodosiou, A., & Mitchell, A. L. (2012). The PRINTS database: a fine-grained protein sequence annotation and analysis resource–its status in 2012. Database: *The Journal of Biological Databases and Curation*, bas019. 10.1 093/database/bas019

Austin, R. S., Hiu, S., Waese, J., Ierullo, M., Pasha, A., Wang, T. T., Fan, J., Foong, C., Breit, R., Desveaux, D., Moses, A., & Provart, N. J. (2016). New BAR tools for mining expression data and exploring Cis-elements in Arabidopsis thaliana. *The Plant Journal: For Cell and Molecular Biology*, 88(3), 490–504. 10.1111/tpj.13261.

Austin, R. S., Vidaurre, D., Stamatiou, G., Breit, R., Provart, N. J., Bonetta, D., Zhang, J., Fung, P., Gong, Y., Wang, P. W., McCourt, P., & Guttman, D. S. (2011). Next-generation mapping of Arabidopsis genes. *The Plant Journal: For Cell and Molecular Biology*, 67(4), 715–725. 10.1111/j.1365-313X.2011.04619.x

Bombarely, A., Menda, N., Tecle, I. Y., Buels, R. M., Strickler, S., Fischer-York, T., Pujar, A., Leto, J., Gosselin, J., & Mueller, L. A. (2011). The Sol Genomics Network (solgenomics.net): growing tomatoes using Perl. *Nucleic Acids Research*, 39(Database issue), D1149–D1155. 10.1093/nar/gkq866

Brian, L., Warren, B., McAtee, P., Rodrigues, J., Nieuwenhuizen, N., Pasha, A., David, K. M., Richardson, A., Provart, N. J., Allan, A. C., Varkonyi-Gasic, E., & Schaffer, R. J. (2021). A gene expression atlas for kiwifruit (*Actinidia chinensis*) and network analysis of transcription factors. *BMC Plant Biology*, 21(1), 121. 10.1186/ s12870-021-02894-x

Bru, C., Courcelle, E., Carrère, S., Beausse, Y., Dalmar, S., & Kahn, D. (2005). The ProDom database of protein domain families: more emphasis on 3D. *Nucleic Acids Research*, 33(Database issue), D212–D215. 10.1093/nar/gki034

Champigny, M. J., Sung, W. W., Catana, V., Salwan, R., Summers, P. S., Dudley, S. A., Provart, N. J., Cameron, R. K., Golding, G. B., & Weretilnyk, E. A. (2013). RNA-Seq effectively monitors gene expression in Eutremasalsugineum plants growing in an extreme natural habitat and in controlled growth cabinet conditions. *BMC Genomics*, 14, 578. 10.1186/1471-2164-14-578

Chen, F., Dong, W., Zhang, J., Guo, X., Chen, J., Wang, Z., Lin, Z., Tang, H., & Zhang, L. (2018). The sequenced angiosperm genomes and genome databases. *Frontiers in Plant Science*, 9, 418. 10.3389/fpls.2018.00418

Choudhuri, S. (2014). *Bioinformatics for Beginners: Genes, Genomes, Molecular Evolution, Databases and Analytical Tools*. New York, NY: Academic Press.

Clevenger, J., Chu, Y., Scheffler, B., & Ozias-Akins, P. (2016). A Developmental transcriptome map for allotetraploid *Arachis hypogaea*. *Frontiers in Plant Science*, 7, 1446. 10.3389/fpls.2016.01446

Conte, M. G., Gaillard, S., Lanau, N., Rouard, M., & Périn, C. (2008). GreenPhylDB: a database for plant comparative genomics. *Nucleic Acids Research*, 36(Database issue), D991–D998. 10.1093/nar/gkm934

Dash, S., Campbell, J. D., Cannon, E. K., Cleary, A. M., Huang, W., Kalberer, S. R., Karingula, V., Rice, A. G., Singh, J., Umale, P. E., Weeks, N. T., Wilkey, A. P., Farmer, A. D., & Cannon, S. B. (2016). Legume information system (LegumeInfo.org):

a key component of a set of federated data resources for the legume family. *Nucleic Acids Research*, 44(D1), D1181–D1188. 10.1093/nar/gkv1159

de Lima Morais, D. A., Fang, H., Rackham, O. J., Wilson, D., Pethica, R., Chothia, C., & Gough, J. (2011). SUPERFAMILY 1.75 including a domain-centric gene ontology method. *Nucleic Acids Research*, 39(Database issue), D427–D434. 10.1093/nar/gkq1130

Dean, G., Cao, Y., Xiang, D., Provart, N. J., Ramsay, L., Ahad, A., White, R., Selvaraj, G., Datla, R., & Haughn, G. (2011). Analysis of gene expression patterns during seed coat development in Arabidopsis. *Molecular Plant*, 4(6), 1074–1091. 10.1093/mp/ssr040

Degenhardt, J., Köllner, T. G., & Gershenzon, J. (2009). Monoterpene and sesquiterpene synthases and the origin of terpene skeletal diversity in plants. *Phytochemistry* 70, 1621–1637. doi: 10.1016/j.phytochem.2009.07.030

Duvick, J., Fu, A., Muppirala, U., Sabharwal, M., Wilkerson, M. D., Lawrence, C. J., Lushbough, C., & Brendel, V. (2008). PlantGDB: a resource for comparative plant genomics. *Nucleic Acids Research*, 36(Database issue), D959–D965. 10.1093/nar/gkm1041

Fasoli, M., Dal Santo, S., Zenoni, S., Tornielli, G. B., Farina, L., Zamboni, A., Porceddu, A., Venturini, L., Bicego, M., Murino, V., Ferrarini, A., Delledonne, M., & Pezzotti, M. (2012). The grapevine expression atlas reveals a deep transcriptome shift driving the entire plant into a maturation program. *The Plant Cell*, 24(9), 3489–3505. 10.1105/tpc.112.100230

Finn, R. D., Bateman, A., Clements, J., Coggill, P., Eberhardt, R. Y., Eddy, S. R., Heger, A., Hetherington, K., Holm, L., Mistry, J., Sonnhammer, E. L., Tate, J., & Punta, M. (2014). Pfam: the protein families database. *Nucleic Acids Research*, 42(Database issue), D222–D230. 10.1093/nar/gkt1223

Geisler-Lee, J., O'Toole, N., Ammar, R., Provart, N. J., Millar, A. H., & Geisler, M. (2007). A predicted interactome for Arabidopsis. *Plant Physiology*, 145(2), 317–329. 10.1104/pp.107.103465

Geng, Y., Wu, R., Wee, C. W., Xie, F., Wei, X., Chan, P. M., Tham, C., Duan, L., & Dinneny, J. R. (2013). A spatio-temporal understanding of growth regulation during the salt stress response in Arabidopsis. *The Plant Cell*, 25(6), 2132–2154. 10.1105/tpc.113.112896

Gonzales, M. D., Archuleta, E., Farmer, A., Gajendran, K., Grant, D., Shoemaker, R., Beavis, W. D., & Waugh, M. E. (2005). The Legume Information System (LIS): an integrated information resource for comparative legume biology. *Nucleic Acids Research*, 33(Database issue), D660–D665. 10.1093/nar/gki128

Goodstein, D. M., Shu, S., Howson, R., Neupane, R., Hayes, R. D., Fazo, J., Mitros, T., Dirks, W., Hellsten, U., Putnam, N., & Rokhsar, D. S. (2012). Phytozome: a comparative platform for green plant genomics. *Nucleic Acids Research*, 40(Database issue), D1178–D1186. 10.1093/nar/gkr944

Hashimoto, K., Goto, S., Kawano, S., Aoki-Kinoshita, K. F., Ueda, N., Hamajima, M., Kawasaki, T., & Kanehisa, M. (2006). KEGG as a glycome informatics resource. *Glycobiology*, 16(5), 63R–70R. 10.1093/glycob/cwj010

Henry, R. J. (2022). Progress in plant genome sequencing. Applied Biosciences, 1, 113–128. 10.3390/applbiosci1020008

Herrbach, V., Chirinos, X., Rengel, D., Agbevenou, K., Vincent, R., Pateyron, S., Huguet, S., Balzergue, S., Pasha, A., Provart, N., Gough, C., & Bensmihen, S.

(2017). Nod factors potentiate auxin signaling for transcriptional regulation and lateral root formation in *Medicago truncatula*. *Journal of Experimental Botany*, 68(3), 569–583. 10.1093/jxb/erw474

Ho, C. L., Wu, Y., Shen, H. B., Provart, N. J., & Geisler, M. (2012). A predicted protein interacted for rice. *Rice (New York, N.Y.)*, 5(1), 15. 10.1186/1939-8433-5-15

Huang, H. H., Xu, L. L., Tong, Z. K., Lin, E. P., Liu, Q. P., Cheng, L. J., & Zhu, M. Y. (2012). *De novo* characterization of the Chinese fir (*Cunninghamia lanceolata*) transcriptome and analysis of candidate genes involved in cellulose and lignin biosynthesis. *BMC Genomics*. 13(1), 648.

Hunter, S., Jones, P., Mitchell, A., Apweiler, R., Attwood, T. K., Bateman, A., Bernard, T., Binns, D., Bork, P., Burge, S., de Castro, E., Coggill, P., Corbett, M., Das, U., Daugherty, L., Duquenne, L., Finn, R. D., Fraser, M., Gough, J., Haft, D., & Yong, S. Y. (2012). InterPro in 2011: new developments in the family and domain prediction database. *Nucleic Acids Research*, 40(Database issue), D306–D312. 10.1093/nar/gkr948

Hyun, T. K., Rim, Y., Jang, H. J., Kim, C. H., Park, J., Kumar, R., Lee, S., Kim, B. C., Bhak, J., Nguyen-Quoc, B., Kim, S. W., Lee, S. Y., & Kim, J. Y. (2012). De novo transcriptome sequencing of *Momordica cochinchinensis* to identify genes involved in the carotenoid biosynthesis. *Plant Molecular Biology*, 79(4-5), 413–427. 10.1007/s111 03-012-9919-9

Ilic, K., Berleth, T., & Provart, N. J. (2004). BlastDigester--a web-based program for efficient CAPS marker design. *Trends in Genetics: TIG*, 20(7), 280–283. 10.1016/j.tig.2004.04.012

Jung, S., Staton, M., Lee, T., Blenda, A., Svancara, R., Abbott, A., & Main, D. (2008). GDR (Genome Database for Rosaceae): integrated web-database for Rosaceae genomics and genetics data. *Nucleic Acids Research*, 36(Database issue), D1034–D1040. 10.1093/nar/gkm803

Kagale, S., Nixon, J., Khedikar, Y., Pasha, A., Provart, N. J., Clarke, W. E., Bollina, V., Robinson, S. J., Coutu, C., Hegedus, D. D., Sharpe, A. G., & Parkin, I. A. (2016). The developmental transcriptome atlas of the biofuel crop Camelina sativa. *The Plant Journal: For Cell and Molecular Biology*, 88(5), 879–894. 10.1111/tpj.13302

Kanehisa, M. (2002). The KEGG database. *Novartis Foundation Symposium*, 247, 91–252.

Kanehisa, M., & Goto, S. (2000). KEGG: kyoto encyclopedia of genes and genomes. *Nucleic Acids Research*, 28(1), 27–30. 10.1093/nar/28.1.27

Kanehisa, M., Araki, M., Goto, S., Hattori, M., Hirakawa, M., Itoh, M., Katayama, T., Kawashima, S., Okuda, S., Tokimatsu, T., & Yamanishi, Y. (2008). KEGG for linking genomes to life and the environment. *Nucleic Acids Research*, 36(Database issue), D480–D484. 10.1093/nar/gkm882

Kanehisa, M., Goto, S., Hattori, M., Aoki-Kinoshita, K. F., Itoh, M., Kawashima, S., Katayama, T., Araki, M., & Hirakawa, M. (2006). From genomics to chemical genomics: new developments in KEGG. *Nucleic Acids Research*, 34(Database issue), D354–D357. 10.1093/nar/gkj102

Kehelpannala, C., Rupasinghe, T., Pasha, A., Esteban, E., Hennessy, T., Bradley, D., Ebert, B., Provart, N. J., & Roessner, U. (2021). An Arabidopsis lipid map reveals differences between tissues and dynamic changes throughout development. *The Plant Journal: For Cell and Molecular Biology*, 107(1), 287–302. 10.1111/tpj.15278

Kersey, P. J., Allen, J. E., Christensen, M., Davis, P., Falin, L. J., Grabmueller, C., Hughes, D. S., Humphrey, J., Kerhornou, A., Khobova, J., Langridge, N., McDowall, M. D., Maheswari, U., Maslen, G., Nuhn, M., Ong, C. K., Paulini, M., Pedro, H., Toneva, I., Tuli, M. A., ... Staines, D. M. (2014). Ensembl Genomes 2013: scaling up access to genome-wide data. *Nucleic Acids Research*, 42 (Database issue), D546–D552. 10.1093/nar/gkt979

Klein, T. E., Chang, J. T., Cho, M. K., Easton, K. L., Fergerson, R., Hewett, M., Lin, Z., Liu, Y., Liu, S., Oliver, D. E., Rubin, D. L., Shafa, F., Stuart, J. M., & Altman, R. B. (2001). Integrating genotype and phenotype information: an overview of the PharmGKB project. Pharmacogenetics Research Network and Knowledge Base. *The Pharmacogenomics Journal*, 1(3), 167–170. 10.1038/sj.tpj.6500035

Krishnakumar, V., Hanlon, M. R., Contrino, S., Ferlanti, E. S., Karamycheva, S., Kim, M., Rosen, B. D., Cheng, C. Y., Moreira, W., Mock, S. A., Stubbs, J., Sullivan, J. M., Krampis, K., Miller, J. R., Micklem, G., Vaughn, M., & Town, C. D. (2015). Araport: the Arabidopsis information portal. *Nucleic Acids Research*, 43(Database issue), D1003–D1009. 10.1093/nar/gku1200

Laverty, K. U., Stout, J. M., Sullivan, M. J., Shah, H., Gill, N., Holbrook, L., Deikus, G., Sebra, R., Hughes, T. R., Page, J. E., & van Bakel, H. (2019). A physical and genetic map of *Cannabis sativa* identifies extensive rearrangements at the *THC/CBD* acid synthase loci. *Genome Research*, 29(1), 146–156. 10.1101/gr.242594.118

Lees, J. G., Lee, D., Studer, R. A., Dawson, N. L., Sillitoe, I., Das, S., Yeats, C., Dessailly, B. H., Rentzsch, R., & Orengo, C. A. (2014). Gene3D: Multi-domain annotations for protein sequence and comparative genome analysis. *Nucleic Acids Research*, 42(Database issue), D240–D245. 10.1093/nar/gkt1205.

Letunic, I., Doerks, T., & Bork, P. (2012). SMART 7: recent updates to the protein domain annotation resource. *Nucleic Acids Research*, 40(Database issue), D302–D305. 10.1093/nar/gkr931

Li, H., Dong, Y., Yang, J., Liu, X., Wang, Y., Yao, N., Guan, L., Wang, N., Wu, J., & Li, X. (2012). *De novo* transcriptome of safflower and the identification of putative genes for oleosin and the biosynthesis of flavonoids. *PLoS One*, 7, e30987.

Li, P., Ponnala, L., Gandotra, N., Wang, L., Si, Y., Tausta, S. L., Kebrom, T. H., Provart, N., Patel, R., Myers, C. R., Reidel, E. J., Turgeon, R., Liu, P., Sun, Q., Nelson, T., & Brutnell, T. P. (2010). The developmental dynamics of the maize leaf transcriptome. *Nature Genetics*, 42(12), 1060–1067. 10.1038/ng.703

Liang, C., Jaiswal, P., Hebbard, C., Avraham, S., Buckler, E. S., Casstevens, T., Hurwitz, B., McCouch, S., Ni, J., Pujar, A., Ravenscroft, D., Ren, L., Spooner, W., Tecle, I., Thomason, J., Tung, C. W., Wei, X., Yap, I., Youens-Clark, K., Ware, D., ... Stein, L. (2008). Gramene: a growing plant comparative genomics resource. *Nucleic Acids Research*, 36(Database issue), D947–D953. 10.1093/nar/gkm968

Lyons, E. H. (2008). *CoGe, a New Kind of Comparative Genomics Platform: Insights Into the Evolution of Plant Genomes*. Ann Arbor, MI: Proquest, Umi Dissertation Publishing.

Mao, X., Cai, T., Olyarchuk, J. G., & Wei, L. (2005). Automated genome annotation and pathway identification using the KEGG Orthology (KO) as a controlled vocabulary. *Bioinformatics (Oxford, England)*, 21(19), 3787–3793. 10.1093/bioinformatics/bti430

Masoudi-Nejad, A., Goto, S., Endo, T. R., & Kanehisa, M. (2007). KEGG bio-informatics resource for plant genomics research. *Methods in Molecular Biology* (Clifton, N.J.), 406, 437–458. 10.1007/978-1-59745-535-0_21

McDonald, A. G., Boyce, S., & Tipton, K. F. (2009). ExplorEnz: the primary source of the IUBMB enzyme list. *Nucleic Acids Research*, 37(Database issue), D593–D597. 10.1093/nar/gkn582

Mehmood, N., Yuan, Y., Ali, M., Ali, M., Iftikhar, J., Cheng, C., Lyu, M., & Wu, B. (2021). Early transcriptional response of terpenoid metabolism to *Colletotrichum gloeosporioides* in a resistant wild strawberry *Fragaria nilgerrensis*. *Phytochemistry*, 181, 112590. 10.1016

Mi, H., Muruganujan, A., Casagrande, J. T., & Thomas, P. D. (2013). Large-scale gene function analysis with the PANTHER classification system. *Nature Protocols*, 8(8), 1551–1566. 10.1038/nprot.2013.092

Mitchell, A., Chang, H. Y., Daugherty, L., Fraser, M., Hunter, S., Lopez, R., McAnulla, C., McMenamin, C., Nuka, G., Pesseat, S., Sangrador-Vegas, A., Scheremetjew, M., Rato, C., Yong, S. Y., Bateman, A., Punta, M., Attwood, T. K., Sigrist, C. J., Redaschi, N., Rivoire, C., & Finn, R. D. (2015). The InterPro protein families database: the classification resource after 15 years. *Nucleic Acids Research*, 43(Database issue), D213–D221. 10.1093/nar/gku1243

Moriya, Y., Itoh, M., Okuda, S., Yoshizawa, A. C., & Kanehisa, M. (2007). KAAS: an automatic genome annotation and pathway reconstruction server. *Nucleic Acids Research*, 35(Web Server issue), W182–W185. 10.1093/nar/gkm321

Mulder, N. J., Apweiler, R., Attwood, T. K., Bairoch, A., Bateman, A., Binns, D., Bradley, P., Bork, P., Bucher, P., Cerutti, L., Copley, R., Courcelle, E., Das, U., Durbin, R., Fleischmann, W., Gough, J., Haft, D., Harte, N., Hulo, N., Kahn, D., & Wu, C. H. (2005). InterPro, progress and status in 2005. *Nucleic Acids Research*, 33(Database issue), D201–D205. 10.1093/nar/gki106

Mustroph, A., Zanetti, M. E., Jang, C. J., Holtan, H. E., Repetti, P. P., Galbraith, D. W., Girke, T., & Bailey-Serres, J. (2009). Profiling translatomes of discrete cell populations resolves altered cellular priorities during hypoxia in Arabidopsis. *Proceedings of the National Academy of Sciences of the United States of America*, 106(44), 18843–18848. 10.1073/pnas.0906131106

Nakabayashi, K., Okamoto, M., Koshiba, T., Kamiya, Y., & Nambara, E. (2005). Genome-wide profiling of stored mRNA in Arabidopsis thaliana seed germination: epigenetic and genetic regulation of transcription in seed. *The Plant Journal: For Cell and Molecular Biology*, 41(5), 697–709. 10.1111/j.1365-313X.2005.02337.x

Nikolskaya, A. N., Arighi, C. N., Huang, H., Barker, W. C., & Wu, C. H. (2007). PIRSF family classification system for protein functional and evolutionary analysis. *Evolutionary Bioinformatics Online*, 2, 197–209.

Patel, R. V., Nahal, H. K., Breit, R., & Provart, N. J. (2012). BAR expressolog identification: expression profile similarity ranking of homologous genes in plant species. *The Plant Journal: for cell and molecular biology*, 71(6), 1038–1050. 10.1111/j.1365-313X.2012.05055.x

Pedruzzi, I., Rivoire, C., Auchincloss, A. H., Coudert, E., Keller, G., de Castro, E., Baratin, D., Cuche, B. A., Bougueleret, L., Poux, S., Redaschi, N., Xenarios, I., Bridge, A., & UniProt Consortium (2013). HAMAP in 2013, new developments in the protein family classification and annotation system. *Nucleic Acids Research*, 41(Database issue), D584–D589. 10.1093/nar/gks1157

Poliakov, A., Foong, J., Brudno, M., & Dubchak, I. (2014). GenomeVISTA—an integrated software package for whole-genome alignment and visualization. *Bioinformatics*, 30, 2654–2655. doi: 10.1093/bioinformatics/btu355

Proost, S., Van Bel, M., Sterck, L., Billiau, K., Van Parys, T., Van de Peer, Y., & Vandepoele, K. (2009). PLAZA: a comparative genomics resource to study gene and genome evolution in plants. *The Plant Cell*, 21(12), 3718–3731. 10.1105/tpc.1 09.071506.

Provart, N., & Zhu, T. (2003) A browser-based functional classification SuperViewer for Arabidopsis genomics. *Current Topics in Computational Molecular Biology*, 271–272.

Provart, N. J., Alonso, J., Assmann, S. M., Bergmann, D., Brady, S. M., Brkljacic, J., Browse, J., Chapple, C., Colot, V., Cutler, S., Dangl, J., Ehrhardt, D., Friesner, J. D., Frommer, W. B., Grotewold, E., Meyerowitz, E., Nemhauser, J., Nordborg, M., Pikaard, C., Shanklin, J., & McCourt, P. (2016). 50 years of Arabidopsis research: highlights and future directions. *The New Phytologist*, 209(3), 921–944. 10.1111/nph.13687

Provart, N. J., Gil, P., Chen, W., Han, B., Chang, H. S., Wang, X., & Zhu, T. (2003). Gene expression phenotypes of Arabidopsis associated with sensitivity to low temperatures. *Plant Physiology*, 132(2), 893–906. 10.1104/pp.103.021261

Ramírez-González, R. H., Borrill, P., Lang, D., Harrington, S. A., Brinton, J., Venturini, L., Davey, M., Jacobs, J., van Ex, F., Pasha, A., Khedikar, Y., Robinson, S. J., Cory, A. T., Florio, T., Concia, L., Juery, C., Schoonbeek, H., Steuernagel, B., Xiang, D., Ridout, C. J., … Uauy, C. (2018). The transcriptional landscape of polyploid wheat. *Science (New York, N.Y.)*, 361(6403), eaar6089. 10.1126/science.aar6089

Schmid, M., Davison, T. S., Henz, S. R., Pape, U. J., Demar, M., Vingron, M., Schölkopf, B., Weigel, D., & Lohmann, J. U. (2005). A gene expression map of *Arabidopsis thaliana* development. *Nature Genetics*, 37(5), 501–506. 10.1038/ng1543

Schmid, R., & Blaxter, M. L. (2008). annot8r: GO, EC and KEGG annotation of EST datasets. *BMC Bioinformatics*, 9, 180. 10.1186/1471-2105-9-180

Seiler, K. P., George, G. A., Happ, M. P., Bodycombe, N. E., Carrinski, H. A., Norton, S., Brudz, S., Sullivan, J. P., Muhlich, J., Serrano, M., Ferraiolo, P., Tolliday, N. J., Schreiber, S. L., & Clemons, P. A. (2008). ChemBank: a small-molecule screening and cheminformatics resource database. *Nucleic Acids Research*, 36(Database issue), D351–D359. 10.1093/nar/gkm843

Sewell, W. (1964). Medical subject headings in MEDLARS. *Bulletin of the Medical Library Association*, 52(1), 164–170.

Sibout, R., Proost, S., Hansen, B. O., Vaid, N., Giorgi, F. M., Ho-Yue-Kuang, S., Legée, F., Cézart, L., Bouchabké-Coussa, O., Soulhat, C., Provart, N., Pasha, A., Le Bris, P., Roujol, D., Hofte, H., Jamet, E., Lapierre, C., Persson, S., & Mutwil, M. (2017). Expression atlas and comparative coexpression network analyses reveal important genes involved in the formation of lignified cell wall in *Brachypodiumdistachyon*. *The New Phytologist*, 215(3), 1009–1025. 10.1111/nph.14635

Sigrist, C. J., de Castro, E., Cerutti, L., Cuche, B. A., Hulo, N., Bridge, A., Bougueleret, L., & Xenarios, I. (2013). New and continuing developments at PROSITE. *Nucleic Acids Research*, 41(Database issue), D344–D347. 10.1093/nar/gks1067

Somerville, C., & Somerville, S. (1999). Plant functional genomics. *Science*, 285, 380–383.

Spannagl, M., Nussbaumer, T., Bader, K. C., Martis, M. M., Seidel, M., Kugler, K. G., Gundlach, H., & Mayer, K. F. (2016). PGSB PlantsDB: updates to the database framework for comparative plant genome research. *Nucleic Acids Research*, 44(D1), D1141–D1147. 10.1093/nar/gkv1130

Su-Fang, E., Zeti-Azura, M., Roohaida, O., Noor, A. S., Ismanizan, I., & Zamri, Z. (2014). Functional characterization of sesquiterpene synthase from *Polygonum minus*. *Scientific World Journal*, 2014, 840592. doi: 10.1155/2014/840592

Sullivan, A., Purohit, P. K., Freese, N. H., Pasha, A., Esteban, E., Waese, J., Wu, A., Chen, M., Chin, C. Y., Song, R., Watharkar, S. R., Chan, A. P., Krishnakumar, V., Vaughn, M. W., Town, C., Loraine, A. E., & Provart, N. J. (2019). An 'eFP-Seq Browser' for visualizing and exploring RNA sequencing data. *The Plant Journal: For Cell and Molecular Biology*, 100(3), 641–654. 10.1111/tpj.14468

Swarbreck, D., Wilks, C., Lamesch, P., Berardini, T. Z., Garcia-Hernandez, M., Foerster, H., Li, D., Meyer, T., Muller, R., Ploetz, L., Radenbaugh, A., Singh, S., Swing, V., Tissier, C., Zhang, P., & Huala, E. (2008). The Arabidopsis Information Resource (TAIR): gene structure and function annotation. *Nucleic Acids Research*, 36(Database issue), D1009–D1014. 10.1093/nar/gkm965

Tanabe, M., & Kanehisa, M. (2012). Using the KEGG database resource. *Current protocols in bioinformatics, Chapter*, 1, 1.12.1–1.12.43. 10.1002/0471250953.bi0112s38

Taylor, J., & Provart, N. J. (2006). CapsID: a web-based tool for developing parsimonious sets of CAPS molecular markers for genotyping. *BMC Genetics*, 7, 27. 10.1186/1471-2156-7-27

Tong, W., Cao, X., Harris, S., Sun, H., Fang, H., Fuscoe, J., Harris, A., Hong, H., Xie, Q., Perkins, R., Shi, L., & Casciano, D. (2003). ArrayTrack--supporting toxicogenomic research at the U.S. Food and Drug Administration National Center for Toxicological Research. *Environmental Health Perspectives*, *111*(15), 1819–1826. 10.1289/ehp.6497

Toufighi, K., Brady, S. M., Austin, R., Ly, E., & Provart, N. J. (2005). The Botany Array Resource: e-Northerns, Expression Angling, and promoter analyses. *The Plant Journal: For Cell and Molecular Biology*, 43(1), 153–163. 10.1111/j.1365-313X.2005.02437.x

Tran, F., Penniket, C., Patel, R. V., Provart, N. J., Laroche, A., Rowland, O., & Robert, L. S. (2013). Developmental transcriptional profiling reveals key insights into Triticeae reproductive development. *The Plant Journal: For Cell and Molecular Biology*, 74(6), 971–988. 10.1111/tpj.12206

Usadel, B., Obayashi, T., Mutwil, M., Giorgi, F. M., Bassel, G. W., Tanimoto, M., Chow, A., Steinhauser, D., Persson, S., & Provart, N. J. (2009). Co-expression tools for plant biology: opportunities for hypothesis generation and caveats. *Plant, Cell & Environment*, 32(12), 1633–1651. 10.1111/j.1365-3040.2009.02040.x

Van Bakel, H., Stout, J. M., Cote, A. G., Tallon, C. M., Sharpe, A. G., Hughes, T. R., & Page, J. E. (2011). The draft genome and transcriptome of Cannabis sativa. *Genome Biology*, 12(10), R102. 10.1186/gb-2011-12-10-r102

Vastrik, I., D'Eustachio, P., Schmidt, E., Gopinath, G., Croft, D., de Bono, B., Gillespie, M., Jassal, B., Lewis, S., Matthews, L., Wu, G., Birney, E., & Stein, L. (2007). Reactome: a knowledge base of biologic pathways and processes. *Genome Biology*, 8(3), R39. 10.1186/gb-2007-8-3-r39

Waese, J., Fan, J., Pasha, A., Yu, H., Fucile, G., Shi, R., Cumming, M., Kelley, L. A., Sternberg, M. J., Krishnakumar, V., Ferlanti, E., Miller, J., Town, C., Stuerzlinger, W., & Provart, N. J. (2017). ePlant: visualizing and exploring multiple levels of data for hypothesis generation in plant biology. *The Plant Cell*, 29(8), 1806–1821. 10.11 05/tpc.17.00073

Waese, J., Pasha, A., Wang, T. T., van Weringh, A., Guttman, D. S., & Provart, N. J. (2016). Gene Slider: sequence logo interactive data-visualization for education and research. *Bioinformatics* (Oxford, England), 32(23), 3670–3672. 10.1093/ bioinformatics/btw525

Waese, J., Provart, N. J., & Guttman, D. S. (2017). Topo-phylogeny: Visualizing evolutionary relationships on a topographic landscape. *PloS One*, 12(5), e0175895. 10.1371/journal.pone.0175895

Wang, L., Czedik-Eysenberg, A., Mertz, R. A., Si, Y., Tohge, T., Nunes-Nesi, A., Arrivault, S., Dedow, L. K., Bryant, D. W., Zhou, W., Xu, J., Weissmann, S., Studer, A., Li, P., Zhang, C., LaRue, T., Shao, Y., Ding, Z., Sun, Q., Patel, R. V., ... Brutnell, T. P. (2014). Comparative analyses of C₄ and C₃ photosynthesis in developing leaves of maize and rice. *Nature Biotechnology*, 32(11), 1158–1165. 10.1038/ nbt.3019

Ward, J. A., Ponnala, L., & Weber, C. A. (2012). Strategies for transcriptome analysis in nonmodel plants. *American Journal of Botany*, 99(2), 267–276. 10.3732/ajb. 1100334.

Waters, M., Stasiewicz, S., Merrick, B. A., Tomer, K., Bushel, P., Paules, R., Stegman, N., Nehls, G., Yost, K. J., Johnson, C. H., Gustafson, S. F., Xirasagar, S., Xiao, N., Huang, C. C., Boyer, P., Chan, D. D., Pan, Q., Gong, H., Taylor, J., Choi, D., & Fostel, J. (2008). CEBS--Chemical Effects in Biological Systems: a public data repository integrating study design and toxicity data with microarray and proteomics data. *Nucleic Acids Research*, 36(Database issue), D892–D900. 10.1093/nar/gkm755

Wheeler, D. L., Barrett, T., Benson, D. A., Bryant, S. H., Canese, K., Chetvernin, V., Church, D. M., Dicuccio, M., Edgar, R., Federhen, S., Feolo, M., Geer, L. Y., Helmberg, W., Kapustin, Y., Khovayko, O., Landsman, D., Lipman, D. J., Madden, T. L., Maglott, D. R., Miller, V., & Yaschenko, E. (2008). Database resources of the National Center for Biotechnology Information. *Nucleic Acids Research*, 36(Database issue), D13–D21. 10.1093/nar/gkm1000.

Winter, D., Vinegar, B., Nahal, H., Ammar, R., Wilson, G. V., & Provart, N. J. (2007). An "Electronic Fluorescent Pictograph" browser for exploring and analyzing large-scale biological data sets. *PloS One*, 2(8), e718. 10.1371/ journal.pone.0000718

Wishart, D. S., Knox, C., Guo, A. C., Cheng, D., Shrivastava, S., Tzur, D., Gautam, B., & Hassanali, M. (2008). DrugBank: a knowledgebase for drugs, drug actions and drug targets. *Nucleic Acids Research*, 36(Database issue), D901–D906. 10.1093/ nar/gkm958

Wu, J., Mao, X., Cai, T., Luo, J., & Wei, L. (2006). KOBAS server: a web-based platform for automated annotation and pathway identification. *Nucleic Acids Research*, 34(Web Server issue), W720–W724. 10.1093/nar/gkl167

Zhang, J., Hu, R., Sreedasyam, A., Garcia, T. M., Lipzen, A., Wang, M., Yerramsetty, P., Liu, D., Ng, V., Schmutz, J., Cushman, J. C., Borland, A. M., Pasha, A., Provart, N. J., Chen, J. G., Muchero, W., Tuskan, G. A., & Yang, X. (2020). Light-responsive expression atlas reveals the effects of light quality and intensity in Kalanchoëfedtschenkoi, a plant with crassulacean acid metabolism. *GigaScience*, 9(3), giaa018. 10.1093/gigascience/giaa018

Zhang, W., Morris, Q. D., Chang, R., Shai, O., Bakowski, M. A., Mitsakakis, N., Mohammad, N., Robinson, M. D., Zirngibl, R., Somogyi, E., Laurin, N., Eftekharpour, E., Sat, E., Grigull, J., Pan, Q., Peng, W. T., Krogan, N., Greenblatt, J., Fehlings, M., van der Kooy, D., ...Hughes, T. R. (2004). The functional landscape of mouse gene expression. *Journal of Biology*, 3(5), 21. 10.1186/jbiol16

Zhao, L., Zhao, X., Francis, F., & Liu, Y. (2021). Genome-wide identification and characterization of the TPS gene family in wheat (*Triticum aestivum* L.) and expression analysis in response to aphid damage. *Acta Physiologiae Plantarum*, 43, 64

Abbreviations

BAR	Bio-Analytic Resource.
InterPro	Protein Sequence Analysis and Classification Database.
CTD	Comparative Toxicogenomics Database.
Phytozome	Plant Comparative Genomics Portal.
KEGG	Kyoto Encyclopedia of Genes and Genomes.
TAIR	The Arabidopsis Information Resource.
IUBMB	ExplorEnz-Enzyme database.
NCBI	National Center for Biotechnology Information.
NGS	Next-Generation-Sequencing.
RNA-Seq	RNA Sequencing.
DNA-Seq	DNA Sequencing.
eFP	electronic Fluorescent Pictograph.

Index

Note: Page numbers in **Bold** refer to tables; and page numbers in *italics* refer to figures

For Product Safety Concerns and Information please contact our EU
representative GPSR@taylorandfrancis.com
Taylor & Francis Verlag GmbH, Kaufingerstraße 24, 80331 München, Germany

www.ingramcontent.com/pod-product-compliance
Lightning Source LLC
Chambersburg PA
CBHW052119230326
41598CB00080B/3869

9 781032 465265